The Upper Palaeolithic Revolution in global perspective

Paul Mellars receiving his Knighthood from Prince Charles, Buckingham Palace, 9 July 2010.
(© Photograph courtesy of British Ceremonial Arts Limited.)

McDONALD INSTITUTE MONOGRAPHS

The Upper Palaeolithic Revolution in global perspective

Papers in honour of Sir Paul Mellars

Edited by Katherine V. Boyle, Clive Gamble & Ofer Bar-Yosef

Published by:

McDonald Institute for Archaeological Research
University of Cambridge
Downing Street
Cambridge, UK
CB2 3ER
(0)(1223) 339336
(0)(1223) 333538 (General Office)
(0)(1223) 333536 (FAX)
dak12@cam.ac.uk
www.mcdonald.cam.ac.uk

Distributed by Oxbow Books
United Kingdom: Oxbow Books, 10 Hythe Bridge Street, Oxford, OX1 2EW, UK.
Tel: (0)(1865) 241249; Fax: (0)(1865) 794449; www.oxbowbooks.com
USA: The David Brown Book Company, P.O. Box 511, Oakville, CT 06779, USA.
Tel: 860-945-9329; Fax: 860-945-9468

ISBN: 978-1-902937-53-3
ISSN: 1363-1349 (McDonald Institute)

Cover design by Dora Kemp.

Edited for the Institute by James Barrett (*Series Editor*) and Dora Kemp (*Production Editor*).

Printed and bound by Short Run Press, Bittern Rd, Sowton Industrial Estate, Exeter, EX2 7LW, UK.

Contents

Contributors vi
Figures viii
Tables ix
Preface xi

Chapter 1 Introduction: Variability, Fate and the Upper Palaeolithic Revolution 1
Katherine V. Boyle, Ofer Bar-Yosef & Clive Gamble

Chapter 2 Paul Anthony Mellars, from Swallownest to Cambridge: the Early Years 9
Pamela Jane Smith

Chapter 3 1970–90: Two Revolutionary Decades 35
Chris Stringer

Chapter 4 Thinking Through the Upper Palaeolithic Revolution 45
Clive Gamble

Chapter 5 The Effect of Organic Preservation on Behavioural Interpretations at the South African Middle Stone Age Sites of Rose Cottage and Sibudu 53
Lyn Wadley

Chapter 6 The Dispersal of Modern Humans into Australia 63
Peter Veth

Chapter 7 Indian Lithic Technology Prior to the 74,000 BP Toba Super-eruption: Searching for an Early Modern Human Signature 73
Michael Haslam, Chris Clarkson, Michael Petraglia, Ravi Korisettar, Janardhana B., Nicole Boivin, Peter Ditchfield, Sacha Jones & Alex Mackay

Chapter 8 The Middle to Upper Palaeolithic Transition in Western Asia 85
Ofer Bar-Yosef & Anna Belfer-Cohen

Chapter 9 The Middle to Upper Palaeolithic Transition in Southern Siberia and Mongolia 103
Anatoly P. Derevianko

Chapter 10 The Transition to Upper Palaeolithic Industries in the Korean Peninsula 115
Kidong Bae

Chapter 11 The Middle to Upper Palaeolithic Transition North of the Continental Divide: Between England and the Russian Plain 123
Janusz K. Kozłowski

Chapter 12 Rethinking the 'Ecological Basis of Social Complexity' 137
Katherine V. Boyle

Chapter 13 Technological Characteristics at the End of the Mousterian in Cantabria: the El Castillo and Cueva Morín (Spain) 153
Federico Bernaldo de Quirós, Granada Sánchez-Fernández & José-Manuel Maíllo-Fernández

Chapter 14 The Demise of the Neanderthals and the Collapse of the Mousterian World 161
Jean-Philippe Rigaud

Index 169

Contributors

Kidong Bae
Department of Anthropology, Hanyang University, Ansan 425-791, Korea.
Email: bkd5374@gmail.com

Ofer Bar-Yosef
Department of Anthropology, Harvard University, Peabody Museum, 11 Divinity Avenue, Cambridge, MA 02138, USA.
Email: obaryos@fas.harvard.edu

Anna Belfer-Cohen
The Institute of Archaeology, The Hebrew University of Jerusalem, Mount Scopus, Jerusalem 91905, Israel.
Email: belferac@mscc.huji.ac.il

Federico Bernáldo de Quiros
Universidad de Leon, Area de Prehistoria, Facultad de Filosofia y letras, 24071 Leon, Spain.

Nicole Boivin
Research Laboratory for Archaeology and the History of Art, University of Oxford, Dyson Perrins Building, South Parks Rd, Oxford, OX1 3QY, UK.
Email: nicole.boivin@rlaha.ox.ac.uk

Katherine V. Boyle
McDonald Institute for Archaeological Research, University of Cambridge, Downing Street, Cambridge, CB2 3ER, UK.
Email: kvb20@cam.ac.uk

Chris Clarkson
School of Social Science, The University of Queensland, St Lucia, Queensland 4072, Australia.
Email: c.clarkson@uq.edu.au

Anatoly P. Derevianko
Institute of Archaeology and Ethnography, Lavrentiev Ave 17, Novosibirsk 630090, Russia.
Email: derev@archaeology.nsc.ru

Peter Ditchfield
Research Laboratory for Archaeology and the History of Art, University of Oxford, Dyson Perrins Building, South Parks Rd, Oxford, OX1 3QY, UK.
Email: peter.ditchfield@rlaha.ox.ac.uk

Clive Gamble
Centre for Quaternary Research, Department of Geography, Royal Holloway University of London, Egham, TW20 0EX, UK.
Email: Clive.Gamble@rhul.ac.uk

Michael Haslam
Research Laboratory for Archaeology and the History of Art, University of Oxford, Dyson Perrins Building, South Parks Rd, Oxford, OX1 3QY, UK.
Email: michael.haslam@rlaha.ox.ac.uk

Janardhana B.
Department of History and Archaeology, Karnatak University, Dharwad, 580 003, India.

Sacha Jones
McDonald Institute for Archaeological Research, University of Cambridge, Downing Street, Cambridge, CB2 3ER, UK.
Email: scj23@cam.ac.uk

Ravi Korisettar
Department of History and Archaeology, Karnatak University, Dharwad, 580 003, India.

Janusz K. Kozłowski
Institute of Archaeology, Jagiellonian University, Gołębia str. 11, 31-007 Cracow, Poland.
Email: kozlowsk@argo.hist.uj.edu.pl

Alex Mackay
Department of Archaeology and Natural History, Australian National University, Canberra, ACT 2600, Australia
Email: alex.mackay@anu.edu.au

José-Manuel Maíllo-Fernández
Departamento de Prehistoria y Arqueología, UNED, c/ Senda del Rey, 7, 28040 Madrid, Spain.
Email: jlmaillo@geo.uned.es

Michael Petraglia
Research Laboratory for Archaeology and the History of Art, University of Oxford, Dyson Perrins Building, South Parks Rd, Oxford, OX1 3QY, UK.
Email: michael.petraglia@rlaha.ox.ac.uk

Jean-Philippe Rigaud
Institut de Préhistoire et de Géologie du Quaternaire, Université Bordeaux I, France.
Email: j.ph.rigaud@wanadoo.fr

Granada Sánchez-Fernández
Departamento de Prehistoria y Arqueología, UNED, c/ Senda del Rey, 7, 28040 Madrid, Spain.
Email: gsanchez@bec.uned.es

Pamela Jane Smith
McDonald Institute for Archaeological Research, University of Cambridge, Downing Street, Cambridge, CB2 3ER, UK.
Email: pjs1011@cam.ac.uk

Chris Stringer
Department of Palaeontology, The Natural History Museum, London, SW7 5BD, UK.
Email: c.stringer@nhm.ac.uk

Peter Veth
National Centre for Indigenous Studies, Faculty of Law, Building # 5, Australian National University, ACT 0200 Australia.
Email: peter.veth@anu.edu.au

Lyn Wadley
Institute for Human Evolution and School of Geography, Archaeology and Environmental Studies, University of the Witwatersrand, PO WITS 2050, South Africa.
Email: lyn.wadley@wits.ac.za

Figures

2.1.	*Paul Mellars, born 29 October 1939, pictured here with his parents.*	9
2.2.	*Paul's maternal grandparents, Ernest and Etheldred Batty photographed during the First World War.*	10
2.3.	*Creswell Crags on 'a cold and miserable' Easter in 1960.*	12
2.4.	*Paul as an undergraduate at Cambridge University, 1959.*	13
2.5.	*The village of Tursac and the Chateau de Marzac where Anny Chanut lived.*	13
2.6.	*Paul in 1965 at the English Lower Palaeolithic site of Purfleet.*	17
2.7.	*Anny Mellars in an old car, 1965.*	18
2.8.	*Paul in 1983 providing scale for a wall of flint nodules at Brandon, Suffolk.*	23
2.9.	*June 1987, Paul relaxing at Natchez-Under-the-Hill, Mississippi.*	28
3.1.	*An extract from a diagram of recent human evolution.*	36
4.1.	*Predicted group sizes for a sample of 145 hominin specimens plotted against geological age.*	48
5.1.	*Middle Stone Age sites mentioned in the text.*	54
6.1.	*Location of landmasses of Sunda and Sahul.*	64
6.2.	*Location of Wallacean islands.*	66
6.3.	*Contemporary Australian landmass and shaded extension of continental shelf reflecting –130 m contour.*	68
6.4.	*Location of the major contemporary desert regions of Australia.*	69
7.1.	*Jwalapuram Localities 3 and 22.*	74
7.2.	*Jwalapuram Locality 3 completed excavation, facing south.*	75
7.3.	*Composite stratigraphy for Jwalapuram Locality 3.*	75
7.4.	*Jwalapuram Locality 22 excavation in progress, facing west.*	76
7.5.	*Jwalapuram Locality 22 north profile.*	77
7.6.	*Spatial distribution of artefacts recovered from the palaeosol, Jwalapuram Locality 22.*	77
7.7.	*Jwalapuram Locality 22 lithic artefacts.*	79
7.8.	*Jwalapuram Locality 22 lithic artefacts.*	79
8.1.	*Map of areas discussed in the text and main sites.*	86
8.2.	*Early Upper Palaeolithic artefacts from the Levant — Emiran and other 'Transitional Industries'.*	89
8.3.	*Ahmarian artefacts from the Levant.*	90
8.4.	*Aurignacian artefacts from the Levant.*	91
8.5.	*Aurignacian bone, horn and teeth artefacts from the Levant.*	92
8.6.	*Upper Palaeolithic artefacts from Dzudzuana Cave.*	95
9.1.	*Map of the Altai showing the locations of Palaeolithic sites.*	104
9.2.	*Early Upper Palaeolithic adornments and implements made of bone, stone, ivory and shell.*	107
10.1.	*Upper Palaeolithic sites in the Korean Peninsula.*	116
10.2.	*^{14}C dates from Palaeolithic sites in the Korean Peninsula.*	119
10.3.	*Blades and tanged points from Gorye-ri and Suyanggae.*	120
11.1.	*Topographic map of Europe and western Russia showing areas mentioned in the text.*	124
11.2.	*Map of the environment during Dansgaard/Oeschger warm events.*	125
11.3.	*Refitted bidirectional volumetric blade cores from Kraków–Księcia Józefa Street.*	126
11.4.	*Szeletian in northern central Europe.*	127
11.5.	*Northernmost Aurignacian sites in western and central Europe.*	128
11.6.	*Most important Lincombian-Ranisian-Jerzmanowician sites.*	129
11.7.	*Lincombian-Ranisian-Jerzmanowician cores.*	130
11.8.	*Lincombian-Ranisian-Jerzmanowician blades.*	130
11.9.	*Early Upper Palaeolithic sites in Russia.*	131
11.10.	*Lithic artefacts from Zaozerie.*	133
12.1.	*Distribution of major sites attributed to the period of the Middle–Upper Palaeolithic transition.*	139
12.2.	*Average species frequencies through the Middle–Upper Palaeolithic transition.*	140
12.3.	*Chart showing change and trend in assemblage diversity during the Middle–Upper Palaeolithic transition.*	142
12.4.	*Chart showing chronological change in mean diversity values through the Upper Palaeolithic.*	143
13.1.	*Central Cantabrian region showing locations mentioned in the text.*	154
13.2.	*Cantabrian Late Mousterian lithic technology.*	156

Tables

3.1. *A personal timeline of discoveries and events concerning modern human origins 1970–90.* 37
4.1. *The major changes between the Middle and Upper Palaeolithic in southwest France.* 45
4.2. *Klein's (1995) traits to recognize, worldwide, fully modern behaviour from 50–40,000 years ago.* 46
4.3. *Carving prehistory at its joints when it comes to the origins of modern humans.* 47
5.1. *Rose Cottage and Sibudu ages.* 55
7.1. *Techno-typological composition of the below-ash Jwalapuram Localities 22 and 3 assemblages.* 78
7.2. *Comparison of size and shape for complete flakes from below-ash Jwalapuram Localities 22 and 3.* 79
7.3. *Comparison of frequencies of different platform types on complete flakes and proximal fragments.* 80
7.4. *Frequencies of platform preparation on complete flakes and proximal fragments.* 80
7.5. *Percentages of raw materials used at Jwalapuram Localities 22 and 3, below-ash.* 80
12.1. *Average large herbivore %NISP frequencies attributable to the 'Transitional' phase.* 140
12.2. *Mean regional diversity values.* 141
13.1. *Cantabrian Late Mousterian sites.* 157

Preface

This book began as an idea in the run up to and during the 'Rethinking the Human Revolution' conference which was held at Corpus Christi College, in Cambridge in September 2005. As so often happens at these events, a few comments were passed and questions asked of and by the 'other organizers'. Chief among these was the question of a volume in tribute to Paul Mellars and his work on the Palaeolithic. By the end of the conference we were discussing possibilities and talks had been held with Professor Graeme Barker, Director of the McDonald Institute for Archaeological Research, over the feasibility of producing a 'Festschrift' which might not be quite a Festschrift in the usual sense of the word.

And so it was that this book began to develop, providing an opportunity for colleagues — including a number of people who had not, for various and numerous reasons, been at the 2005 conference — to write for another tome little realizing that not long before its publication the book would be paying tribute to Sir Paul Mellars's work.

Paul Mellars's research career has spanned forty years, during which he has established a reputation as arguably the most senior and internationally respected figure in the cultural and behavioural emergence, evolution and geographical dispersal of anatomically and behaviourally fully 'modern' humans. His work has been characterized by a combination of penetrating new interpretative insights and a wide-ranging interdisciplinary approach to evolutionary issues, which has kept him continuously at the cutting edge of international research in these fields. For this reason it was essential that the papers in the present volume cover a wide range of topics, many of which supplement the papers from the 2005 conference (now published as *Rethinking the Human Revolution*, McDonald Institute Monographs, 2007).

As new disciplines and techniques come into play, furnishing key information which archaeology and physical anthropology alone could not provide, so the subject diversifies and become yet more complex. Paul, however, despite an unrelenting focus on stone-tool assemblages and radiocarbon dates, has never viewed the subject as a simple one and, despite an everlasting love affair with southwestern France, has certainly considered the Human Revolution in its wider geographical context. For this reason, if for no other, the papers in this volume look beyond the classic region of the Périgord to areas such as Korea and Australasia, and cover the complete chronological span of the Middle to Upper Palaeolithic Transition in the Old World.

As always there are numerous people whom the editors would like to thank. In particular these include Pamela Jane Smith who doggedly persevered in order to be able to furnish us with a fascinating account of Paul's 'academic development' and research; Dora Kemp who has not only done us proud in the production of the volume but, despite regular contact with Paul, has kept all information to herself; Graeme Barker for agreeing that there should indeed be a 'volume for Paul', without whom Palaeolithic archaeology in Cambridge would not be what it is today; and James Barrett for continuing to support a volume which had already begun to develop before he joined the community in Cambridge; Manuel Arroyo-Kalin for helping with the editorial work on the paper by Federico Bernaldo de Quirós and colleagues; and finally, the authors for bearing with us through what must have seemed like interminable, and often inexplicable, delays and, we hope, for failing to mention the book to Paul.

Katherine V. Boyle, Clive Gamble and Ofer Bar-Yosef
January 2009

Chapter 1

Introduction: Variability, Fate and the Upper Palaeolithic Revolution

Katherine V. Boyle, Ofer Bar-Yosef & Clive Gamble

The opportunity to celebrate the achievements of one scholar in a field as diverse and contentious as human origins comes with challenges. The scale must be broad to reflect his impact on debates and the analysis of data. But the focus must be sharp enough to avoid a familiar criticism that such celebrations miss the importance of the target by either ascending into the ether of generalization or descending into dull parochial detail.

The career of Sir Paul Mellars, as Pamela Jane Smith so elegantly shows in her oral-historical account (Chapter 2), presents particular opportunities to meet these challenges head-on and turn them into an advantage. At an early stage of planning this volume we decided that our focus would be Paul's contribution to the great debate of our earliest prehistoric times: the Upper Palaeolithic Revolution. This decision allowed us to invite papers about the archaeology of every continent, except the Americas, and here we present the results of research from Australia, Korea, Siberia, India, South Africa, the countries of western Asia and Europe, including where, for Paul, it all began: southwest France. We very much hope that other volumes — and Paul certainly merits them — will deal in greater depth with his contributions to the Mesolithic, cognitive archaeology (although there is coverage of this topic in many of our contributions) and economic prehistory generally.

Two lasting themes, two great debates

The themes that drove our selection were variability and fate. These twin themes have been played out in two seminal debates among archaeologists — debates which have spilled over into other disciplines and out into public archaeology.

The debate about variability had as its principals Lewis Binford, François Bordes and Paul Mellars. It pitted the new against the traditional when it came to explaining the consistent patterning in stone-tool technology and typology. It built on Bordes's frequency analysis of many Mousterian assemblages from sites in southwest France and in particular his excavations at Combe Grenal (Dordogne). What had surprised Bordes (1972) was that a technological and typological analysis revealed repeated clusters in artefact assemblages rather than continuous variation. The interpretation he favoured for the five Mousterian variants was tribal. Southwest France was viewed as a favoured region that saw competition between five Neanderthal tribes that won temporary rights to inhabit the rockshelters. The history of Neanderthal occupation as recorded in the main variants, Denticulate, Typical, Quina, Ferrassie and Mousterian of Acheulean Tradition A and B, was therefore a summary of their changing regional fortunes.

Lewis and Sally Binford disagreed. Their interpretation of variability in stone-tool assemblages also acknowledged mobility but placed it within a wider environmental setting. These were not tribal signatures but tool kits (Binford & Binford 1966). Their variation reflected the distribution of resources in the environment of the Neanderthals which meant that different tools were needed at different times of the year and in different places. In the Binfords' view, supported by statistical correlation rather than cumulative frequency, the variation in tool kits was a result of the diverse activities that took place at base and hunting camps.

Paul Mellars provided a third view. His PhD at Cambridge had examined the stratigraphic ordering of Bordes's Mousterian variants (Mellars 1969). The law of superimposition clearly demonstrated a stratigraphic ordering in those same variants. Time, as well as space, had to be factored into the explanation. His data showed that Binford had collapsed time to reach an explanation, while Bordes's insistence on the iron hand of tradition in stone-tool making, backed by some largely erroneous interpretations of the cave sediments, had also written it out of the story. In other words that most archaeological of dimensions — time — had been ignored.

Of course such temporal variability still needed to be explained, which brings us to our second theme: fate. The variability debate had far-reaching significance for the development of archaeology as a scientific discipline but it mostly concerned other archaeologists. A much older debate about the fate of the Neanderthals drew in a larger audience, as Chris Stringer documents in Chapter 3, when it re-surfaced in the early 1970s just as the intense interest in variability was beginning to wane. The key moment was a conference on explaining culture change, held in Sheffield and organized by Colin Renfrew (1973). Bordes, Mellars and Binford all attended, and the exchanges are part of archaeological legend. However, the papers that Mellars and Bordes published moved on from the Mousterian to consider the Upper Palaeolithic. For Bordes (1973) it was business as usual with the Upper Palaeolithic showing comparable tribal variability to the Middle Palaeolithic. However, Mellars (1973) opened up an altogether different vista by setting out the archaeology of an Upper Palaeolithic Revolution (Gamble, Chapter 4).

The fate of the Neanderthals

Both variability and fate are fully discussed in *The Neanderthal Legacy* (Mellars 1996) and there is no need to rehearse their history further. What concerns us here is their impact and in particular the central role that Paul played in clarifying the Palaeolithic sequences of southwest France. His injection of time, via stratigraphy, into the archaeological debate was a turning point for Palaeolithic studies since it opened the door to evolutionary approaches that examined cultural adaptation through environmental selection. What were the capabilities of the Neanderthals and the people who replaced them? How did the differences in the archaeology of these two hominins play out in an evolutionary scenario of competitive advantage?

Furthermore, faced with explaining change through time archaeologists were also forced to consider the historical dimensions of their data. Here the issues passed beyond archaeological concerns to wider reflections about human nature by novelists, artists and commentators, for example Jean Auel's *Earth's Children* which takes the reader across Europe during the Early Upper Palaeolithic. These remain one of archaeology's most potent reference points to contemporary concerns about the uses of the past.

Therefore, it is the second theme we want to explore further in order to set the stage for this volume of papers. In the 35 years since Paul defined an Upper Palaeolithic Revolution there has been a wealth of new data and in particular the chronology has changed out of all recognition. In 1973 the dating of Neanderthal occupations depended for the most part on a four ice-age model, transferred ingeniously to rockshelters (Laville *et al.* 1980), and a few radiocarbon dates whose errors now would not be tolerated. The dating revolution was just starting with U-series, ESR, TL and OSR, while the ocean record would soon sweep all previous chronologies before it like a cleansing tidal wave. Paul has been a pioneer in promoting these approaches, believing quite rightly that archaeologists are lost without a firm chronology. In particular he has championed the use of accelerator dating (Mellars & Bricker 1986; Gravina *et al.* 2005) and has used the results to argue his case for an Upper Palaeolithic Revolution that points, in his view unequivocally, to the replacement of Neanderthals by modern humans in Europe. He dismissed the alternative, that Neanderthals had their own revolution late in the day just prior to their extinction, as 'an impossible co-incidence' (Mellars 2005).

However, that is not the full story. By pursuing so vigorously the themes of variability and fate we now have a very different understanding of Neanderthals. Today these once shambolic cavemen are seen instead as nuanced, intelligent hominins who, we feel sure, were we able to replay the tape-of-history, might well be editing this volume in our place and commenting instead on the evolutionary fate of the graciles!

The Neanderthal's changing world

The Neanderthals, a successful European population of large-brained hominins, dispersed during the first half of the last glacial cycle into western and northern Asia, reaching the Altai in central Asia, the southern Caucasus, the Zagros Mountains and the Levant.

They became extinct within a few thousand years sometime during Marine Isotope Stage (MIS) 3. The fluctuating climate of this long interval, which lasted from *c.* 60 ka cal. BP until the onset of the Last Glacial Maximum *c.* 24 ka cal. BP (van Andel & Davies 2003), was characterized by increasingly colder and drier conditions, and is considered by several scholars to be the culprit for the extinction of a population that had survived more than one previous glacial cycle (Gamble *et al.* 2004). While the detailed European climate record allows elaborate simulations (Banks *et al.* 2008) most archaeologists avoid explaining what the 'adaptive system' of the Neanderthals entailed or why it failed. Others view the dispersal of modern humans during the latter part of this period into this vast territory as triggering the demise of the local native populations through competitive exclusion (e.g. Banks *et al.* 2008; Bar-Yosef 2006; Shea 2008). But once again detailed explanations are rare.

As a result, both hypotheses which address the ultimate fate of the Neanderthals — climate and competition — lack testable models. For example, those who view abrupt climatic shifts as the main cause for extinction should tell us how did it happen? Most studies simply imply that harsh climatic conditions affected hominins and other large mammals in similar ways by reducing the availability of plant food for herbivores and the numbers of prey species for the guild of predators.

Faced with climatic deterioration Neanderthals had two options; to become locally extinct or migrate into more favourable areas south of the continental divide (Gamble 2009). In practice this meant moving to the three peninsula refuges of the Mediterranean; Iberia, Italy and Greece (Stewart & Cooper 2008). It is suspected that the patchiness of resources in the northern latitudes forced Neanderthals to become hyper-carnivores (Soffer 2000) — an adaptation supported by stable isotope studies (Richards *et al.* 2000). However, this would not explain their disappearance from the Mediterranean landscapes where they survived only one or two thousand years later than their relatives who inhabited temperate Europe, and where they probably successfully exploited plants as much as animal tissues. This extinction also applied to the fourth refugium of the Mediterranean region, the Levant.

To illustrate the challenge that still faces explanation and which is the subject of this volume we pose six simple, but challenging questions that archaeological research still needs to address:

1. Why did a population who successfully inhabited the ecological niches of temperate Europe during colder climate such as MIS-4 become extinct in MIS-3?
2. Does their extinction mean that the Neanderthals lacked the ability to adapt to new environmental conditions?
3. Was their technology and range of tool types too inflexible so that it could not be changed to assist their survival?
4. Were the archaeologically invisible tools and objects (made of organics that are not preserved) also inadequate when faced with new economic challenges?
5. Was it, for example, their inability to understand that bone and antler can be shaped into new tool forms and not just retouched as if they were hard rocks?
6. Did Neanderthals lack the kind of social organization that characterized small groups of modern humans as demonstrated by the organization of their 'living floors'?

We remain some way from providing adequate answers to any or all of these basic questions and opinion is often divided. For example, according to Farizy (1994), Neanderthals were adapted to particular environments in what she referred to as *locational fidelity*. They differed from modern humans who had a sense of a *homeland* which took in a large territory that for successful exploitation required greater social and economic flexibility. Stringer & Gamble (1993) also argued that the difference lay in the different scale of their social networks as indicated by the transfer distances of raw materials. Support for locational fidelity comes from stable-isotope studies of a Neanderthal tooth from Lakonis, Greece (Richards *et al.* 2008), that point to a lifetime range of possibly only 20 km for that individual. However such findings have to be compared with recent work on raw materials in France (Slimak & Giraud 2007) and Europe (Féblot-Augustins 2009) that indicate larger circulation territories for Middle Palaeolithic raw materials than was once believed. Moreover, the demonstration that Neanderthals were regular and efficient hunters of prime-aged animals (Adler *et al.* 2006; Blasco Sancho 1995; Gaudzinski 1996; Gaudzinski & Roebroeks 2000) has now overwhelmed opposing views (e.g. Klein 2008), raising the possibility that technological advantage may not have been the decisive factor between the two hominins.

Such convergence in their adaptive skills raises other possibilities. For example, did Neanderthals succumb instead to unrecorded failures such as diseases or did they simply lack the ability to think ahead and create stores to buffer against famine? Storage, as Wadley shows in Chapter 5 for the South African evidence, remains a great archaeological unknown until the appearance of pits containing bones at Upper Palaeolithic sites such as Kostenki and Avdeevo (Soffer & Praslov 1993). However, even then it is difficult to demonstrate storage practices with any frequency until the Neolithic. In short the material differences between Neanderthals and modern humans do not appear to be that great.

Equifinality and replacement

The climatic conditions of MIS-3 cannot be ignored in the search for satisfactory answers to the six questions listed above and the general issue of what caused Neanderthal extinction. It is perhaps noteworthy that among the European large mammals there were two main extinction periods either side of the Last Glacial Maximum. The first which lasted between 30 and *c.* 26 ka cal. BP included cave bear, hyaena and Neanderthals (Stuart & Lister 2007; Tzedakis *et al.* 2007). The later extinction period saw the demise in Europe of the mammoth, woolly rhino, cave lion and giant deer. This extinction process started in the late glacial interstadial and ended in the Younger Dryas/early Holocene (Stuart *et al.* 2004; Stuart & Lister 2007). Neanderthals were attuned, like these other species, to a repeated pattern of local extinctions in their European rather than their western Asian range. And what links all of these animals is that their principal refuge and source of re-colonization lay to the east rather than the south (van Kolfschoten 1992).

As a consequence, what now lies at the heart of investigations of the Upper Palaeolithic Revolution is not parallel or convergent evolution, as so often debated (d'Errico 2003; Mellars 2005; Zilhão & d'Errico 1999), but rather equifinality. For example, the adaptive strategies of Neanderthals and modern humans, as Boyle discusses in Chapter 12, are different but the outcomes, for example hunting fleet-footed prime-aged animals, seem identical for indicating hunting success and by inference reproductive advantage through an enhanced dietary niche. Moreover, as Brumm & Moore (2005) discuss using Australian data, there are more ways to be symbolic than making stone blades or decorating the body with thousands of beads. If the standards of the Upper Palaeolithic Revolution in Europe were applied to Australia then the conclusion must be that this continent never benefited from a human revolution. Such a patently ludicrous conclusion, although advocated in some quarters for all modern hunters and gatherers (Renfrew 2007), has to be rejected. But applying a similar logic to Europe and the fate of the Neanderthals not only raises the issue of equifinality but makes Mellars's case for out-and-out replacement compelling.

But how do we investigate the process of such replacement? Our suggestion is to model the different social and ecological roles of the new forager-colonizers who dispersed into the world occupied by the Neanderthals (Bernaldo de Quirós *et al.*, Chapter 13; Rigaud, Chapter 14), as well as into new territories such as northern Siberia (Derevianko, Chapter 9) and Australia (Veth, Chapter 6). The significance of dispersal into uninhabited continents lies in the demonstration of the social and ecological adaptations unique to *Homo sapiens* that permitted such a major extension of their range. In the case such as Europe and western Asia where a resident population of hunter-foragers was encountered, interaction may have varied regardless of their respective technological ability and social organization. We identify three social and economic options that characterize this interaction:

1. newcomers, moving into an unknown territory may ignore the presence of the indigenous population and continue to move on;
2. friendly encounters may take place and could lead to a certain degree of hybridization and acculturation through mate and object (and idea) exchange; or
3. infrequent conflicts between the two populations may result either in inter-group homicide or forceful occupation of centrally located sites from which control is exerted by the newcomers over the resource-rich niches.

The recently published Neanderthal genome (Green *et al.* 2010) should go some way towards resolving several of these issues but only archaeology, and improved chronologies will shed light on the qualitative aspects of the encounters.

Furthermore these considerations can be framed as testable hypotheses. We must remember that the incoming modern humans dispersed into Europe when the climate, for warm-adapted humans, was not in their favour. What it means, even if we choose the peaceful scenario, is that the colonizers had technical and social advantages that facilitated their acts and enabled them to occupy, for ever, the land of the 'others'.

The process of competitive exclusion meant that the territorial continuity of the Neanderthals was divided into smaller areas so that a geographical pattern of islands was produced. Colonizers, owing to their ability to maintain social networks across distances (Gamble 2009), moved into the optimum habitats. The Neanderthals then were cut off from their wider mating system that had preserved their meta-population during previous periods. Past 'bottlenecks' of this meta-population were overcome when climatic conditions improved, but at that time *c.* 45–38 ka cal. BP the 'foreigners' disrupted the demographic human cline across temperate Europe and parts of the Mediterranean lands.

The fate of the Neanderthals may therefore have been to be the wrong hominin in the wrong place at the wrong time.

A global perspective

The scale of these questions, hammered out with data from the Neanderthal world, therefore needs a global perspective. The background to the Upper Palaeolithic Revolution and the debates that Mellars's model has provoked are covered in detail by Pamela Jane Smith (Chapter 2) and Chris Stringer (Chapter 3). More than thirty years is a long time for an academic debate and it is necessary, as both authors remind us, to situate the discussions both at the level of an individual's scholarly development, in this instance Paul Mellars, and that of the wider community. These papers are examples of the intersecting scales that any intellectual history needs to encompass. Theories and ideas are never independent of person, time and context and by weaving together the micro (Paul) and the macro (the global community of palaeoanthropologists) the process of understanding is enriched. Archaeologists are all too reluctant to examine the subjective in their analyses of academic enquiry. When the 'micro' is as 'macro' as the subject of this volume is, we feel that such distinctions are easily over-ridden. There may be some who still consider that archaeology must be a scientific discipline and where the individual is driven out in the interests of objectivity. That, however, can never be the case and how better to illustrate this point than by these two papers combining personal histories and the testimony of those who lived and created the events.

The framework established, the next paper by Clive Gamble (Chapter 4) sounds a note of caution based on current readings of the Palaeolithic mind. Once again a sense of history is advocated. Why are Palaeolithic archaeologists still wedded to only one view of how the mind works?; a stance which not only informs interpretation but actively closes off other avenues of enquiry. Gamble's conclusion is that the Upper Palaeolithic Revolution needs expanding to include not only its traditional concerns of diet, technology and symbolism but also the different imaginations — social, geographical and material — that can be traced through the varied metaphorical engagements with the world of things by worldwide hominins.

Using data from South Africa, Lyn Wadley (Chapter 5) scrutinizes the proxies for symbolism. The issue revolves around the extent to which symbolism only exists when it is stored externally to the body. She builds on Merlin Donald's (1998) model that stresses the importance of mimesis but adds to this by arguing for the importance of social relationships when it comes to the process by which hominins became humans. Drawing on her Middle Stone Age excavations at Sibudu Cave, and comparing these with results from Rose Cottage, she shows that the pattern of change in South Africa was incremental rather than revolutionary. Such changes at Sibudu were well established prior to 60,000 years ago.

Africa was the continent of change for humans. In his discussion of the evidence for Australia, Peter Veth (Chapter 6) reminds us that the archaeology of this continent remains a benchmark for establishing the inception of a global human diaspora, that occurred within the last 60,000 years. The overwatering capabilities of the humans that first reached the Pleistocene continent of Sahul were matched by their ability, once established, to colonize the arid interior. Australia is therefore the archaeologist's laboratory to test the relative importance of social networks and technology in the colonization of foragers. It is also, as Veth shows, central to interpreting unexpected discoveries such as *Homo floresiensis* from the wider region. What was the fate of this creature? Is this an opportunity to replay the Neanderthal–modern human tape in a different setting?

It is now possible to trace a southern route for the global diaspora that led to the settlement of Australia. Central to this, as Michael Haslam and his co-authors (Chapter 7) show, are the state-of-the-art excavations in Andhra Pradesh, India. Much has been written about the possible significance of the Toba eruption 74,000 years ago for the creation of bottleneck conditions that could lead to the genetic pattern of humans. Very few archaeological investigations are available to test such models. When that is possible, as reported here, the pattern that emerges is far more complex than the model of a bottleneck supposes. The main significance of Toba may turn out to be as a synchronous marker to compare stone-tool assemblages rather than the perfect storm which blew us into existence.

The interest then turns to the particulars of the Upper Palaeolithic Revolution as the human diaspora flowed in different directions. As Ofer Bar-Yosef and Anna Belfer-Cohen (Chapter 8) show, much hinges on the transition in the key area of western Asia. This region is critical because of its cross-roads position. By turning both east and west humans entered both arms of the Neanderthal world and their fate was sealed. Bar-Yosef and Belfer-Cohen's analysis of the transition uses lithic technology and typology, and points to the considerable variability between areas and sites. Of prime concern is whether this variation arose either from adaptation to local conditions or is the signature of different expanding populations. How replacement occurred then becomes a central issue for future research and of especial interest is the role of disease as a barrier and agent for change by replacement.

The volume then takes the eastern arm of dispersal with Anatoly Derevianko (Chapter 9) examining the transition to the Upper Palaeolithic in Siberia and Mongolia and Kidong Bae (Chapter 10) continuing this theme in the Korean peninsula. Both papers highlight the variability that characterizes this transition. This is expressed both in the timing and in the form of the artefacts and their technology. Derevianko traces links to western Asia while Bae has to be more cautious when making connections owing to the more limited research in the Korean peninsula. But even so the possibility that the hallmark blades of the Upper Palaeolithic Revolution developed there from an elongated flake technology point to more complex patterns than simple replacement.

We then turn west towards Europe. Janusz Kozłowski (Chapter 11) examines the significance of cultural geography on the transition to the Upper Palaeolithic. Instead of following the traditional axis of the Balkans and the Danube he makes the case for the importance of the European lowlands in the process of continental scale change. What this view accentuates, from what many regard as the periphery of Palaeolithic Europe, is a different history for the transition and one that challenges the orthodoxy of lower latitudes.

The traditional heartland of the Upper Palaeolithic Revolution is addressed in three papers. Katherine Boyle (Chapter 12) takes as her theme Paul Mellars's interest in cultural ecology. She examines the evidence for that link between the exploitation of resources and the greater complexity of material culture, and by inference of society, among humans rather than Neanderthals. What she finds is abundant evidence for blurring. Social complexity and subsistence change did not go hand in hand.

Federico Bernaldo de Quirós and his co-authors (Chapter 13) use one of the most important archives for the study of the transition; El Castillo cave from Cantabria. Here they draw comparisons with Cueva Morín to examine Mousterian variability during the process of transition. They remind us of the catch-all nature of the Mousterian and argue on the basis of their evidence for a more fluid approach to the concept of transitions.

Finally, but by no means least in the context of this volume, we arrive in the Dordogne (Chapter 14). Jean-Philippe Rigaud focuses on the key points in the occupation of this region by Neanderthals and humans. He stresses the role of variability in social organization and presents evidence to show how this was nuanced at different times and places. There were many different fates involved and the outcomes are unpredictable rather than certain. He concludes by comparing the diversity of social organization today with that of the transition and, by so doing, warns us against adopting a single cause as sufficient explanation for such a complex historical process.

* * *

Palaeolithic archaeology is world history writ large. As Paul Mellars's mentor Sir Grahame Clark argued in his seminal *World Prehistory*:

> Prehistory is not merely something that human beings passed through a long time ago: it is something which properly apprehended allows us to view our contemporary situation in a perspective more valid than that encouraged by the study of our own parochial histories. (Clark 1961, 3)

The global perspective on one such major prehistoric concern — how we became human — owes a great debt to the example of Paul Mellars and his insistence that the Palaeolithic past is, to paraphrase a famous football manager, not only a matter of life and death but much more important than that.

References

Adler, D.S., G. Bar-Oz, A. Belfer-Cohen & O. Bar-Yosef, 2006. Ahead of the game: Middle and Upper Palaeolithic hunting behaviors in the southern Caucasus. *Current Anthropology* 47, 89–118.

van Andel, T.H. & W. Davies (eds.), 2003. *Neanderthals and Modern Humans in the European Landscape during the Last Glaciation: Archaeological Results of the Stage 3 Project*. (McDonald Institute Monographs.) Cambridge: McDonald Institute for Archaeological Research.

Banks, W.E., F. d'Errico, A.T. Peterson *et al.*, 2008. Human ecological niches and ranges during the LGM in Europe derived from an application of eco-cultural niche modeling. *Journal of Archaeological Science* 35(2), 481–91.

Bar-Yosef, O., 2006. Neanderthals and modern humans: a different interpretation, in *When Neanderthals and Modern Humans Met*, ed. N. Conard. (Tübingen Publications in Prehistory.) Tübingen: Kerns Verlag, 467–82.

Binford, L.R. & S.R. Binford, 1966. A preliminary analysis of functional variability in the Mousterian of Levallois facies. *American Anthropologist* 68(2), 238–95.

Blasco Sancho, M.F., 1995. *Hombres, fieras y presas: estudio arqueozoológico y tafonómico del yacimiento del Paleolítico Medio de la Cueva de Gabasa 1 (Huesca)*. (Monografías arqueológicas 38.) Zaragoza: Universidad de Zaragoza.

Bordes, F., 1972. *A Tale of Two Caves*. (Harper's Case Studies in Archaeology.) New York (NY): Harper and Row.

Bordes, F., 1973. On the chronology and contemporaneity of different Palaeolithic cultures in France, in *The Explanation of Culture Change: Models in Prehistory*, ed. C. Renfrew. London: Duckworth, 217–26.

Brumm, A. & M.W. Moore, 2005. Symbolic revolutions and the Australian archaeological record. *Cambridge Archaeological Journal* 15(2), 157–75.

Clark, J.G.D., 1961. *World Prehistory in New Perspective*. Cambridge: Cambridge University Press.

Donald, M., 1998. Mimesis and the executive suite: missing links in language evolution, in *Approaches to the*

Evolution of Language: Social and Cognitive Bases, eds. J.R. Hurford, M. Studdert-Kennedy & C. Knight. Cambridge: Cambridge University Press, 44–67.

d'Errico, F., 2003. The invisible frontier: a multiple species model for the origin of behavioral modernity. *Evolutionary Anthropology* 12, 188–202.

Farizy, C., 1994. Behavioral and cultural changes at the Middle to Upper Paleolithic transition in western Europe, in *Origins of Anatomically Modern Humans*, eds. M.H. Nitecki & D.V. Nitecki. New York (NY): Plenum Press, 93–100.

Féblot-Augustins, J., 2009. Revisiting European Upper Paleolithic raw material transfers: the demise of the cultural ecological paradigm?, in *Lithic Materials and Paleolithic Societies*, eds. B. Adams & B.S. Blades. Oxford: Blackwell Publishing, 25–46.

Gamble, C., 2009. Human display and dispersal:a case study from biotidal Britain in the Middle and Upper Pleistocene. *Evolutionary Anthropology* 18(4), 144–56.

Gamble, C., W. Davies, P. Pettitt & M. Richards, 2004. Climate change and evolving human diversity in Europe during the last glacial. *Philosophical Transactions of the Royal Society, London* Series B 359, 243–54.

Gaudzinski, S., 1996. On bovid assemblages and their consequences for the knowledge of subsistence patterns in the Middle Paleolithic. *Proceedings of the Prehistoric Society* 62, 19–39.

Gaudzinski, S. & W. Roebroeks, 2000. Adults only: reindeer hunting at the Middle Palaeolithic site Salzgitter Lebenstedt, northern Germany. *Journal of Human Evolution* 38(4), 497–521.

Gravina, B., P. Mellars & C. Bronk Ramsey, 2005. Radiocarbon dating of interstratified Neanderthal and early modern human occupations at the Chatelperronian type-site. *Nature* 438, 51–6.

Green, R.E., J. Krause, A.W. Briggs *et al.*, 2010. A draft sequence of the Neandertal genome. *Science* 328(5979), 710–22.

Klein, R.G., 2008. Out of Africa and the evolution of human behavior. *Evolutionary Anthropology* 17(6), 267–81.

van Kolfschoten, T., 1992. Aspects of the migration of mammals to northwestern Europe during the Pleistocene, in particular the reimmigration of *Arvicola terrestris*. *Courier Forschungsinstitut Senckenberg* 153, 213–20.

Laville, H., J.-P. Rigaud & J. Sackett, 1980. *Rock Shelters of the Perigord*. New York (NY): Academic Press.

Mellars, P., 1969. The chronology of Mousterian industries in the Périgord region of south-west France. *Proceedings of the Prehistoric Society* 35, 134–71.

Mellars, P., 1973. The character of the Middle–Upper Palaeolithic transition in south-west France, in *The Explanation of Culture Change: Models in Prehistory*, ed. C. Renfrew. London: Duckworth, 255–76.

Mellars, P., 1996. *The Neanderthal Legacy: an Archaeological Perspective from Western Europe*. Princeton (NJ): Princeton University Press.

Mellars, P., 2005. The impossible coincidence: a single-species model for the origins of modern human behavior in Europe. *Evolutionary Anthropology* 14(1), 12–27.

Mellars, P., 2006a. A new radiocarbon revolution and the dispersal of modern humans in Eurasia. *Nature* 439, 931–5.

Mellars, P., 2006b. Why did modern human populations disperse from Africa *ca.* 60,000 years ago? A new model. *Proceedings of the National Academy of Sciences of the USA* 103(25), 9381–6.

Mellars, P. & H. Bricker, 1986. Radiocarbon accelerator dating in the earlier Upper Palaeolithic, in *Archaeological Results from Accelerator Dating*, eds. J.A.J. Gowlett & R.E.M. Hedges. (Monograph 11.) Oxford: Oxford University Committee for Archaeology, 73–80.

Renfrew, C. (ed.), 1973. *The Explanation of Culture Change: Models in Prehistory*. London: Duckworth.

Renfrew, C., 2007. *Prehistory: Making of the Human Mind*. London: Weidenfeld & Nicolson.

Richards, M.P., P.B. Pettitt, E. Trinkaus, F.H. Smith, M. Paunović & I. Karavanić, 2000. Neanderthal diet at Vindija and Neanderthal predation: the evidence from stable isotopes. *Proceedings of the National Academy of Sciences of the USA* 97(13), 7663–6.

Richards, M., K. Harvati, V. Grimes *et al.*, 2008. Strontium isotope evidence of Neanderthal mobility at the site of Lakonis, Greece using laser-ablation PIMMS. *Journal of Archaeological Science* 35(5), 1251–6.

Shea, J.J., 2008. Transitions or turnovers? Climatically-forced extinctions of *Homo sapiens* and Neandertals in the east Mediterranean Levant. *Special issue: The Coastal Shelf of the Mediterranean and Beyond: Corridor and Refugium for Human Populations in the Pleistocene. Quaternary Science Reviews* 27(23–4), 2253–70.

Slimak, L. & Y. Giraud, 2007. Circulations sur plusieurs centaines de kilomètres durant le Paléolithique moyen. Contribution à la connaissance des sociétés néanderthaliennes. *Comptes Rendus Palevol* 6(5), 359–68.

Soffer, O., 2000. The last Neanderthals. *ERAUL* 92, 139–45.

Soffer, O. & N.D. Praslov (eds.), 1993. *From Kostenki to Clovis: Upper Paleolithic–Paleo-Indian Adaptations*. New York (NY): Plenum Press.

Stewart, J.R. & A. Cooper, 2008. Ice Age refugia and Quaternary extinctions: an issue of Quaternary evolutionary palaeoecology. *Special issue: Ice Age Refugia and Quaternary Extinctions: an Issue of Quaternary Evolutionary Palaeoecology. Quarternary Science Reviews* 27(27–8), 2423–48.

Stringer, C. & C. Gamble, 1993. *In Search of the Neanderthals: Solving the Puzzle of Human Origins*. London: Thames & Hudson.

Stuart, A.J. & A.M. Lister, 2007. Patterns of Late Quaternary megafaunal extinctions in Europe and northern Asia. *Courier-Forschungsinstitut Senckenberg* 259, 287–97.

Stuart, A.J., P.A. Kosintsev, T.F.G. Higham & A.M. Lister, 2004. Pleistocene to Holocene extinction dynamics in giant deer and woolly mammoth. *Nature* 431, 684–9.

Tzedakis, P.C., K.A. Hughen, I. Cacho & K. Harvati, 2007. Placing late Neanderthals in a climatic context. *Nature* 449, 206–8.

Zilhão, J. & F. d'Errico, 1999. The chronology and taphonomy of the earliest Aurignacian and its implications for the understanding of Neandertal extinction. *Journal of World Prehistory* 13(1), 1–68.

Chapter 2

Paul Anthony Mellars, from Swallownest to Cambridge: the Early Years[1]

Pamela Jane Smith

When asked to describe Paul Mellars, James Sackett replied 'He's the kind of guy I would want standing next to me in an Anglo-Saxon shield wall.'[2]

Childhood

'My father had an allotment garden in our small village of Swallownest near Sheffield' Paul Mellars recalled in our first interview in 2006. 'One evening, he came back with a Charles II copper coin found in the potato patch. That was about the most exciting thing I had ever seen ... I became really interested and started reading Grahame Clark, including *Prehistoric Europe, the Economic Basis* and *Archaeology and Society*.'[3]

Paul was fortunate in his parents. His father, Herbert Mellars, was a member of the religious sect known as the Plymouth Brethren and began his working life as a coal miner, a penalty imposed for taking a religious stance as a conscientious objector to fighting in the Second World War. Although Paul broke from his father's strict religious faith as a young man, he nevertheless inherited a steadfast commitment to basic truths. Within his father's faith, there was clear right and clear wrong. Paul grew up with an elementary sense of basic definitive choices; this seemed to contribute to forming a character often described by colleagues as straightforward and remarkably staunch and consistent in interests and behaviour.

Paul matured into an uncomplicated 'straight bloke' with robustly defined and steadfast views in

Figure 2.1. *Paul Mellars, born 29 October 1939, pictured here with his parents; Paul was sturdy, cautious, determined and bright. 'My mother would say that I never took unnecessary risks. I would carefully consider all options'; and then presumably look for 'the big picture',[4] the '64 thousand dollar question',[5] light a cigar and synthesize the available literature. (Photograph courtesy of Mr Herbert Mellars.)*

Figure 2.2. *Paul's maternal grandparents, Ernest and Etheldred Batty photographed during the First World War. In addition to contributing natural cleverness, his family supported Paul's academic pursuits. For example, Harvey Bricker, Professor emeritus at Tulane University, remembers Paul's family loaning their car for a summer expedition to Scottish Obanian sites in 1963.*[6] *Years later, Paul used the knowledge thus gained when choosing to excavate Oronsay. (Photograph courtesy of Mr Herbert Mellars.)*

which there is little room for intense, ill-defined subtleties. Our species would be seen by Paul as upright and definitively intelligent with clear boundaries, rather than essentially nuanced. With a clean definition, we are rational, thinking, moral beings. As an academic, Paul would support strong arguments and stick with the steady accumulative development and expansion of certain central themes of knowledge. His father clearly believed in God; Paul, a 'fundamentalist agnostic'[7] would be equally passionately, simply and solidly committed to definite archaeological views. Paul seems to have inherited his father's strong character.

Coming from a modest family, Paul attended the village school. Growing up surrounded by a variety of social classes, he was still aware of those who were less fortunate. Paul realized that some could barely afford proper clothing. He was one of the brightest boys in his school and sometimes bullied and, therefore, delighted, after the 'eleven plus' examination, to qualify for and move up to Woodhouse, an excellent West Riding County Council Grammar School where he thrived. Founded in 1909 to supply free, quality education, Woodhouse had 'a good mix of social classes all judged to be capable of an academic education'.[8]

Paul's mother, exceptional for her era, had also attended Woodhouse when women seldom had that opportunity. Although Paul speaks less frequently of his mother than his father, she also was a strong supporter of his education. In 1965, she was the one who spotted the newspaper advertisement for what became Paul's research fellowship in 1965 at Sheffield University where archaeology had just begun with Warwick Bray as the first Lecturer.

Paul enjoyed Woodhouse, had inspiring and well-educated teachers, spent his teenage years specializing in science and maths and was eventually appointed Head Boy or Head Prefect which is a position of distinction and considerable responsibility within the British school system. He was certainly the first in his family to be interested in archaeology, 'I don't think they knew the word!';[9] with fellow Woodhouse student and long-term friend, Jeff Radley, Paul collected flints at the local Mesolithic site of Hail Mary Hill and with the Headmaster's encouragement, he studied A-Level archaeology without a tutor. None other than Leslie Armstrong examined his essay on the Hail Mary Hill site and came to his school to examine the finds; Armstrong was, at that time, famous for his excavations at the well-known Upper Palaeolithic site of Creswell Crags just 20 miles from Paul's home.

And, with the Latin Master, Paul founded the first Woodhouse Archaeology Society, took part in weekend excavations and at the age of 17 published the following mission statement extolling 'Archaeology as a Hobby' in the school magazine, *Woodnotes*. 'Even the richest of unearthed relics is, to the archaeologist, only as important as the new light which it throws on the life and times of the people who made it.' And when discussing the value of amateur discoveries, he wrote, 'I have found truth in the statement "Seek and ye shall find".' (Mellars 1957/8, 27–8).

By the time Paul left Woodhouse in 1958, he had already discovered Palaeolithic and Mesolithic archaeology, had conducted his first excavations, had published his first article and was obviously 'mad keen on archaeology'.[10] He would go on to become the first in his family to attend university.

> People had said that I would never make a living in archaeology and advised me to do something useful. "Get a job and then do archaeology in your spare time." I thought that the pure sciences sounded a little bit boring; my mother's cousin was married to an engineer so I thought, "Why don't I do civil engineering?"[11]

Paul applied to Leeds University and UCL, was offered places in civil engineering at both and started attending engineering lectures at UCL in October 1958. 'They were the most boring things I had ever heard.' Every afternoon he would skip classes in drawing and go 'around all the second-hand bookshops, buying archaeology books', reading avidly. And, to add 'to the depression ... in order to get to the Engineering Department, I had to walk daily past the Institute of Archaeology in Gordon Square which had recently been built'.[12] Within weeks, Paul had left engineering and returned to Sheffield.

During his 'gap' year, Paul took a temporary post as a schoolteacher, learning a skill and trade that served him well throughout his later career. 'I then spent about six months teaching nine- to eleven-year-olds in a mining village near Sheffield. I was teaching these little mining kids in an all-age school. That was probably the hardest work that I have ever done.'

The experience paid off. Paul's ability as a teacher is now legendary at Cambridge. Among numerous other accomplishments, he was nominated twice for a University Teaching Prize, has supervised over 40 PhD candidates, has been an external examiner at 18 British and foreign universities, has been Director of Studies for Corpus Christi College for over 25 years and is one of the few professors in Britain with a dedicated Facebook, 'The Professor Paul Mellars Big Love Society' for 'those who appreciate the archaeology god/genius that is Professor Mellars'.[13] This site is set up by undergraduates who admire Paul's charms and who are thankful for his abilities to present organized, judicious, easy-to-grasp clear lectures. As Tim Reynolds, Paul's first 'home-grown' Cambridge PhD student, recently stated, 'As a lecturer for undergraduates, Paul was brilliant and had excellent technique; he stated what he was going to say, said it and then summarized what he said; everyone understood his points.'[14]

During 1958–59, Paul also worked as a temporary archaeological assistant at the Sheffield Museum, excavating at Ash Tree Cave, and was encouraged by the Deputy Director, Stanley West, to apply to Cambridge to read archaeology; at that time, Cambridge was one of the few universities offering an undergraduate degree in the subject. West wrote to Glyn Daniel who suggested that Paul 'bombard the colleges for entry now [this was July/early August], I might just get a place. I wrote out a screed about who I was, why I was interested in archaeology and what I had done, addressed to every Senior Tutor in Cambridge' and was miraculously offered an interview at Fitzwilliam in early September, three weeks before term. 'So, that is how I got into Cambridge ... by the skin of my teeth.'[15]

The early Cambridge years

'There were two kinds of people who came up to Cambridge in 1959,' Paul remembers,

> There were those who had done National Service. They were the men. And there were those who came up from school and they were the boys. Also, public school men dominated Cambridge and, having come from a northern grammar school, I found this all rather intimidating. I was pretty low-key. I must have been rather inward-looking and felt a bit gauche.[16]

When Paul came up to Fitzwilliam House in Cambridge University in October, 1959, he entered what was to become one of the most illustrious classes ever to graduate from Archaeology and Anthropology; Barry Cunliffe 'Mortimer Wheeler's protégé',[17] Charles Higham, 'the great rugger player' and Paul were awarded Firsts in Part I and Charles Higham, Colin Renfrew, who joined the class after National Service and a Part I in Natural Sciences, Barry Cunliffe, Gavin Brown and Paul all earned Firsts in the final examinations.[18]

A young Ray Inskeep, who later became an acclaimed Africanist, taught Palaeolithic archaeology to Paul in the first year. In his first supervision, along with Charles Higham, Paul was set an essay on C.D. Forde's *Habitat, Economy and Society: a Geographical Introduction to Ethnology* to examine the relevance of hunter-gatherer ethnography to the understanding of the Palaeolithic. Paul remembers this as a most challenging undergraduate essay which, more than any other, absolutely convinced him of the importance of having anthropological and ethnographic perspectives on the Palaeolithic. 'That stayed with me throughout my life.'

Ethnographically-informed interpretations have featured consistently in Paul's research. This interest was reinforced by the Man the Hunter conference held in 1966 and the subsequent widely read volume (Lee & DeVore 1968). In 1970, after arriving at Sheffield, Paul suggested that one of his first students, Gillian Drinkwater, 'collect systematic data on population

Figure 2.3. *Creswell Crags on 'a cold and miserable' Easter in 1960, left to right, Wilfred Shawcross, Robert Soper, Colin Breese, Warwick Bray, Paul Mellars, unknown,*[19] *Glynn Isaac. 'The highlight of every day was the trip to the pub after a meal which I have NO memories of. The contrast between the dirty archaeologists and the spic-and-span miners, who had emerged from the bowels of the earth to hot showers and hot high teas, made an impression', remembers Nic David who was then a Cambridge undergraduate.*[20] *(Photograph courtesy of Barbara Isaac.)*

densities, group sizes and seasonal mobility on a wide range of hunter-gatherers to see if she could find patterns in that data that might be applicable to archaeology'.[21] Forty years on, Paul argued that New Guinea cargo cults illustrate how technology can be copied without a simultaneous transfer of all aspects of cultural, symbolic and cognitive patterns (Mellars as quoted by d'Errico *et al.* 1998)

Also in his first year at Cambridge, in Easter 1960, Charles McBurney took a group of now-famous students to an excavation at Mother Grundy's Parlour, Creswell Crags. Paul recalls:

> Charles never forgot his participation in Montgomery's campaign in the western desert. He thought that everything should be done in a military way. The camping was all about making us men. We all had to take turns cooking dinner. People had never even fried an egg. My friend, Desmond Collins, was told to cook pigeons. That was the worst gastronomic experience ever. Barbara also went on that excavation and that is how she and Glynn Isaac met. Things blossomed.[22]

In the following year, McBurney taught Palaeolithic archaeology. Paul remembers McBurney, as

> a god-like figure. We treated his book (1960), the *Stone Age of Northern Africa*, almost as the *Koran*, or *Holy Writ*, in the first year. But, by the second year, those ahead of me, people like Nic David, had worked in the summers at the Abri Pataud with Hallam Movius and Desmond Collins had worked with Francois Bordes at Combe-Grenal; and they were exposed to the very sophisticated excavations that were going on in France.

Figure 2.4. *Paul as an undergraduate at Cambridge University, 1959. 'I didn't go on major excavations until the summer I graduated in 1962 when I went to work at Clark Howell's Lower Palaeolithic site, Ambrona' and subsequently at Harvard's Professor Hallam Movius's invitation, as an assistant to Nic David who was analysing the Noaillian levels at the Abri Pataud in Les Eyzies, Dordogne, France. Paul would also meet James Sackett there and later Harvey Bricker and Anny Chanut, friends for life. (Photograph courtesy of the Master and Fellows of Fitzwilliam College.)*

Figure 2.5. *The village of Tursac and the Chateau de Marzac where Anny Chanut lived. The Vézère River runs beneath it. 'I wonder if it's true that Paul swam down the Vézère and Anny let down her hair so that he could climb up to her.'*[23] *A more realistic rumour is that Nic David himself introduced Paul to his future wife in 1963.(Photograph by courtesy of Nic David.)*

Also, shortly after, there was an intellectual shift in the Cambridge Department when Grahame Clark received funds from the British Academy for the British Academy Major Research Project in the Early History of Agriculture and appointed Eric Higgs as his Research Assistant, in charge of that project. Eric attracted many PhD students, including Mike Jarman, Derek Sturdy and Paul Wilkinson. Students became excited by catchment analyses and economic prehistoric approaches.

Although Paul never felt that he challenged McBurney on excavation technique as other students began to do, he did write an essay in which he argued,

> the French Châtelperronian developed from the Mousterian. I came to the conclusion that the first Upper Palaeolithic communities in France were derived from the last Mousterian, which was totally against what Charles believed. We used to have supervisions, in his very cold, cavernous study with books all over the place. Charles would sit on one side of the fireplace and the students would sit on the other side. And you had to read out your essay. So, I read out this essay to Charles and he sat in stony silence; then there were a few moments of pause and he snorted, 'I don't believe a word of it'.[24]

Paul remained consistently faithful to this interpretation of the Châtelperronian, explaining it in detail 35 years later in *The Neanderthal Legacy*.

When remembering his PhD work at the Abri Pataud in 1962, Nic David states,

> I started what Movius expected would be a de Sonneville-Bordian analysis of the Proto-Magdalenian materials, However, I soon realized that I was doing violence to the data by shoehorning tools into imposed categories. I began to experiment with what became known as descriptive attribute analysis [using means and standard deviations] to get at the social units.[25]

At the same time, Jim Sackett was finishing his dissertation on the Aurignacian culture in the Dordogne. His was a pioneering attempt 'to use contingency table analysis, introduced into archaeology by Albert Spaulding. It may have been the first time anyone attempted analysing multi-variable interaction on a real sample of archaeological data,' Jim Sackett stated in an interview.

> I was supposed to write a dissertation on Abri Pataud; then Hallam Movius kicked me off the project when it became apparent to him that the problems he was attempting to solve were, in my opinion, out-of-date. So Francois Bordes gave me my thesis topic, which entailed studying old Aurignacian collections. I then came up with the idea of re-doing artefact typology within a new statistical frame.[26]

The resulting article (Sackett 1966), 'Quantitative Analysis of Upper Paleolithic Stone Tools' is now considered a classic.

Nic remembers discussing typology 'ad infinitum' that summer. 'Ideas were flowing back and forth between Jim, Paul and me'. Paul and Jim had dinner together nightly and 'we talked archaeology, archaeology, archaeology' and although Paul (Mellars 1996, 344–5) never engaged in the 'style debates' of the 1970s and '80s in which Jim Sackett played a key role, he greatly admires Jim's capacity for subtle thought. 'I regard those conversations with Jim in 1962 as my main education in Palaeolithic archaeology. He introduced me to the ethnologist Julian Steward's research and to the principles of what became known as cultural ecology.' Steward's arguments were set out in his article 'Ecological aspects of southwestern society' (1937), in a paper with F.M. Setzler (1938) and in *Theory of Culture Change* (1955). Steward maintained that archaeologists should use their data to study changes in subsistence economies, population size and settlement patterns.[27] Paul continued:

> Almost everything I have done since is cultural ecology, within a strongly ecological, adaptive framework, looking at the relationships between humans and their environment. I liked ecological thinking because it put the emphasis on the kinds of things that I felt archaeology can get at, technology, environments, subsistence, seasonality; this approach did then and still does lead to understanding long-term change. That summer in '62, talking to Jim Sackett, made me a Palaeolithic archaeologist.[28]

It was an exciting time; young researchers were re-evaluating and refining the Bordes and de Sonneville-Bordes systems. Paul added:

> My school background in the sciences influenced me and I became increasingly convinced that descriptive attribute analysis was the way to go. I was obsessed with the idea that we needed to put the studies of technology and typology on a more scientific basis. We shouldn't just classify a tool. We should conduct an analysis, defining what was meant by a Levallois flake with metrical results, length, breadth, thickness, and numbers of flakes, character of the striking platform. Since the others were studying the Upper Palaeolithic, I thought I must do this sort of analysis on the Mousterian, the Middle Palaeolithic. So that is what I decided to do.[29]

Paul's PhD project was registered as the 'Metrical and Statistical Analysis of Middle Palaeolithic Artefacts'. In looking back to statistical analyses in that era, Derek Roe, who was one year ahead of Paul at Cambridge, states, 'It is worth remembering that Paul and I were working in pre-computer days. What we did, we did by hand though David Clarke was about to change all that in his own thesis.' Derek Roe processed 38,000 items of metrical data, one at a time, using a slide rule.[30]

Paul remembers first measuring lithics excavated by Charles McBurney from La Cotte de St Brelade in Jersey. 'I spent a lot of time trying to figure out exactly how you measure various dimensions of the flakes and how to record striking platforms and flake scars. It became problematic. It was difficult to get consistent measurements that really meant something.' Shortly after he arranged to return to the Abri Pataud to dig with Movius in the summer of 1963.

In the meantime, Paul read everything he could find on the Mousterian and was particularly influenced by Maurice Bourgon's (1957) *Les industries mousteriennes et pré-mousteriennes du Perigord*. In this publication, Bourgon systematically analysed a number of Mousterian industries from the Perigord region using the quantitative techniques which Francois Bordes eventually perfected.

Bourgon's research heralded a new era for Paul, who went on to become intricately enmeshed in the now famous 'Francois Bordes/Lewis Binford' debate. For decades, this debate exerted broad academic influence on the nature and course of Palaeolithic archaeology as it was taught in the UK and America. As Derek Roe pointed out in a recent letter, 'It wasn't long before Cambridge archaeology and anthropology students were being asked to compare, contrast and evaluate the views of Bordes, Binford *and Mellars* on the nature of the Mousterian in South-west France.'[31] The debates placed industrial variability 'front and centre', making it a key disciplinary focus around which archaeology continues to revolve even today. People interviewed, who lived through the era, view that period as a time of great change that gave birth to 'the New Archaeology, processualism, new interests in theoretical archaeology'.[32] Jim Sackett noted that Paul Mellars came 'at a time when the rules of the game were changing, indeed, when nobody quite knew what the rules were'.[33] As Paul commented in *The Neanderthal Legacy* (1996, 315), the debates on the significance of European and Near Eastern industrial variability were to dominate 'studies of the Middle Palaeolithic' for 30 years. In 1997, Sackett discussed the on-going relevance of this debate. In his review of *The Neanderthal Legacy*, he (Sackett 1997, 149) writes,

'Mellars has broken the log-jam created by the classic debate.' By giving 'the Mousterian a time dimension, function and style, activity and ethnicity, appear as complementary reflections of hominid behaviour that must be wrestled with.'

This debate was Paul's first experience with a major disciplinary, archaeological controversy. His life history would intersect with several controversies at turning points in method, theory and policy. Paul went on to be involved openly in many key issues, such as the debate as to whether South African delegates should be invited to the World Archaeological Congress in Southampton in 1986. He is also currently entrenched in a painful disagreement on how best to excavate and to preserve the famous British Mesolithic site, Star Carr. Although very difficult for those involved, Paul's career has added value for an historian precisely because he participates fully and forcefully in these 'turning-point' controversies.

As will be discussed fully later, Paul's contribution to the Mousterian variability debate, the chronological model, was eventually vindicated by Hélène Valladas's (Valladas *et al.* 1986) TL dating of burnt flint samples from Le Moustier published in *Nature*. Chronology and dating remained absolutely central themes in all of Paul's subsequent work.

It is therefore well worth recounting exactly how Paul developed his early thinking. Also, although readers of this biography will have studied and be aware of the functional versus cultural versus tool reduction explanations for variability, few know the personal events which led Paul to his chronological conclusions. It is best then to let Paul reconstruct his own story in his own words.

When Paul arrived in 1963, Bordes had been excavating the site of Combe-Grenal since the 1950s where he had discovered an incredible stratigraphy of 55 different Mousterian levels over nine Acheulean levels and had already published the basic sequence of the different Mousterian industries. This was not done in detail but the stratigraphic layers were listed and on the basis of this enormous sequence, and considering Bordes strong intellectual reaction to the terribly simplistic unilinear nineteenth-century French evolutionary sequences, he had concluded that the five major industrial variants of the Mousterian demonstrated 'a mosaic of different cultures and different cultural variants, more or less contemporary' (Bordes 1973, 221).

Bordes's work 'did indeed show some interstratification of what Bordes called Denticulate and Typical Mousterian' Paul noted. 'There was no question of making the Denticulate/Typical Mousterian into a simple chronological sequence'.[34] But Bordes had also found at Combe-Grenal a sequence of Ferrassie Mousterian layers overlaid by Quina Mousterian layers with Mousterian of Acheulean Tradition in the top three levels.

Paul spotted this sequence immediately and began to check all other published stratigraphies on Mousterian sites. Paul stated in an interview in 2006:

> It started to dawn on me, I could begin to see that at several sites the Mousterian of Acheulean Tradition was stratified above the Quina Mousterian. By digging through all old reports, the idea began to take hold that there might be a chronological sequence. These were the most typologically, technologically distinctive variants, the most clearly characterized. The more I looked, the more impressive it was. I thought that this emerging sequence could be highly significant! That is how the idea of this crucial chronological sequence developed. From that point on I was scanning literature all the time and trying to find every known stratigraphy I could.

Paul stayed in touch with excavators digging all major sites, checking the stratigraphy.

> Every time I checked a new source I was on tenterhooks as to whether the next site might contradict the sequence suggested. Only a single well-documented occurrence where the Mousterian of Acheulean occurred under the Quina Mousterian would effectively shatter my whole scheme.[35]

Then, while digging at the Abri Pataud in 1963, Paul met Bordes's illustrator Pierre Laurent with his wife, Jocelyne, who were living and working as part of the team and Laurent arranged for Paul to spend a few days digging at Combe-Grenal. Paul remembers that Bordes received him very warmly.

> I only spent three or four days there but he put me to work at one of the metre squares with a little notebook. He explained how he recorded the flints and the bones by giving those numbers, square numbers and measuring the '3D' co-ordinates. Bordes was doing high-quality excavation and he was very helpful and friendly.

Then Pierre Laurent also got permission from Bordes to let Paul work in Bordes's laboratory located in an eighteenth-century chapel on the grounds of Bordeaux University.

Paul remembers:

> Laurent asked Bordes to allow me to look at all of the analyses of the different assemblages from Combe-Grenal. Bordes had the full analysis of each assemblage on a large card and he recorded each of his 63 different types. He had them all in a box by layer numbers, layers 1 to 55. I spent a few days together with Desmond Collins transcribing nearly all the information from those sheets and I got all the data from all Bordes's sub-layers at Combe-Grenal on the understanding that I would never use that without

> receiving permission. Bordes was immensely generous to let me have access to that material. So, I had all that data from Combe-Grenal and the sequence I was beginning to formulate was beautifully represented in the 55 levels at Combe-Grenal. Amazing!!

Paul was surprised that Bordes had never spotted this; that such an obvious chronological scheme could have been missed was puzzling. 'It never struck him', Paul stated with a certain amount of frustration during an interview. Nic David remembers that Paul, as a young Englishman, was a bit nervous about going 'up against Bordes'.[36] But, he also realized, and as Sackett (Laville *et al.* 1980) later explained in painstaking detail, that Bordes primarily trained as a soft-rock geologist, was completely convinced by the complex scheme of intercorrelations of sites based on sedimentological evidence; a climatic sequence had been reconstructed from the stratigraphy and sedimentology rather than from the typology.[37] These results provided Bordes with what he thought was a detailed, accurate chronostratigraphy for southwestern France.

In addition to looking for a chronology, Paul was also interested in applying Sackett's and Philip Phillips's idea of seriation changes within particular traditions. Paul thought he could put the different industries into a sequence according to gradually changing frequencies of different attributes.

> I had the idea of seriation very firmly in mind; the Quina Mousterian occurred immediately above the Ferrassie Mousterian. So again I started gathering all the data on percentages of transverse side-scrapers, percentages of pieces which retained no cortex and of straight-edged forms, etc. I plotted the results.

With the information that Bordes had given him, Paul discovered evidence of seriation, typological trends, 'a gradual decrease in the frequencies of Levallois flakes, faceted striking platforms and blades from Ferrassie to Quina Mousterian'.

> About this time, I became vaguely aware that someone in America was working on the French Mousterian; my first reaction was that he would come to exactly the same conclusion as I did so I rushed out the paper in *Nature* in 1965, pointing out a clear sequence of chronological patterning with seriation within traditions. As history has revealed, Lewis Binford had come to exactly the opposite conclusion. The obvious explanation to Binford was that the same people were doing different things, at different sites, either in different seasons or different activities or different groups of people, males or females, doing different activities using different tools at different frequencies. That could explain the five variants.[38]

'Time passed with virtually no one noticing my 1965 *Nature* article', Paul continued in a later interview.[39]

> A few people saw the subsequent 'The chronology of Mousterian industries in the Perigord region of south-west France' published in the *Proceedings of the Prehistoric Society* in 1969, and my 1970 *World Archaeology* article 'Some comments on the notion of functional variability in stone-tool assemblages'. I was still subdued in my whole approach and almost apologetic about what I was doing. I felt diffident. I felt that the French totally ignored everything I did.[40]

This reaction of the Bordeaux University researchers did not surprise observers. As Harvey Bricker explained,

> Paul's arguments were rejected because they could not be made to fit with the inter-site geological chronology being developed by Bordes's student, Henri Laville. When it came to a choice between accepting Paul's view of things versus Henri's view, it was a no-brainer. The Bordes [Francois and Denise de Sonneville-Bordes] were so convinced of the validity of Henri's approach that they seemed to regard Paul's model as almost an affront. Consider what they said on pages 65 and 66 in their (1970) paper 'We shall try here to dispose, quite definitely we hope, of the antiquated hypothesis that the different types of Mousterian represent an evolution, a hypothesis recently brought forward once again in the face of the most flagrant contradiction by stratigraphical data.'

Harvey Bricker continued:

> This quote does not speak directly to Paul's model, but it shows a total impatience with his sort of approach. I don't think they saw anything methodologically unacceptable in Paul's work; it was his results that could not be allowed. Paul's chronological model, or at least much of it, was not accepted until Henri's approach was blown out of the water by chronometric dating.[41]

Jim Sackett also suggests that non-scientific factors underlie the Mousterian debate and that Paul was somewhat naive and failed to understand fully those factors. It was possible, for example, that the Bordeaux school 'did not appreciate a young Brit who had little experience in rockshelters telling them they were wrong'.[42] In addition, Jim observes:

> The extent Bordes really wanted parallel phyla is questionable. Between Art Jelinek and myself, Art feels he really did believe in 'tribes', whereas I believe that one has to understand the subtler background of French thinking about systematics. I'm not so sure. Bordes told me he originally assumed that the four Mousterian types represented seasonal variants of one and the same culture. However, the 'debate' with Binford hardened Bordes's attitude, as Binford was aggressive and combative and Bordes always took the opposite side whenever anyone disagreed with him.

Also

> Bordes's assignment of individual strata assemblages to one of the four Mousterian industries was often

pretty iffy, given either assemblage size, assemblage make-up itself, or stratigraphic ambiguities.[43] Ultimately, he realized, perhaps correctly, that there weren't four industries at all and that his quartet had simply schematicized a much more complex scene of industrial variation.[44]

However, Jim continued:

> the debate made Bordes famous in the USA, which was very important to him. He occasionally suggested to me where USA might put its missiles but he nonetheless always wore a cowboy hat and bola string tie and hugely enjoyed the admiration of American students. He was known to say that the debate made him well known in a country where people couldn't tell an end-scraper from a handaxe.

It is significant that Bordes rarely talked or wrote about the debate in France.[45] There was a general feeling among others whom I interviewed that, although the Bordes/Binford debate could not be said to have been 'staged', nevertheless Bordes and the Bordeaux centre greatly benefited; the Bordeaux traditional concern with industrial variability was successfully projected onto a world stage.

Jim Sackett also suspects that some continental scholars in the 1960s may have held a slight natural antipathy toward the 'Anglo-Saxon' who could be viewed as insular, culturally blinkered and unaware of the European research. It was also well known that Francois Bordes and Charles McBurney, Paul's PhD supervisor, thoroughly disliked each other.[46] Bordes's silent reaction to Paul's chronological results therefore seems, in retrospect, to be understandable.

In contrast to Bordes's reaction, Binford quite quickly and publicly recognized and debated Mellars's chronological arguments. He (Binford 1973, 231) accepted that there was evidence for temporal sequence but suggested that he did not 'anticipate an exclusive sequential arrangement of all the variability' and did not like the evolutionary suggestion that one form of assemblage was ancestor to another. When Colin Renfrew finally introduced them in 1971 at the University of Sheffield 'Explanation of Culture Change' conference, Lewis Binford and Paul enjoyed each other's company and have remained great friends ever since.

The Sheffield years, 1970–1980

Paul was awarded his PhD in 1967 with John Coles and Roy Hodson, who had helped with the statistics, as Examiners. From 1965 to 1968, Paul had a Research Fellowship at Sheffield University and then took up a Sir James Knott Research Fellowship at Newcastle University, after which he was offered a newly-established position to teach Palaeolithic archaeology at Sheffield. This started in October 1970.

Figure 2.6. *Paul in 1965 at the English Lower Palaeolithic site of Purfleet, pointing out evidence of cryoturbation deformation at the top of the chalk. (Photograph courtesy of Harvey Bricker.)*

> It was Cambridge in the coalfields [remembers Paul]. All the prehistorians at Sheffield were ex-Cambridge, Warwick Bray, the first, helped to set up the teaching, then Colin Renfrew, Andrew Fleming, Colin's wife, Jane and myself came in quick succession. It was still a broad ranging department with some prehistorians, some classical archaeologists and classical historians. I started teaching courses, first-, second-, third-year courses on the Palaeolithic and offering tutorials. The number of students doing archaeology at that time was small; we could teach them in groups of three or four much as the Cambridge system. There were girls and boys from working-class backgrounds, rather timid, grey compared to Cambridge students. But, when you got to know them, they brightened up.

Paul had excellent PhD students while at Sheffield. According to Professor Keith Branigan,[47] who was at

Figure 2.7. *Anny Mellars in an old car, 1965. 'At the beginning of Paul's career, he and Anny did not have a lot of money and made do with used cars.'[48] (Photograph courtesy of Harvey Bricker.)*

Sheffield at the time and as we know from university records, some of these were Mike Wilkinson who worked on otoliths from the Oronsay middens, Margaret Deith who investigated seasonality as revealed by Oronsay limpets, Christine Williams who produced the thesis *Palaeoecological Studies of a Mesolithic Landscape in the Central Pennines*, David Jones whose PhD examined shellfish remains at Oronsay, Richard Nolan who wrote a PhD thesis on the spatial analysis of the Mesolithic shell midden, Cnoc Coig and Derek Sloan, who came to Cambridge to work with Paul on the re-excavation of the Star Carr area in the late 1980s. Sloan was awarded a Cambridge PhD for work on Scottish shell-midden economies in 1993.

Paul also remembers Alan Turner, Nicholas Ralph, Sally Reinhardt and Katie [Katherine] Boyle. Katie later migrated to Cambridge to become one of Paul's first Cambridge PhD students. And, in 1970, 'there was Clive Bonsall who was incredibly bright'[49] as Paul's first Sheffield PhD student. 'He had originally gone to Cambridge to study geography', Paul remembers. 'When he got there, he decided on archaeology.' But his request to transfer was declined, 'so he gave up the place at Cambridge and applied to Sheffield to read prehistory and archaeology'. Half way through his PhD, Clive, who was one of the Sheffield students who remained successfully in archaeology, was offered a good position at the British Museum; he remembers Paul strongly advising him to take it. Before leaving for that job, Clive stated that he acted as site supervisor at the Oronsay excavation for two seasons in 1971 and 1973.[50] In remembering that work together, Paul continued

> That takes the story up to the start of the Oronsay work. I did the first excavation there in collaboration with Sebastian Payne whom I met in Cambridge. Sebastian had worked with Colin Renfrew and had been a member of the Eric Higgs team. He was especially interested in how much extra small material could be recovered from fine wet sieving. He was obsessive about that.[51]

> One day, we were talking about the possibility of doing wet sieving on shell-midden sites; I knew about the shell-midden sites on Oronsay; I had, when at Newcastle, published an article on an antler harpoon-head from County Durham which suggested a link to the Scottish Obanian (Mellars 1970b). That fired my interest, [Paul continued] so, we decided to have a look. In Easter of 1970, Sebastian, Sebastian's wife, Rosemary, baby Polly and I drove to western Scotland; somehow we got across to Oronsay at low tide. We re-discovered the famous middens that had been dug in the nineteenth century and found a new

> midden; it was immediately obvious that there was potential for new work. Sebastian thought it would be a fantastic site for sieving because we might find fish bones.

This was a prophetic thought. Oronsay is now remembered by generations of students for its innovative analysis of fish otoliths as indicators of seasonality.

At that time, Paul was involved with teaching at a summer school in archaeology for American students organized by a Cambridge-based body, the Association for Cultural Exchange. The residential courses were held in one of Oxford's Colleges and since 1967 Paul had been running the course.

> After we had given the students three weeks of theoretical archaeology we sent them off on digs. So I suggested that I take 10 or 12 students to Oronsay. The Association of Culture Exchange therefore funded the first excavation in July and August 1970. That is how the Oronsay operation started.[52]

Sebastian Payne remembers that first summer,

> I only dug with Paul once, and only for one season at Oronsay, in 1970. Paul was a very careful excavator, very conscious of the need to dig as little as possible in order to leave as much as possible for future archaeologists and was careful about not going beyond the available evidence.[53]

To this day, Paul prefers to excavate small square test pits, to take column samples, to examine intensely the material recovered and to preserve the site whenever possible.

The Oronsay excavation ran in alternate years until 1979 and was conducted against a backdrop of expansive debate and innovation in British archaeology paralleled in the USA by the New Archeology. Three volumes, in which Paul's articles appeared, published during the 1970s, testify to this perceived change. 'Archaeology today is changing with astonishing rapidity' Colin Renfrew (1973, ix) observed. Again in 1974, Renfrew wrote:

> Britain's past has changed in the past few years almost beyond recognition: the new datings, new discoveries and new assessments have come so fast that any survey written more than five years or so ago is inevitably out of date. (Renfrew 1974, xi)

In 1978, Paul (1978a, vii) commented, 'In common with most areas of prehistory, research into the earlier stages of postglacial occupation in northern Europe has advanced at an impressive — almost alarming — rate during the last decade.' 'To speak of a "crisis" in archaeology', Renfrew (1973, ix) stated, 'or a "methodological revolution" is now a commonplace.'

When considering the grand shifts in excavation technique which occurred in the 1960s and 1970s, Robin Dennell commented:

> The days of big English excavations with scores of workmen in foreign climes were over by the 1960s so there was an enormous incentive to get more bits of data per pound per cubic metre; that was the tremendous incentive that led to the development of the seeds and bones revolution.[54]

Sebastian Payne built a version of the water-sieving machine that he, David French, Andrew Sherratt and John Watson had developed and introduced at Can Hasan 3 in Turkey in 1969.[55] Bags of sieved remains and column samples were brought back from Oronsay to Sheffield. Over several years, Anny Mellars meticulously sorted the washed material in her kitchen.[56]

Paul published the first results of the Oronsay work immediately (Mellars & Payne 1971) and then added further analyses over the following years (1978b; 1979; 1981; Andrews *et al.* 1985; Mellars & Wilkinson 1980), culminating in the interdisciplinary and multi-authored, team-produced, *Excavations on Oronsay: Prehistoric Human Ecology on a Small Island* in 1987. 'The remarkable concentration of Mesolithic shell middens on Oronsay' turned out to be 'unique within the context of the British Mesolithic and can be paralleled from only two or three other localities in Europe' (Mellars 1978b, 371). The material was well excavated and properly preserved so that the remains are still available for on-going investigation. The skeletal material has recently been reanalysed and may show evidence of ritualistic behaviour (Meiklejohn *et al.* 2005).

In retrospect, Sebastian Payne notes the Oronsay shell middens were especially suited to Paul's archaeological background and to the era. Paul had learned to excavate on sites which could be contained and could be treated as bounded entities and where restrained ecological interpretations were especially productive.[57]

In a recent interview, Paul reaffirmed his indebtedness to Jim Sackett's 1962 tutorials, stating that the work at Oronsay and the title of the publication were influenced not only by Jim but also by Julian Steward's concept of cultural ecology, the 'rage in the early 1960s'. According to Paul, Steward had explained

> the differences in some aspects of hunter/gatherer society and behaviour in terms of adaptation to the environment. I think that everything that I have done since then is in one form or another cultural ecology. Ecological thinking was a formative influence on the 'New Archeology'.[58]

During the late 1960s and early 1970s, Paul was exposed to the 'New Archaeology' through reading Binford.

> I became a committed processualist during the 1970s; it put emphasis on the kind of information I think we

> have access to — technology, subsistence, seasonality, food supply, demography, settlement, environment and the relationship between society and environment. These are the things that archaeology can do well. And, processualists also emphasize the explanation of change. It seems to me, if archaeology can get at anything, it is change.[59]

The 'Explanation of Culture Change' conference, held in 1971 in Sheffield, was Colin Renfrew's creation.

> Its aim was to bring together archaeologists, anthropologists, geographers and contributors from other fields ... in order to discuss afresh both the aims of archaeology and ... new approaches and methods of research. (Renfrew 1973, ix)

The conference was one of a series founded by Peter Ucko and Peter's constructive suggestion that the papers be pre-circulated[60] made the conference especially exciting and productive, resulting in a large and ever-popular textbook. Many observers have vivid memories of the combustive, almost violent, confrontation between Edmund Leach, who argued that archaeologists could not reconstruct the internal organization of prehistoric social systems and Binford who optimistically claimed he could.[61] But Colin Renfrew also remembers Bordes and Binford as lively 'interlocutors'.

'Binford, accompanied by his wife, MaryAnn, came to stay with Jane and me during the conference' remembers Colin. 'I had come to know him quite well during my semester at UCLA in 1967.'[62] Shortly after their arrival and before the conference, the Mellars invited the Binfords and Renfrews to tea at their beautifully restored home in Eyam in the Peak District.

> It was a rather special occasion [Colin wrote in a recent letter]. [It] set the scene for a very pleasant social atmosphere. There is no doubt that the trip into the country from Sheffield, and Anny's excellent tea, made this into a very agreeable excursion for all concerned and helped keep the later Mousterian interlocutions on a cheerful level.[63]

Colin further recalls, 'Paul had a good control of his data and was in a position to talk seriously to Binford.' When Lewis Binford became too polemical, Paul could hold 'him up short with a firm statement. Binford quite rapidly realized that Paul knew more about the stratigraphical sequences than he did.' Both Colin and Paul suggest that the 'Explanation of Culture Change' conference offered Paul, for the first time, a platform and opportunity personally to present his hard-earned knowledge before a large, international audience.

One of the conference sessions was entitled the 'Explanation of Artifact Variability in the Palaeolithic' with Bordes, Binford and Paul speaking. Although Bordes and Binford continued their Mousterian variability debate, Paul (1973) decided to present his first synthetic, overview paper on a new subject, 'The Character of Middle–Upper Palaeolithic Transition in South-west France'. In hindsight, if Bordes had been more receptive to Paul's chronological analysis, Paul may have remained stuck into the Mousterian debate but, given the circumstances, he was free to move quickly to new intellectual territory.

> I didn't think that I had much else to say about the Mousterian. I used to make lists of possible future papers and noticed that everyone was paying lip service to the Middle/Upper Palaeolithic transition as a major dividing point.[64]

It may be difficult to believe today, but in 1970, Paul states, 'There were all these stock generalizations in the textbooks but no one had analysed the archaeological evidence in detail.'[65] No one had yet questioned the idea of a major change and there was no thorough analysis or summarizing documentation. Paul found this to be surprising

> in view of the universally acknowledged importance of these developments. [My] primary aim ... is therefore an empirical one: to examine closely the evidence from one particular area in an attempt to define what the true pattern of cultural innovations at the time of the Middle–Upper Palaeolithic transition was. (Mellars 1973, 255)

Paul's paper went beyond mere summary of existing opinion and scholarship; he searched for the gaps in the current evidence, adding the results of his own research and analysis. For example, textbook generalizations in 1970 generally stated that Middle Palaeolithic implements were manufactured from flakes and that the Upper Palaeolithic tools were made on blades. But, on close reading, Paul found that this was an oversimplification and misleading. 'The most original feature of Upper Palaeolithic stone-tool technology lies in the shapes of the finished tools rather than in the techniques of manufacture' (Mellars 1973, 257). He also looked at comparative evidence on fauna specialization, number of sites, gross area of the sites and seasonality, each time pushing the limits of what was then the received knowledge. When considering Paul's research, Colin observed, 'He has a good sense of problem and feels things in a strong way. He broods over the problem, thoroughly and carefully reads around it and redefines it for himself.'[66]

Paul considers this work to be a turning point in his career which has defined his interests ever since. This 1973 Middle–Upper Palaeolithic transition summary analysis is now seen as an iconic paper which expresses the received knowledge of its day. 'The first real synthesis of a broad range of data pertaining to the Middle–Upper Paleolithic transition was that of Mellars (1973) for southwestern France,' wrote Randall White (1982, 169). 'Mellars article is well organized and cogently written. It therefore makes an effective baseline from which debate can proceed.' Paul's article is viewed as an intellectual watershed which marks the beginning of the ongoing debate about the nature of this transition.

The painstaking presentation of all available evidence and literature in analytical and synthetic articles is now Paul's trademark. As one of his former PhD students, Laura Basell, recently wrote, 'He writes excellent big picture papers with full credit to the relevant authors and has always kept up to date with the latest developments. The criticism is that anyone can do that but in fact they don't. Clarity in thinking and writing makes the reading easy but it is a real skill.'[67]

Paul believes that synthetic, analytical articles, more than books, are the best tools to further the development and spread of knowledge. The best scholarly way to create knowledge is to produce sharply focused descriptive and detailed, concise analytical studies. Paul does this, for example, in his ever-expanding list of articles on the Middle–Upper Palaeolithic transition. The theme and purpose of these papers remain the same but the data presented changes and expands. In these 'review' articles, Paul follows an historical model of explanation.

'I wrote quite a few general, synthetic papers in the 1970s on industrial variability, chronology, new radiocarbon dating methods, British Mesolithic settlement patterns and industrial variability and fire ecology' (Mellars 1974; 1975; 1976a,b; Jacobi *et al.* 1976; Mellars & Reinhardt 1978). Some of these were especially important in the era. The 1974 volume, *British Prehistory: a New Outline*, for example, edited by Colin Renfrew, was 'the first time a calibrated radiocarbon chronology became available for Britain.'[68] Several of these articles still figure prominently in undergraduate reading lists.

Considering the important intellectual changes that took place at the end of the 1970s and in the early 1980s, with the introduction of what became known as 'postprocessual' approaches,[69] it is well worth looking in more detail at one of these papers. Paul's research into fire ecology illustrates the dominance yet complexity of ecological thinking in the era immediately before the literary turn of the early 1980s.

'I was invited by Gene Sterud to spend one semester in 1974 at The State University of New York at Binghamton', Paul remembers.[70] Rhys Jones, who was a year behind Paul at Cambridge, had already introduced him to the Australian ethnographic evidence of 'fire-stick farming' which Aboriginal people used to encourage the growth of new vegetation and attract game. Ian Simmons (1969) had already published his 'Evidence for vegetation changes associated with Mesolithic man in Britain' in which he found evidence of burning. G.W. Dimbleby, (1962) Professor of Human Environment at the Institute of Archaeology, London, a specialist in forest ecology and soil formation, had published his classic study *The Development of British Heathlands and their Soils*. Dimbleby argued that the present heathlands were not natural, that they had all originally been forested and that the deforestation was an artefact of human activity during the Mesolithic. He suggested that fire could have been a major factor in this.

Coincidentally, Paul had access, at Binghamton, to the library at Cornell, a university known for its strength in forestry and ecology; Paul spent considerable time in an 'incredible library' and read widely an enormous amount of work on the impact of forest fires on the regeneration of vegetation and the beneficial impact on game populations. 'I suddenly discovered this massive literature. I read that avidly. It was very exciting, all relevant to what was being discussed in Britain concerning the Mesolithic.'[71] In his resulting article, 'Fire ecology, animal populations and Man', (Mellars 1976a, 41), Paul went beyond description and summary of the available evidence. After a detailed analysis of the ethnographic and Mesolithic pollen evidence, he argued,

> You could see how the use of burning might have turned hunting strategies into husbandry, herding, pastoralist economies. Burning allowed an increase in the health, numbers and birth rate of the animals and allowed humans to control the distribution of the animals and how they moved. It allowed people to a put a territorial stamp on the environment; if you had burned a bit of land, you had invested work in improving that land. You could perhaps claim ownership. I suggested that this could lead to social and economic, structural change.[72]

The processualism of the 1970s is characterized as functionalist and behaviourist and concerned only with problems of economic subsistence, settlement patterns and demography but with Paul's human ecology work from that period, we have an example of an attempt to explain the emergence of new social values and the importance of human agency, themes that would become popular during the 1980s.[73]

With good students, research opportunities, a pleasant Department milieu, promise of promotion and a beautiful home, there were few incentives to leave Sheffield. As Colin Renfrew later recalled,

> I do remember great discussions with Paul as to whether he should come to Cambridge or not because I was very keen that he should. It was all very open to doubt because he was well placed in Sheffield and then, of course, he did decide to come and that all worked out extremely well.[74]

Cambridge, 1981–1990

The late 1970s and early 1980s were an exciting time in Cambridge. Ian Hodder organized a series of postgraduate 'think-tank' seminars to consider alternatives to the New Archaeology; this was in reaction to what was perceived as Binford's polemical, reductionistic and oversimplified application of functionalist and behaviourist approaches. Hodder and others were responding as well to the naïve use of the deductive-nominological (DN) model of explanation. As Hodder (1982, 14) stated, 'I have tried to show that the New Archaeology can be extended by reconsideration of the issues outlined by traditional and historical archaeologists, and that culture, ideology and structure must be examined as central concerns.' A conference in 1980 on symbolism and structuralism in archaeology resulted in the edited volume, *Symbolic and Structural Archaeology* (Hodder 1982); this book now marks the beginning of postprocessualism.

Henrietta Moore, who attended that conference and who was then writing her PhD thesis, 'Men, Women and the Organization of Domestic Space among the Marakwet of Kenya' (1983), recently remembered those years.

> I was a student with, amongst others, Danny Miller and Christopher Tilley. It was a time of immense intellectual optimism in Cambridge. We were discussing a kind of social theory that created space for human agency, for interpretation, for a creative practice. The feminist activism that I was involved in was an enormously important part of the way I approached my work ... I was interested in the questions of political economy, what difference systems of production and reproduction make to the way people lived their lives and how men and women are differentially placed in their system. Of course, we had Tony Giddens on our doorstep; we talked about human intention, motivation and much about the great problem of social theory, the relation between structure and agency. I developed an interest in Pierre Bourdieu [who was by then moving beyond French structuralism].[75]

Alison Wylie, who also attended the Cambridge graduate seminars, remembered the

> extraordinary group of graduate students, a pantheon of people who have since become key players in archaeology. Hodder was critiquing what he referred to as functionalist, adaptationist conceptions of material culture. He was insisting that material culture couldn't be understood only in terms of manifest behaviours and technical functions of material artefacts. In order to understand cultural process, it is necessary to attend to the symbolic structures. If you are going to make sense of cultural process, you must understand the inside of actions, the cognized worlds, the motivations, the symbolically-rendered environments in which people move. [It was obvious] that the New Archaeology was ill served by its positivist rhetoric. Attention to the symbolic dimensions of cultural action in the past [would require a realist philosophy of science].[76]

Throughout this lively and at times derisive debate, Paul remained faithful to his preferred ecological stance, arguing that it was more productive than postprocessual approaches for Palaeolithic evidence. Although he agreed with Wylie's argument that realist philosophy best described how science proceeds, he suspected that the social theories discussed by postprocessualists could be applied more effectively to recent periods of prehistory. Paul, with characteristic caution, felt that early postprocessual approaches relied too heavily on speculation and imagination and that people were taking too many risks with interpretation.

> We would all love to know the meaning of things [he explained], but we don't have direct access to that! The greater part of variation between Palaeolithic assemblages can be understood by reference to the environment.[77]

Paul often reiterated in interviews, 'Palaeoecological approaches put emphasis on demography, the sizes of groups, seasonality, mobility, data we can get access to.' He would ask with a hint of frustration, 'Why did processualism become treated as a swear word!?'[78] Paul seemed concerned by what he saw as intellectual intolerance and snobbery driven by the new post-processual approaches.

Although concerned, therefore, about post-processualism, Paul was obviously more impressed by the scientific innovations which became available at the same time as postprocessualism emerged. Robert Hedges, who has just been awarded the Royal Society's Royal Medal for his work on accelerator mass spectrometry C14 dating, received his first large SERC grant to build the Oxford Radiocarbon Accelerator Unit in 1978.[79] The major advantage of the new technique was its ability to measure extremely small samples allowing large numbers of closely stratified samples to be dated. The ability to extract specific chemical components of the samples also reduced the possibility of contamination (Mellars *et al.* 1987).

Paul had always accepted the clear value and necessity of radiocarbon dating. As early as 1969, he sought and published dates of bone collagen samples

Figure 2.8. *Paul in 1983 providing scale for a wall of flint nodules at Brandon, Suffolk. (Photograph courtesy of Harvey Bricker.)*

from a new Cresswellian site (Mellars 1969b). Hedges remembered 'Paul as being a strong (!) champion of the Unit in its early days'.[80] Thus, in the early 1980s, Paul suggested to Colin Renfrew, who was then Chair of the grant's steering committee, that the Unit date the Upper Palaeolithic in southwestern France. The Abri Pataud was ideal in that it was well stratified and excavated; also the whole sequence had previously been thoroughly dated by conventional radiocarbon methods during the 1960s at the Groningen Laboratory. The Unit offered to measure a series of 30 to 40 samples to find out how the two sets of results compared. 'To get those samples I collaborated with Harvey Bricker who had been the site director.'[81] When informed, Harvey responded, 'With 30 to 40 dates we can date the hell out the site,' which they did.

During a frigid December and January 1984, Harvey and Paul obtained samples from the Abri Pataud and eventually also from other Upper Palaeolithic sites which produced 'a beautiful series of dates'. The radiocarbon accelerator dates from the Abri Pataud fitted well with the Groningen dates, confirming both; these samples together with those from other sites provided a 'secure radio carbon chronology of the French Upper Palaeolithic'.[82] 'The Upper Palaeolithic project was an early and important demonstration of where the new method could be powerfully applied',[83] commented Hedges in a recent message. Paul hoped that the new scientific techniques would resolve the 'complex issues surrounding the chronology of the Chatelperronian and the earliest Aurignacian' (Mellars *et al.* 1987, 132); however, these issues would continue to haunt him.

At about the same time, in the summer of 1981, Paul had been a Visiting Research Fellow at The Australian National University in Canberra, just after Peter Ucko left the Institute of Aboriginal Studies to take up his position as Professor at Southampton. At ANU, Paul heard many stories of Peter Ucko's 'radical', abrasive, confrontational behaviour as an advocate for Aboriginals and their control of native archaeology. Immediately upon his return to England, Peter was invited to be National Secretary of the British Congress of the International Union of Pre- and Protohistoric Sciences (IUPPS). According to Peter,

> I agreed to organize the conference on condition that it would take as its most serious commitment the full participation of the countries of the Third World [and indigenous peoples], to end the European domination of previous Congresses. (Ucko 1987, 3)

By all accounts, Peter's proposal to broaden the IUPPS' mandate and expand the inclusiveness of the next congress, to be held in England in 1986, was greeted with approval and considerable enthusiasm. There was general agreement that recent IUPPS' congresses may have been poorly organized and that the IUPPS was losing touch with a growing international archaeological community. The value of Peter's vision was well acknowledged and he felt that there was a 'genuine desire to make the next IUPPS Congress into a truly world Congress' (Ucko 1987, 8).

One of the sub-themes of the forthcoming World Archaeological Congress was 'Archaeology and the Origins and Dispersal of Modern Man' organized by Paul; this theme had interested Paul since his early 1970s research on the Middle to Upper Palaeolithic transition. In 1982 and 1983, however, when Paul first suggested the topic, it was only beginning to attract the attention of British and North American colleagues. As Paul pointed out in his 1973 article, there had been some work by French scholars, but, when considering the English-speaking audience, Randall White (1982, 169) could state, 'Despite its importance, the Middle/Upper Paleolithic transition in Western Europe has been the subject of very little informed debate.' At this early stage, Paul and Randall White thought that the emergence of modern man was best documented by evidence of the transition to the Upper Palaeolithic in southwestern Europe.

Peter Ucko (1987, 34) remembered that Paul seemed ponderous and cautious at first but 'From 1983 onwards, he did much more than just the required letter-writing and took the whole enterprise [of the WAC] very seriously; he and I were to become close friends.' For his part, Paul remembered Peter as 'so charismatic, arguing his ideas with such force. He was an incredible entrepreneur, had incredible energy, drive and passion for whatever he did.' Paul admired Peter and found him 'a good man to work with'. 'Ucko used to send me off as his delegate ... to Paris once to talk to Henry de Lumley about a Lower Palaeolithic session.'[84]

'Everything was going swimmingly', when in late 1983, the Pan African Association on Prehistory and Related Studies passed a motion censuring all contact with colleagues and institutions in South Africa. Then, as a State of Emergency was declared in South Africa in 1985, and hundreds were detained and many were murdered, the worldwide Anti-Apartheid movement gained momentum. The UN, Commonwealth and EEC called for sanctions. The ANC proposed that South Africa be excluded from cultural and sporting events, UNESCO banned cultural interaction with South Africans and the regime and then, after a complicated, emotionally intense series of events, the British Executive Committee voted to support an academic boycott of South Africa and exclude South African delegates from the upcoming WAC.

'All Hell broke loose.' Paul remembered intense media coverage and appalled, passionate letters from would-be participants, especially Americans 'How can you do this! You should disassociate yourself! Support the principle of academic freedom!!' Paul recalled 'personal attacks', 'immense psychological stress' and then 'people began pulling out under the pressure'. Ezra Zubrow, who lived through the McCarthy era in the States and who witnessed academic lives destroyed by political interference, explained, during an interview, that most Americans were passionately against banning South Africans because they interpreted this as a dangerous move against individual and academic freedoms. Thurstan Shaw, on the other hand, who had worked for years in Africa, accepted the ANC's, UNESCO's and the Pan African Association's request for support. He argued that the higher hope of freedom for all South Africans must take priority over European and North American conceptualizations of individual freedom. Ironically, if Thurstan had advocated on behalf of his South African colleagues who might have wished to attend, his students and friends from the University of Ibadan would have refused or been unable to attend. The world archaeological community was thus clearly, dramatically and *very* publicly split and, in interviews, painful memories persist.

As participation in his session dwindled and after long discussions with his co-organizer, Clive Gamble, and a memorable, agonizing meeting with Peter,[85] Paul decided to withdraw and to plan an independent conference on the emergence of modern humans in 1987 in Cambridge. According to Peter (1987, 132), Paul and Clive remained helpful to the organization of what became the first WAC. This was held successfully, independently of the IUPPS, in Southampton from 1 to 6 September 1986.

Paul's decision to hold a separate, larger interdisciplinary conference was not politically or personally motivated. Paul did not participate in the 'freedoms' debate. His over-riding concern was that discussion on the emergence of modern humans should proceed and he realized that he would need a proper venue and broader stage.

It should also be remembered that Allan Wilson's (Cann *et al.* 1987) revolutionary research on mitochondrial DNA (mtDNA) as 'a source of new perspectives concerning the evolutionary history of our species and the genetic relatedness of human populations' was then, in 1985, just becoming available and known.

> We proposed that all mtDNAs in modern human populations are descended from a single common ancestor who lived in Africa some 200,000 years ago (Stoneking & Cann 1989, 17).

Chris Stringer remembers

> When things went pear-shaped at Southampton, it was Paul's idea to rerun the sessions now wrecked by

> withdrawals ... I was already in correspondence with [Allan Wilson] about his latest results in early 1986, having heard about them from Roger Lewin and Peter Andrews, who had attended the Cold Spring Harbor meeting with Wilson. However, he [Allan Wilson] refused to participate in the Cambridge plans because he accepted [the] view that taking part would undermine the Anti-Apartheid movement. Nevertheless, he said that he would not oppose my proposal to invite Mark Stoneking and Becky Cann. I told Peter Ucko that the Cambridge meeting might be of great benefit to science and to Africa.[86]

'The new line of evidence [was] extremely healthy for science', stated Chris in Roger Lewin's report on the Cambridge meeting for *Science* (Lewin 1987, 1292). 'We decided to do this [meeting] on a worldwide basis. We had everybody,' Paul remembered. 'I was very grateful to Paul and Chris for their courage in inviting me to the first conference in the political climate of academic boycotts' wrote Hilary Deacon, Professor of Archaeology at Stellenbosch University.[87] Paul concluded, 'The story we had to tell to the world about modern human origins was tremendously important for all people.'[88]

Stan Ambrose, in a recent message, wrote, that the 'Origins and Dispersal of Modern Humans' meeting, which took place from 22 to 26 March 1987,

> happened at a crucial time when most scholars of human evolution were first confronting serious scientific evidence which challenged the idea that modern humans and modern human behaviour developed first in Europe.

Academics began to realize, 'that we should look to Africa ... it set the tone for a new understanding'.[89]

Roger Lewin (1987) agreed that the conference offered an unusual opportunity. In his review of the conference proceedings, Roger noted,

> Until relatively recently there was a strong sentiment among anthropologists in favor of extensive local continuity. In addition, Western Europe tended to dominate discussions ... In fact ... if it is the origin of modern humans you are interested in, then Western Europe is something of a backwater. One of the things that is becoming clear is that the real action was elsewhere (Fred Smith as quoted by Lewin 1987, 1293).

Lewin (1987, 1295) concluded, 'Overall, the Cambridge meeting probably further tilted opinion toward the idea of replacement' and an African origin.

To some Africanists, however, the tilt was insignificant when considering the archaeological as opposed to the interdisciplinary papers. Hilary Deacon wrote,

> The archaeological portion of the first conference was really about the replacement of Neanderthals by *Homo sapiens* and how dumb the Neanderthals were and how Upper Palaeolithic newcomers were seen as the first undoubtedly modern people. Another colonial remarked that the ideas she had heard *x* number of years ago as a student had not changed at all at the first conference. She and perhaps other non-Europeans were a little jaded at hearing about the wonders of the Upper Palaeolithic without any explanation of what that meant for other areas of the globe where the Upper Palaeolithic was not recognized as a leap in human achievement. It was a very much a Eurocentric conference focused on the Middle to Upper Palaeolithic transition. I was totally on a limb stating heretically that at more than 100 ka, people in Africa were in all respects modern ... The fixed idea was that the Mousterian and the Middle Stone Age were non-modern equivalents because they share a Levallois technology and, therefore, it needed an African Late Stone Age to initiate an Upper Palaeolithic. Paul was concerned with the dating of the Upper Palaeolithic showing it appeared earlier in the east than the west and famously saw the Upper Palaeolithic as symbolizing the revolution.[90]

The European focus of the archaeological presentations, published as *The Emergence of Modern Humans: an Archaeological Perspective* (Mellars 1990), was also noted by Michael J. Mehlman, who wrote in his review:

> This book, with its strong Eurocentric bias in subject matter, is really not about the emergence of modern humans; it is mostly about a regional changeover from Middle to Upper Palaeolithic that seems to be broadly penecontemporary with replacement of Neanderthals by anatomically modern humans in Europe. (Mehlman 1992, 731)

To be fair, it should be noted that both Paul and Chris Stringer disputed the charge of Eurocentrism, pointing out that six of the papers presented at the meeting were devoted specifically to the African evidence, four to the Australian evidence and that several other African and Asian specialists were invited. Chris wrote in a letter (25 August 2008):

> Maybe [these] comments are justified from the archaeological contributions but I would take a different view, given the people I invited on the biological side, and what they presented and published in volume I (e.g. all the genetic contributions favoured OOA). In fact I was accused by some of bias towards an African origin. It may be that the biological side was more open to the idea of a recent African origin, though there was fierce opposition from the likes of Milford Wolpoff, of course.

Scholars certainly did find that the broader interdisciplinary and international nature of the discussions opened intellectual venues. Marcel Otte noted that the conference, made me discover the English thinking and speaking world. It is totally different than the one in which we are accustomed to working in Europe! No doubt this is thanks to Paul who, in a way, belongs to both worlds. The first meeting opened my mind in a drastic way and, yes, provoked an inflexion in my research.[91]

'I don't think we could claim that this kind of meeting was the first,' remembered Chris Stringer. 'Wenner-Gren had run a number of conferences on that scale and breadth'. Also, Erik Trinkaus (1989) had organized a closed workshop in Santa Fe in 1986, which brought together Binford, Wolpoff and others.

But, I think that it was the first major international meeting that focused purely on modern human origins from all perspectives: behaviour; fossils; genetics; what we now call evolutionary psychology; and other related disciplines ... and it ... had a great effect on my thinking.[92]

Other attendees also remembered the conference as marking a crossroads. Hilary Deacon suggested that, since the conference, 'there has been a remarkable shift in thinking about where Paul's symbolic revolution fits into the picture. Paul too has changed his ideas.'[93] Geoffrey Clark states in his forthcoming tribute,

Attendance at the conference was a career-altering event for me. After the early 1980s, I shifted my geographical focus from Iberia to the other end of the Mediterranean (Jordan) and my theoretical interests from the Mesolithic back into 'deep time' ... [this] also resulted in a growing critique of Paul's construal of the Middle–Upper Paleolithic transition.[94]

And, John Shea commented:

I attended the 1987 conference as a grad student. I think that I was possibly the only pre-PhD on the speaker list, and thus both very honored to be invited and fairly intimidated by the company. Dr Mellars was a most gracious host who made this very junior researcher very comfortable ... being invited ... was a turning point in my professional career.[95]

In addition to the momentous DNA research which became available just before the conference was convened, Valladas and colleagues (1987; 1988) released the ground-breaking dating results from two critical Levant sites before the papers were published. A robust Neanderthal fossil from Kebara Cave was dated *c.* 65,000 bp (Valladas *et al.* 1987) and

Burned flints from strata containing the remains of anatomically modern-looking humans at the caves of Skhul and Qafzeh ... yielded dates between 80,000–100,000 bp ... Having early modern humans dating some 15,000 to 35,000 years earlier than their purported ancestors understandably causes some difficulty for unilinear models of human evolution. (Shea 1992, 81)

Thermoluminescence dating therefore revolutionized the then received view that the Neanderthal were necessarily ancestral to modern humans.

Looking back on the 1987 conference, Paul stated:

It was organized from Corpus Christi College, through my room [Paul recalled]. We had hardly any assistance administratively. Chris and I did everything ourselves, xeroxing papers, organizing coffees etc. It was incredibly hard work. We were both utterly exhausted and I hardly remember the conference itself ... but it all went off successfully.[96]

The conference resulted in a massive series of 55 papers and CUP said it was too big. So I approached Edinburgh Press who had published the excavations at Oronsay. Chris and I decided to make the big red volume the interdisciplinary one. I remember debating what to put on that cover of *The Human Revolution: Behaviour and Biological Perspectives in the Origins of Modern Humans* (1989). The late Neanderthal from St Cesaire in western France had been dated, as had the skull from Qafzeh, an early modern human. So I put these two iconic images on the cover. The spine was so thick, we decided to add the Venus of Willendorf; she filled the space very nicely.[97]

According to Paul, the big article he published in *Current Anthropology* in 1989 was, in hindsight

a very important article for me because it came in the wake of that conference. As I started to write the introduction to *The Human Revolution*, I realized that I wasn't writing an introduction but instead a massive article. So, I submitted the larger version to *Current Anthropology*. Adam Kuper was then Editor. I was dreading the review process. Having almost killed myself writing this incredibly comprehensive review, I would have to let a dozen smart alicks read it but Adam made a decision to get it out fast.[98]

The article, 'Major issues in the emergence of modern humans' (Mellars 1989) has since been widely used by students and other colleagues. Its content reveals that, as Marcel Otte and Hilary Deacon suggest, Paul had broadened and changed his thinking. In the article's first section, 'Population dispersal and replacement', Paul went way beyond any of his previous work on the subject. For example, he agreed with some Africanists that, 'the situation' in 'the extreme western fringes of Eurasia ... could no doubt be seen in certain respects as ... highly peripheral and atypical' (Mellars 1989, 354). Certainly Paul's concentration on origins and his detailed inclusion of material from Africa reflected the conference agenda. He accordingly expanded and greatly refined his now well-known definition of the 'human revolution' (Mellars 1989, 370–72). He included observations about 'behaviour adaptations' in southern Africa and also theorized as to why a period of 'essential equilibrium' could have been followed by such apparent strong changes in Europe, western Asia, and northeast Africa. When considering new evidence for behavioural change, Paul demonstrated a maturing recognition of the complexities involved.

> At the outset, it must be recognized that these problems are ... largely dependent on different theoretical approaches to the interpretation of the archaeological data. For example, exactly what criteria can be used to discriminate objectively between the products of 'hunting' and those of 'scavenging' in faunal assemblages? (Mellars 1989, 355)

Also, his augmented stress on the central importance of language, as a 'single development in human evolution which could, potentially, have revolutionized the whole spectrum of human culture and behaviour' and his questions about 'whether there were some significant shifts in the basic mental or cognitive abilities of human groups' anticipated his work with Kathleen Gibson in the early 1990s (Mellars 1989, 375, 377). Finally, his statement that a period of coexistence between the Aurignacian and Châtelperronian populations 'led to a significant degree of acculturation' was to remain central to his thinking.

As mentioned briefly earlier in this essay, immediately before the 'human revolution' conference, there had been a major development in Mousterian research. In the July issue of *Nature* 1986, Helene Valladas (1986), 'the finest chronometric worker in France ... an excellent physicist who had developed 'TL' dating methods'[99] published, with her co-authors, the now famous thermoluminescence dates on burnt flint samples from all the levels at Le Moustier. Reviewing the impact of this paper on the long-standing Mousterian chronology debate, Paul stated in an interview:

> Henri Laville believed, based his chronostratigraphy, that he could correlate the main Mousterian sequences in southwestern France, particularly the sequence at Le Moustier, with the much longer and more detailed 55-layer sequences that Bordes had excavated at Combe Grenal which had MTA at the top. He claimed that he could correlate the sequences of MTA at Le Moustier with Combe Grenal and that Le Moustier could be synchronized with the whole of the Ferrassie Mousterian and Quina sequences at Combe Grenal.
>
> When *Nature* published the Valladas *et al.* article, the editors asked me to contribute a 'news and views' article commenting on the significance of the dating results. What was absolutely amazing about this dating was that it showed that the entire sequence at Le Moustier only spanned about 15,000 years. Whereas according to Laville's and Bordes's interpretation it spanned the entire Mousterian going back to 70,000 or more years, exactly the same time length as Combe Grenal. In fact, the earliest layers at Le Moustier dated from 55,000 at the bottom of the MTA; the top layers dated about 40,000. This was a 'God send' to me! Of course, I hammered home the point.
>
> This proved that the chronology I had proposed in my PhD thesis was right!! That was a flashpoint!! That dating led to the collapse of Laville's scheme. It was a long time to keep faith with my early work but there was no way that so many stratified sites could be wrong! I had been fighting that battle for 20 years and it was over.[100]

Unfortunately as a result of time and space constraints, this biography must end here with the conclusion of the 1980s but Paul's life continued productively, beginning the new decade of the 1990s with a flourishing cluster of postgraduate researchers which included Mark White, Paul Pettitt, Dimitra Papagianni, William Davies and Nathan Schlanger. Referred to by students as 'Oncle', Paul began work on his *magnum opus*, *The Neanderthal Legacy: an Archaeological Perspective from Western Europe* (1996) which was to be brilliantly reviewed. His ferocious appetite for new material and solid, very difficult debates would lead him to the intense Neanderthal acculturation exchanges as well as the seriously taxing controversy concerning excavation methods at Star Carr and the vast over-arching issues of the origins of biologically and behaviourally modern human populations in Africa and their dispersal over all regions of Europe and Asia. In 2008, Paul was awarded the prestigious Grahame Clark Medal for Prehistory by the British Academy and, in 2010, a Knighthood 'for services to scholarship' in the New Year's Honours List, making his 95-year-old father proud and representing a long journey from the coal mines of Yorkshire.

When asked to contribute a concluding comment and fair evaluation of the early decades of Paul's life portrayed here in 'From Swallownest to Cambridge', his colleague, Clive Gamble, stated:

> Paul's biggest contribution has been that he [has often been] right. He was right about the vexed Mousterian variability question; he was right about the origins of modern humans; he was right to resign from WAC; and he was right about the significance of science in archaeology. It is an impressive record and most of us just follow his lead.[101]

Figure 2.9. *June 1987, Paul relaxing at Natchez-Under-the-Hill, Mississippi, following a visit to an archaeological site on the Natchez Trace. (Photograph courtesy of Harvey Bricker.)*

Acknowledgements

Thank you to Sir Paul and to the dozens of friends and colleagues who generously contributed their time and memories. It was my great pleasure to work with such a pleasant group!! The piece is dedicated to my husband, Thurstan Shaw, to commemorate his 96th year and to our great friend Jim Sackett.

Notes

1. *Preface:* For those unfamiliar with oral-historical work, a biography based on oral-historical evidence specifically values and concentrates on the individual's own narrative. This narrative is then used to augment written sources. This essay, therefore, offers a personal explanation of how and why Sir Paul Mellars produced his well-known publications. It fills in the details of his life and illuminates questions of motivation, influence and intention. This type of biography is not expected to be utterly objective and does not summarize or analyse Mellars's written material only. This is an intellectual biography based on detailed, new oral-historical interviews conducted over a number of years. The story has not been publicly available before and hopefully you, the reader, will find this type of history enjoyable, educational and valuable. Through Paul's experiences, we see aspects of the development of archaeology over 55 years. It is assumed that the reading audience has some archaeological background and some knowledge of Paul's published research. It should be mentioned that all quotations have been approved.
2. James Sackett, correspondence with author, 20 April 2006.
3. Paul Mellars discussing his Yorkshire childhood with author, 27 October 2006.
4. 'Paul Mellars has a knack for grasping and clearly articulating the "big picture": John Shea, correspondence with author, 31 March 2008.
5. When asked to name a memorable 'Paulism' Laura Basell responded 'And the 64,000 dollar question is ...': correspondence with author, 14 March 2008.
6. Harvey Bricker, correspondence with author, 6 April 2008.
7. Anny Mellars describing her husband's religious views, 25 March 2008.
8. Woodhouse Grammar School ex-student (1946–54), Mr Jon Layne, correspondence with author, 3 April 2008.
9. Paul Mellars, interview with author, 27 October 2006.
10. Paul Mellars, interview with author, 27 October 2006.
11. Paul Mellars, interview with author, 27 October 2006.
12. At that point, the Institute offered only a postgraduate diploma. As John Evans stated in an interview with the author, 17 August 2000, 'It wasn't until 1969 we were able to establish a first degree course. Some people on the staff were very much against it, like Kathleen Kenyon and Wheeler himself.' Archaeology was considered to be a postgraduate activity. Paul could not therefore have studied for a degree in prehistoric archaeology at UCL or the Institute in 1958.
13. The site includes poems as well as tributes, 'Paul Mellars was my DOS's name, yes grey bearded, of Neanderthal fame, Bumbling of walk, middling of gait. But rather wonderful and simply great.' copyright L.A. Tyrrell, 27 January 2006.
14. Tim Reynolds, interview with author, 17 March 2008.
15. Paul Mellars, interview with author, 27 October 2006.
16. Paul Mellars, interview with author, 3 November 2006.
17. 'Barry used to refer to him as Rik; the rest of us would call him Sir Mortimer Wheeler. We were enormously impressed by that.' Paul Mellars, interview with author, 3 November 2006.
18. The Cambridge undergraduate degree is a three-year course with a Part I and II; a perfect GPA in the North American system would be the closest equivalent to a First in the British system; in Cambridge, Firsts are rare indeed.
19. Derek Roe, who was also there, suspects that the unknown undergraduate is 'Edward' whom I cannot trace. (Roe, correspondence with author, 2 June 2008).
20. Nic David (described by former classmates as 'very dashing'), correspondence with author, 22 November 2006.
21. Paul Mellars, interview with author, 11 January 2007.
22. Paul Mellars, interview with author, 27 October 2006.
23. Nic David, correspondence with author, 15 May 2006.
24. Paul Mellars, interview with author, 3 November 2006.
25. Nic David, correspondence with author, 27 October 2006, 29 November 2006.
26. Jim Sackett, correspondence with author, 28 May 2007.
27. For an explanation of the beginnings of ecological archaeology in the 1930s, see my PhD thesis (Smith 2009) and Trigger (2006).
28. Paul Mellars, interview with author, 6 May 2008 and 20 June 2008.
29. Paul Mellars, interview with author, 3 November 2006. This was a hard decision for Paul. Since his school days the British Mesolithic had fascinated him and, as an undergraduate, he had investigated Deepcar, a Mesolithic site in Yorkshire. During 1962 and 1963, Paul was writing up these results which successfully challenged the long-held view that the Maglemosean was found only in eastern lowland England (Radley & Mellars 1964). Paul further explored this in 1974, finding that the broad blade and narrow blade assemblages were chronological phrases of an earlier and later Mesolithic.
30. Derek Roe, correspondence with author, 3 April 2008.
31. Derek Roe, correspondence with author, 3 April 2008.
32. Derek Roe, correspondence with author, 3 April 2008.
33. Jim Sackett, correspondence with author, 31 May 2007.
34. Paul Mellars, interview with author, 24 November 2006.
35. Paul Mellars, interview with author, 24 November 2006.
36. Nic David, interview with author, 15 May 2006.
37. This was Laville's mistake. 'In some parts of the world at certain times, typology is a LOT more capable of sensitive dissection of the past than geoarchaeology,' wrote Nic David in a letter to the author, 7 July 2008.
38. Paul Mellars, interview with author, 24 November 2006.
39. Paul Mellars, interview with author, 11 January 2007.

40. Paul Mellars, interview with author, 11 January 2007.
41. Harvey Bricker, correspondence with author, 6 May 2008.
42. Jim Sackett, correspondence with author, 5 March 2008.
43. 'The real problem was that he put novice excavators to work and gave them very little guidance, and Combe Grenal was way too complex to excavate without constant supervision.' Jim Sackett, correspondence with author, 10 August 2008.
44. Jim Sackett, correspondence with author, 8 July 2008.
45. Jim Sackett, correspondence with author, 5 March 2008.
46. Derek Roe, correspondence with author, 3 April 2008.
47. Branigan, correspondence with author, 15 March 2008. Branigan remembers Marmaduke taking Paul for walks and recalls the persistent rumour that Paul was an admirer of Dolly Parton's 'music rather than the lady herself'.
48. Harvey Bricker, correspondence with author, 6 April 2008.
49. Paul Mellars, interview with author, 11 January 2007.
50. Clive Bonsall, interview with author, 29 November 2006 and correspondence, 4 May 2008.
51. Paul Mellars, interview with author, 11 January 2007.
52. Paul Mellars, interview with author, 11 January 2007.
53. Sebastian Payne, correspondence with author, 25 May 2008.
54. Robin Dennell speaking at the 2006 Personal Histories in Archaeological Theory and Method panel, 23 October 2006.
55. The first water sieve was, 'a square tank with a sieve sitting on a ledge inside the tank, water coming in through a sort of bubbler and then bringing all the water out over a weir so that the floating stuff went into a second sieve that caught it. This machine was later further developed at Siraf by David Williams.' Sebastian Payne, interview with author, 23 May 2008.
56. Paul Mellars, interview with author, 13 March 2008.
57. Sebastian Payne, interview with author, 2 July 2008.
58. Paul Mellars, interview with author, 20 June 2008.
59. Paul speaking at the Personal Histories Panel, 23 October 2006.
60. Colin Renfrew, interview with author, 31 March 2008.
61. It was at this conference that Edmund Leach (1973, 762) made his oft-quoted, prophetic statement, 'The paradigm which is currently high fashion among the social anthropologists, namely that of structuralism, has not as yet caught up with archaeologists. Don't worry, it will!'
62. Colin Renfrew, interview with author, 31 March 2008. For additional information about Colin Renfrew and other former 'New Archeologists', see the film, Personal-Histories in Archaeological Theory and Method; the Beginnings of Processualism http://www.arch.cam.ac.uk/personal-histories/video.html
63. Colin Renfrew, correspondencewith author, 12 July 2008.
64. Paul Mellars, interview with author, 16 April 2008.
65. Paul Mellars, interview with author, 16 April 2008.
66. Colin Renfrew, interview with author, 31 March 2008.
67. Laura Basell, correspondence with author, 14 March 2008.
68. Colin Renfrew, interview with author, 31 March 2008.
69. For documentation of these shifts, see 'Personal-Histories; the Beginnings of Post-processualism.' On YouTube at http://middlesavagery.wordpress.com/2008/03/28/personal-histories-at-cambridge/
70. Paul Mellars, interview with author, 16 April 2008.
71. Paul Mellars, interview with author, 16 April 2008.
72. Paul Mellars, interview with author, 16 April 2008.
73. Despite the fact that Paul wrote fluidly and clearly and published repeatedly in major journals, he suffered from intense writer's block throughout the 1970s and still remains always cautious about his results.
74. Colin Renfrew, interview with author, 31 March 2008.
75. Henrietta Moore, speaking at the Personal Histories Panel, Cambridge, 22 October 2007.
76. Alison Wyllie, speaking at the Personal Histories Panel, Cambridge, 22 October 2007.
77 Paul Mellars, interview with author, 20 June 2008.
78. Paul Mellars, interview with author, 20 June 2008.
79. Robert Hedges, correspondence with author, 24 July 2008.
80. Robert Hedges, correspondence with author, 21 July 2008.
81. Paul Mellars, interview with author, 16 May 2008.
82. Paul Mellars, interview with author, 16 May 2008.
83. Robert Hedges, correspondence with author, 21 July 2008.
84. Paul, speaking, 6 May 2008. 'And, Peter was so utterly disheveled. I remember when we went to Paris, I met him at the airport. Peter typically arrived; he wasn't wearing a jacket, his trousers were falling down, half his shirt tail was out and he was carrying an enormous bag with every conceivable file that possibly related to anything we were doing, a massive weight ... and that is the way Peter operated.'
85. 'We did respect each other ... Peter said, 'I got to talk to you.' He took me to a little restaurant outside Waterloo Station and spent an hour and a half trying to persuade me. There were tears streaming down his face. This was terrible and that's how it ended': Paul Mellars, interview with author, 6 May 2008.
86. Chris Stringer, correspondence with author, 4 May 2008.
87. Hilary Deacon, correspondence with author, 8 May 2008.
88. Paul Mellars, interview with author, 6 May 2008.
89. Stan Ambrose, correspondence with author, 28 April 2008.
90. Hilary Deacon, correspondence with author, 8 May 2008.
91. Marcel Otte, correspondence with author, 29 April 2008.
92. Chris Stringer, correspondence with author, 27 April 2008.
93. Hilary Deacon, correspondence with author, 8 May 2008.
94. Geoffrey Clark, manuscript.
95. John Shea, correspondence with author, 31 March 2008.
96. Paul Mellars, interview with author, 6 May 2008.
97. Paul Mellars, interview with author, 6 May 2008.
98. Paul Mellars, interview with author, 20 June 2008.
99. Paul Mellars, interview with author, 16 May 2008.
100. Paul Mellars, interview with author, 16 May 2008.
101. Clive Gamble, correspondence with author, 25 March 2008.

References

Andrews, M.V., D.D. Gilbertson, M. Kent & P.A. Mellars, 1985. Biometric studies of morphological variation in the intertidal gastropod *Nucella lapillus* (L): environmental and palaeoeconomic significance. *Journal of Biogeography* 12, 71–87.

Binford, L., 1973. Interassemblage variability: the Mousterian and the 'functional' argument, in *The Explanation of Culture Change: Models in Prehistory*, ed. C. Renfrew. London: Duckworth, 227–54.

Bordes, F., 1973. On the chronology and contemporaneity of different Palaeolithic cultures in France, in *The Explanation of Culture Change: Models in Prehistory*, ed. C. Renfrew. London: Duckworth, 217–26.

Bordes, F. & D. de Sonneville-Bordes, 1970. The significance of variability in Palaeolithic assemblages. *World Archaeology* 2(1), 61–73.

Bourgon, M., 1957. *Les industries moustériennes et pré-moustériennes du Périgord*. (Archives de l'Institut de Paléontologie Humaine, Mémoire 27.) Paris: Masson.

Cann, R.L., M. Stoneking & A.C. Wilson, 1987. Mitochondrial DNA and human evolution. *Nature* 325, 31–6.

Clark, J.G.D., 1939. *Archaeology and Society*. London: Methuen.

Clark, J.G.D., 1952. *Prehistoric Europe: the Economic Basis*. London: Methuen.

Clark, J.G.D., 1954. *Excavations at Star Carr: an Early Mesolithic Site at Seamer near Scarborough, Yorkshire*. Cambridge: Cambridge University Press.

Dimbleby, G.W., 1962. *The Development of British Heathlands and their Soils*. Oxford: Oxford University Press.

d'Errico, F., J. Zilhão, M. Julien, D. Baffier & J. Pelegrin, 1998. Neanderthal acculturation in western Europe? A critical review of the evidence and its interpretation. *Current Anthropology* 39, S1–S44.

Forde, C.D., 1934. *Habitat, Economy and Society: a Geographical Introduction to Ethnology*. London: Methuen.

Hodder, I. (ed.), 1982. *Symbolic and Structural Archaeology*. (New Directions in Archaeology.) Cambridge: Cambridge University Press.

Jacobi, R.M., J.H. Tallis & P.A. Mellars, 1976. The Southern Pennine Mesolithic and the ecological record. *Journal of Archaeological Science* 3(4), 307–20.

Laville, H., J.-P. Rigaud & J. Sackett, 1980. *Rock Shelters of the Perigord: Geological Stratigraphy and Archaeological Succession*. New York (NY): Academic Press.

Leach, E.R., 1973. Comments, in *The Explanation of Culture Change: Models in Prehistory*, ed. C. Renfrew. London: Duckworth, 762.

Lee, R.B. & I. DeVore (eds.), 1968. *Man the Hunter*. Chicago (IL): Wenner-Gren Foundation for Anthropological Research.

Lewin, R., 1987. Africa: cradle of modern humans *Science* 237(4820), 1292–5.

Mehlman, M.J., 1992. Review of *The Emergence of Modern Humans: an Archaeological Perspective* by P.A. Mellars. *American Anthropologist* 94, 730–31.

Meiklejohn,C., D.C. Merrett, R.W. Nolan, M.P. Richards & P.A. Mellars, 2005. Spatial relationships, dating and taphonomy of human bone from the Mesolithic site of Cnog Coig, Oronsay, Argyll, Scotland. *Proceedings of the Prehistoric Society* 71, 85–105.

Mellars, P.A., 1957/8. Archaeology as a hobby. *Woodnotes* 1957/8, 27–8.

Mellars, P.A., 1965. Sequence and development of Mousterian traditions in south-western France. *Nature* 205, 626–7.

Mellars, P.A., 1969a. The chronology of Mousterian industries in the Périgord region of south-west France. *Proceedings of the Prehistoric Society* 35, 134–71.

Mellars, P.A., 1969b. Radiocarbon dates for a new Creswellian site. *Antiquity* 43, 308–10

Mellars, P.A., 1970a. Some comments on the notion of 'functional variability' in stone-tool assemblages. *World Archaeology* 2(1), 74–89.

Mellars, P.A., 1970b. An antler harpoon-head of Obanian affinities from Whitburn, County Durham. *Archaeologia Aeliana* 48, 337–46

Mellars, P.A., 1973. The character of the Middle–Upper Palaeolithic transition in south-west France, in *The Explanation of Culture Change: Models in Prehistory*, ed. C. Renfrew. London: Duckworth, 255–76.

Mellars, P.A., 1974. The Palaeolithic and Mesolithic periods in Britain, in *British Prehistory: a New Outline*, ed. A.C. Renfrew. London: Duckworth, 41–100, 268–79.

Mellars, P.A., 1975. Ungulate populations, economic patterns, and the Mesolithic landscape, in *The Effect of Man on the Landscape: the Highland Zone*, eds. J.G. Evans, S. Limbrey & H. Cleere. (Research Report 11.) London: Council for British Archaeology, 49–56.

Mellars, P.A., 1976a. Fire ecology, animal populations and Man: a study of some ecological relationships in prehistory. *Proceedings of the Prehistoric Society* 42, 15–46.

Mellars, P.A., 1976b. Settlement patterns and industrial variability in the British Mesolithic, in *Problems in Economic and Social Archaeology*, eds. I.H. Longworth, G. Sieveking & K.E. Wilson. London: Duckworth, 375–400.

Mellars, P.A., 1978a. Preface, in *The Early Postglacial Settlement of Northern Europe: an Ecological Perspective*, ed. P.A. Mellars. London: Duckworth, vii–ix.

Mellars, P.A., 1978b. Excavation and economic analysis of Mesolithic shell middens on the Island of Oronsay (Inner Hebrides), in *The Early Postglacial Settlement of Northern Europe: an Ecological Perspective*, ed. P.A. Mellars. London: Duckworth, 371–96.

Mellars, P.A., 1979. Excavation of Mesolithic shell middens on the Island of Oronsay (Hebrides), in *Early Man in the Scottish Landscape*, ed. L.M. Thoms. Edinburgh: Edinburgh University Press, 43–61.

Mellars, P.A., 1981. Cnoc Coig, Druim Harstell and Cnoc Riach: problems of the identification and location of shell middens on Oronsay. *Proceedings of the Society of Antiquaries of Scotland* 111, 516–18.

Mellars, P.A., 1986. A new chronology for the French Mousterian period. *Nature* 322, 410–11.

Mellars, P.A. (ed.), 1987. *Excavations on Oronsay: Prehistoric Human Ecology on a Small Island*. Edinburgh: Edinburgh University Press.

Mellars, P.A., 1989. Major issues in the emergence of modern humans. *Current Anthropology* 30(3), 349–85.
Mellars, P.A. (ed.),1990. *The Emergence of Modern Humans: an Archaeological Perspective*. Edinburgh: Edinburgh University Press.
Mellars, P.A., 1996. *The Neanderthal Legacy: an Archaeological Perspective from Western Europe*. Princeton (NJ): Princeton University Press.
Mellars, P.A. & S. Payne, 1971. Excavation of two Mesolithic shell middens on the Island of Oronsay, Inner Hebrides. *Nature* 231, 397–8.
Mellars P.A. & S.C. Reinhardt, 1978. Patterns of Mesolithic land-use in southern England: a geological perspective, in *The Early Postglacial Settlement of Northern Europe: an Ecological Perspective,* ed. P.A. Mellars. London: Duckworth, 243–93.
Mellars, P.A. & C.B. Stringer (eds)., 1989. *The Human Revolution: Behavioural and Biological Perspectives on the Origins of Modern Humans*. Edinburgh: Edinburgh University Press.
Mellars, P.A. & M.R. Wilkinson, 1980, Fish otoliths as evidence of seasonality in prehistoric shell middens: the evidence from Oronsay (Inner Hebrides). *Proceedings of the Prehistoric Society* 46, 19–44.
Mellars, P.A., H.M. Bricker, J.A.J. Gowlett, & R.E.M. Hedges, 1987. Radiocarbon accelerator dating of French Upper Palaeolithic sites. *Current Anthropology* 28(1), 128–33.
Radley, J & P. Mellars, 1964, A Mesolithic structure at Deepcar, Yorkshire, England and the affinities of its associated flint industries. *Proceedings of the Prehistoric Society* 30, 1–24.
Renfrew, C., 1973. Preface, in *The Explanation of Culture Change: Models in Prehistory*, ed. C. Renfrew. London: Duckworth, ix–x.
Renfrew, C., 1974. Preface, in *British Prehistory: a New Outline*, ed. C. Renfrew. London: Duckworth, xi–xiv.
Sackett, J., 1966. Quantitative analysis of Upper Paleolithic stone tools. *American Anthropologist Special Publication* 68(2), 356–94.
Sackett, J., 1997. Neanderthal behaviour: the archaeological evidence. Review of *The Neanderthal Legacy: an Archaeological Perspective from Western Europe*, by Paul Mellars. *Cambridge Archaeological Journal* 7(1), 148–9.
Shea, J.J., 1992. Review of *The Emergence of Modern Humans: Biocultural Adaptations in the Later Pleistocene,* edited by E. Trinkaus. *Journal of Field Archaeology* 19, 80–83.
Simmons, I.G., 1969. Evidence for vegetation changes associated with Mesolithic Man in Britain, in *The Domestication and Exploitation of Plants and Animals,* eds. P.J. Ucko & G.W. Dimbleby. London: Duckworth, 113–19.
Smith, P.J., 2009. *A Splendid Idiosyncrasy: Prehistory at Cambridge, 1915–50.* (British Archaeological Reports, British Series 485.) Oxford: Archaeopress.
Steward, J.H., 1937. Ecological aspects of southwestern society. *Anthropos* 32, 87–104.
Steward, J.H., 1955. *Theory of Culture Change: the Methodology of Multilinear Evolution*. Urbana (IL): University of Illinois Press.
Steward, J.H. & F.M. Setzler, 1938. Function and configuration in archaeology. *American Antiquity* 4(1), 4–10.
Stoneking, M. & R.L. Cann, 1989. African origin of human mitochondrial DNA, in *The Human Revolution: Behavioural and Biological Perspectives on the Origins of Modern Humans*, eds. P.A. Mellars & C.B. Stringer. Edinburgh: Edinburgh University Press, 17–30.
Trigger, B.G., 2006. *A History of Archaeological Thought*. Cambridge: Cambridge University Press.
Trinkaus, E. (ed.), 1989 *The Emergence of Modern Humans: Biocultural Adaptations in the Later Pleistocene.* (School of American Research Advanced Seminar Series.) Cambridge: Cambridge University Press.
Ucko, P., 1987. *Academic Freedom and Apartheid: the Story of the World Archaeological Congress.* London: Duckworth.
Valladas, H., J.M. Geneste, J.L. Joron & J.P. Chadelle, 1986. Thermoluminescence dating of Le Moustier (Dordogne, France). *Nature* 322, 452–4.
Valladas, H., J.L. Joron, G. Valladas *et al.*, 1987. Thermoluminescence dates for the Neanderthal burial site at Kebara in Israel. *Nature* 330, 159–60.
Valladas, H., J.L. Reyss, J.L. Joron, G. Valladas, O. Bar-Yosef, & B. Vandermeersch, 1988. Thermoluminescence dating of Mousterian 'Proto-Cro-Magnon' remains from Israel and the origin of modern man. *Nature* 331, 614–16.
White, R., 1982. Rethinking the Middle/Upper Paleolithic transition. *Current Anthropology* 33(1), 85–108.

Manuscripts
Woodnotes 1957/8 printed by the Woodhouse, West Riding County Council Grammar School, Sheffield, in possession of Pamela Jane Smith.
Paul Anthony Mellars – an Appreciation by Geoffrey A. Clark in possession of Clark.
Moore, Henrietta Louise, 1983. Men, Women and the Organization of Domestic Space amoung the Marakwet of Kenya. Cambridge PhD thesis in archaeology. CUL Manuscripts D53366/85.

Correspondence
Stan Ambrose, 28 April 2008
Laura Basell, 14 March 2008
Clive Bonsall, 4 May 2008
Keith Branigan, 15 March 2008
Harvey Bricker, 6 April 2008, 6 May 2008
Nic David, 27 October 2006, 29 November 2006, 7 July 2008
Hilary Deacon, 8 May 2008
Clive Gamble, 25 March 2008
Robert Hedges, 21 July 2008, 24 July 2008
Jon Layne, 3 April 2008
Marcel Otte, 29 April 2008
Sebastian Payne, 25 May 2008
Colin Renfrew, 12 July 2008
Derek Roe, 3 April 2008, 2 June 2008
James Sackett, 20 April 2006, 28 May 2007, 31 May 2007, 5 March 2008, 8 July 2008, 10 August 2008
John Shea, 31 March 2008
Chris Stringer, 27 April 2008, 4 May 2008, 25 August 2008

Oral-historical interviews
Clive Bonsall, 29 November 2006
Nic David, 15 May 2006
Robin Dennell speaking at the 2006 Personal Histories in Archaeological Theory and Method panel, 23 October 2006
John Evans, 17 August 2000
Barbara Isaac, 17 May 2007
Anny Mellars, 25 March 2008
Paul Mellars, 23 October 2006 Personal Histories Panel, 27 October 2006, 3 November 2006, 24 November, 2006, 11 January 2007, 13 March 2008, 16 April 2008, 24 April 2008, 6 May 2008, 16 May 2008, 20 June 2008
Henrietta Moore, 22 October 2007. Speaking at the Personal Histories in Archaeological Theory and Method, the Beginnings of Post-processual approaches
Sebastian Payne, 23 May 2008, 2 July 2008
Colin Renfrew, 31 March 2008
Tim Reynolds, 17 March 2008
Thurstan Shaw, 12 July 2008
Alison Wylie, 22 October 2007, speaking at the Personal Histories in Archaeological Theory and Method, the Beginnings of Post-processual approaches
Ezra Zubrow, July 2008

Web sources
The Professor Paul Mellars Big Love Society, FACEBOOK, L.A. Tyrrell, 27 January 2006
2006 Personal-histories in Archaeological Theory and Method; the Beginnings of Processualism http://www.arch.cam.ac.uk/personal-histories/video.html
2007 Personal-Histories; the Beginnings of Post-processual approaches. On YouTube at http://middlesavagery.wordpress.com/2008/03/28/personal-histories-at-cambridge/

Chapter 3

1970–90: Two Revolutionary Decades

Chris Stringer

In my paper for *Rethinking the Human Revolution* (Stringer 2007), I reviewed some of the developments in palaeoanthropology that followed in the twenty years after the 1987 'Human Revolution' meeting in Cambridge. Every paper in the 2007 volume has something valuable to add to our rapidly evolving perspectives on modern human origins, and I will refer to many of them informally in the paper that follows. Here, however, I also want to look further back, to the period from 1970–90, a time that encompassed the graduate or postgraduate careers of Paul Mellars, myself, and a number of other contributors to this volume honouring Paul. Those of us studying or researching during those two decades were highly privileged to witness a period of dynamic, even revolutionary, growth in our knowledge of modern human origins. Table 3.1 presents a personal selection of some of the main discoveries and developments in research during this period. Here I would like to review what we really knew at that time, what we thought we knew (but which seems to have been erroneous), and something of what we have learnt since.

It is very difficult, now, to transport ourselves back to the time and intellectual environment around 1970 when unilinear gradualism was the dominant model for recent human evolution, when the term *Homo sapiens* usually encompassed fossils as morphologically diverse as those from Crô-Magnon, La Chapelle-aux-Saints, Broken Hill and Ngandong, and when material culture and human biology were often envisaged as linked in a close and intimate evolutionary dance through time. In addition, there was a major limitation that few of us even appreciated — there was little meaningful way to reach beyond the indirect application of conventional radiocarbon dating to the fossil and archaeological records. With its inevitable (but then unrecognized) compression of time, this limitation blocked the provision of a realistic chronology for the fossil human and archaeological records of the middle–late Pleistocene.

Chronological limitations

The accompanying Figure 3.1 provides my interpretation from 1976 of the dating of the later stages of human evolution, and it is immediately apparent from that figure why I had little or no basis to argue for an African origin of modern humans at that time, despite my early placement of the Omo Kibish material. The only continent that seemed to have a reasonably stretched post-*erectus* chronology was Europe, based on relative dating derived from its glacial–interglacial sequence, but I was already fairly confident from my and Bill Howells's work that Europe contained Neanderthals and their ancestors, not those of modern humans. However, the enigmatic early Upper Palaeolithic Châtelperronian industry in France seemed to have evolved out of the locally preceding Mousterian of Acheulean tradition, leaving open the possibility that a parallel *in situ* evolutionary transition from Neanderthal to moderns had occurred, that undiscovered modern humans were in fact responsible (as suggested by François Bordes), or that Neanderthals were actually the manufacturers. Elsewhere there seemed to me to be a dearth of informative material in the appropriate time-frame, with no plausible alternative ancestors from Africa, or anywhere else for that matter. I (mistakenly, in hindsight) placed Broken Hill and Saldanha as about the same date as Omo, while Jebel Irhoud was considered younger still.

The European transition from Neanderthals and their associated Middle Palaeolithic tools to modern *Homo sapiens* (the Cro-Magnons) and the associated Upper Palaeolithic was believed from radiocarbon dating of fauna and charcoal to have occurred about 35,000 years ago. But because of dating limitations, the nature and extent of evolution or overlap between these taxonomic and archaeological units could not be determined. In the Levant, a similar transition period from Neanderthals to *Homo sapiens* was believed to lie only slightly beyond this 35,000-year-old watershed. Even as late as 1985 it was believed by most workers,

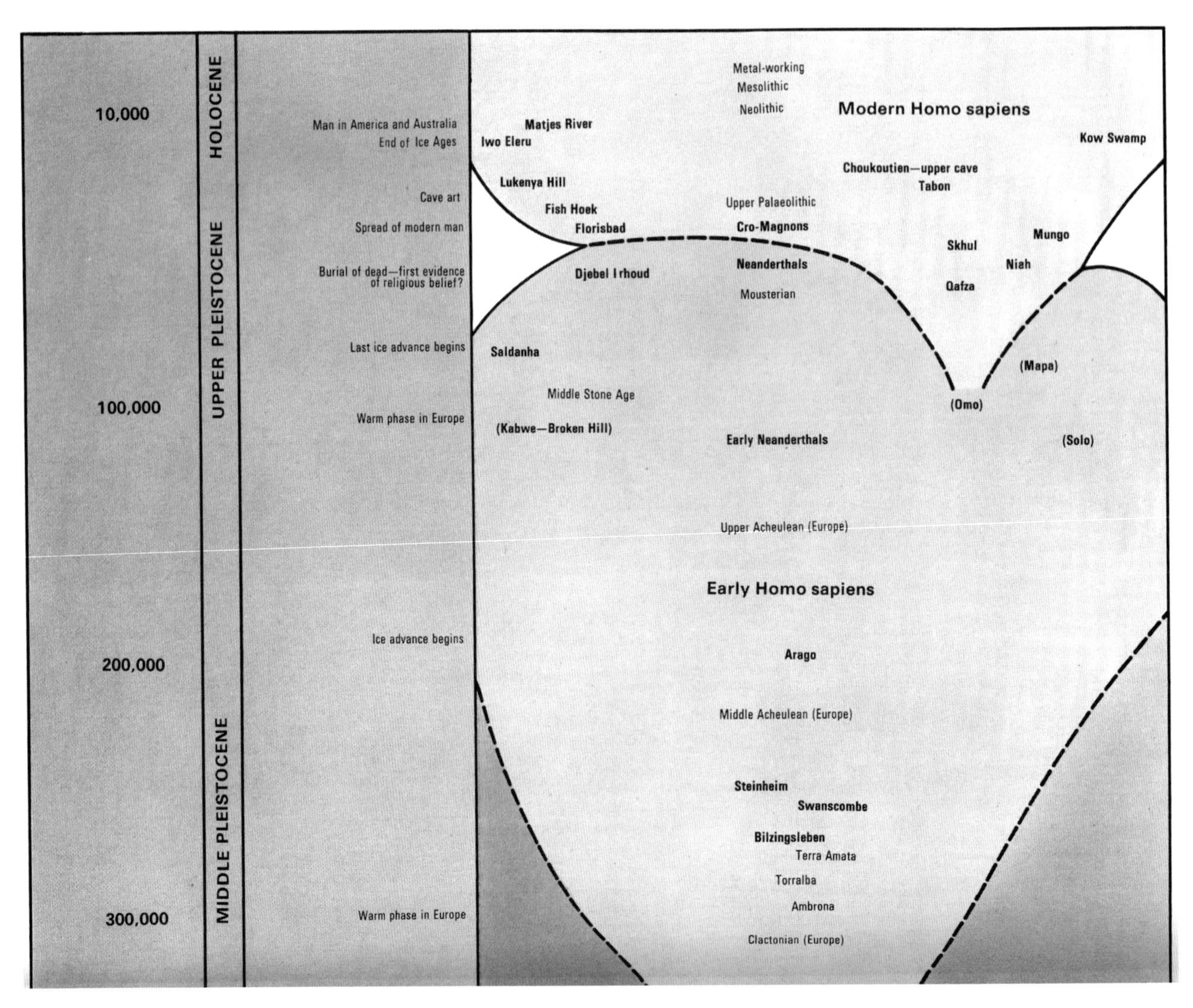

Figure 3.1. *An extract from a diagram of recent human evolution, which I produced in 1976 (Clapham 1976, 53). Note the early placement of the Omo material, but the late dates assigned to a range of African fossils, as well as the Levantine early moderns from Qafzeh (here spelt Qafza) and Skhul.*

including me, that the pattern of population change in this area followed that of Europe, or rather preceded it, by only a small amount of time. Thus Neanderthals at Israeli sites such as Tabun and Amud evolved into, or gave way to, early modern humans such as those known from Skhul and Qafzeh by about 40 kya (see e.g. Trinkaus 1984). For some workers (following Clark Howell's ideas: Howell 1957) there had been interlinked technological and biological changes, leading to the evolution of modern humans in the region from 'generalized' Neanderthals, and it was postulated that these early moderns then migrated into Europe, giving rise to the Cro-Magnons.

In Africa, the Lower Palaeolithic was generally believed to have continued until about 50,000 years ago, while the local transition from the Middle Stone Age (technologically equivalent to the Middle Palaeolithic) to the Later Stone Age (technologically equivalent to the Upper Palaeolithic) was generally believed to date from an even younger time than further north, perhaps as recent as 12,000 years ago. Hence African cultural and physical evolution was thought to have lagged considerably behind that of Europe and western Asia, and this was reinforced by the belief (from radiocarbon dating) that archaic humans such as Florisbad and Djebel Irhoud dated from only about 40,000 years ago.

In the Far East, the pattern of human evolution was even more difficult to discern in the 1970s. Some Asian fossils such as those from Zhoukoudian (and their associated stone industries) were dated via correlation with European glacial–interglacial sequences

Table 3.1. *A personal timeline of discoveries and events concerning modern human origins 1970–90. (Modified from: Neanderthals meet modern humans,* Athena Review *2(4) (2001)).*

1965–80	Jebel Qafzeh Cave, Israel: Bernard Vandermeersch and his team re-excavate the site, discovering skeletal material from at least ten more individuals. Vandermeersch concludes that the Skhul and Qafzeh material is not Neanderthal, but 'proto-Cro-Magnon', yet is associated with the Middle Palaeolithic.
1971	David Brose and Milford Wolpoff argue that Neanderthals evolved into modern humans. They support Loring Brace's contention that toolkits changed gradually from the Middle Palaeolithic to the Upper Palaeolithic, with associated evolution to modernity.
1974	Mungo 3 red-ochre burial discovered in the Willandra Lakes region of Australia.
1974	William Howells and Chris Stringer conclude that Neanderthals were too different to be human ancestors, based on quantitative studies of their cranial form in comparison to modern humans, including Cro-Magnon fossils.
1975	Erik Trinkaus's study of Neanderthal postcrania confirms that they were highly evolved and walked like modern humans.
1975	Desmond Clark publishes *Africa in Prehistory: Peripheral or Paramount?*
1975–78	Rainer Protsch, Günter Bräuer and Peter Beaumont present early versions of the 'Out of Africa' hypothesis.
1976	Fred Smith's study of the Krapina Neanderthals concludes that they were our direct ancestors.
1978	Jean-Jacques Hublin's thesis on Pleistocene crania concludes modern humans could not have evolved from Neanderthals.
1979	St Césaire, France: discovery of a Neanderthal burial with Châtelperronian tools (often argued to have been manufactured by *Homo sapiens*), dated at about 35,000 BP.
1979–81	Bernard Vandermeersch and Ofer Bar-Yosef suggest that the Qafzeh early modern human remains may be older than Neanderthals in the region, supported by William Farrand and Eitan Tchernov's geological and faunal studies.
1982	Michael Day and Chris Stringer restudy the Omo Kibish remains from Ethiopia and confirm Omo 1 is a modern-looking, but ancient, human.
1983	Trinkaus publishes *The Shanidar Neanderthals.*
1983	Kebara Cave Israel: discovery of the burial of an adult male Neanderthal.
1984	Frank Spencer and Smith publish *The Origins of Anatomically Modern Humans* volume, including the first detailed presentation of Multiregional Evolution by Milford Wolpoff, Alan Thorne and Wu Xin Zhi.
1986	Modern Human Origins Research Seminar organized by Trinkaus in Santa Fe.
1987	From genetic data Rebecca Cann, Mark Stoneking and Allan Wilson propose a Recent African Origin for modern humans in the journal *Nature*, based on a study of mtDNA haplotypes.
1987–88	New thermoluminescence dates in the Levant place Neanderthal levels at Kebara at *c.* 60,000 BP and modern humans at Qafzeh at about 90,000 BP. These dates are subsequently supported by ESR dates, and by ESR and TL dates that also place Skhul earlier than Kebara (Vandermeersch 1989; Mercier *et al.* 1991).
1987	'Human Revolution' meeting organized by Paul Mellars and Stringer in Cambridge.
1988	Stringer and Peter Andrews publish their review paper on modern human origins in *Science.*
1989	Publication of Santa Fe and *Human Revolution* volumes .

as far back as 400,000 years, but the Braceian view that fossils such as Maba and Ngandong were representatives of a Neanderthal stage in our evolution meant that they, too, were often dated to the equivalent of the last glaciation, while fossils like Wajak were seen as transitional or 'Neanderthaloid'. However, there was apparently a significant gap before the appearance of unequivocally modern humans in the Far East. Until the first radiocarbon dating of the Mungo material in the 1970s, it was believed that the arrival of humans in Australia was a very late event, dated at perhaps 10–15,000 years. With the possible age of Mungo moving back towards 30,000 years, and the Niah Cave material tentatively dated at about 40,000 years, the apparent gap between archaics and moderns narrowed. But even accepting Omo Kibish 1 as the oldest known modern human, I was unable to construct a credible model for the evolution of modern humans during the 1970s, because of confused intercontinental chronologies and the lack of potential ancestors in the right time frame.

The impact of the St Césaire discovery

Many important fossil discoveries were made in the period 1970–90, but one of the most significant was the partial skeleton found in Châtelperronian (early Upper Palaeolithic) levels at St Césaire, France, in 1979. It was briefly mentioned as a Neanderthal in a French report, news then picked up by Arthur ApSimon, who wrote a commentary on its significance for the journal *Nature* the following year (ApSimon 1980). In the 1974 summary of my thesis results (Stringer 1974) I had postulated that the Châtelperronian could well have been made by Neanderthals, echoing the views of Richard Klein (1973), but ApSimon's interpretation of the find seemed odd to me, since he appeared to argue that it complicated the Neanderthal–modern transition, and provided support for population continuity. In contrast, I thought the discovery clarified the situation in Europe by demonstrating polyphyletic origins for the Upper Palaeolithic, and a clear demarcation between late Neanderthals and the Cro-Magnons who had apparently produced the contemporaneous Aurignacian. Accordingly, together with Robert Kruszynski and Roger Jacobi, I wrote a critical reply to ApSimon's piece, and this was published, together with a quite different reply by Milford Wolpoff, and ApSimon's response in 1981 (Wolpoff *et al.* 1981).

This discovery really catalysed the debate about events in Europe around the Neanderthal–modern transition, and I corresponded with a number of other researchers over its significance, exchanging views as different as Wolpoff's (who argued that the find strongly supported local technological and morphological evolution from Neanderthal–modern populations) and Bordes's (who argued that the skeleton represented a victim of the modern manufacturers of the Châtelperronian, who had considerately given him a decent burial!). And I began to discuss the significance of the find with Paul, who was working through its implications for his own research on the Middle–Upper Palaeolithic transition. Reading archaeological commentaries around the time is both instructive and sobering as many took the position summarized by White (1982): 'if there is a relationship between culture and biology across the [MP–UP] boundary, cultural developments ... are stimulating biological change rather than vice versa'. To my surprise, then, rather than challenging the fundamentals of biological continuity, the association of a Neanderthal with the Châtelperronian seemed for a while to instead reinforce the models of Brace and Brose and Wolpoff that technological change (or perhaps social change) was catalysing evolutionary trajectories towards modern humans. My response, influenced both by an analysis I had recently completed with Erik Trinkaus on the Shanidar crania (Stringer & Trinkaus 1981) and by my frustration with the debate, was to write a polemical and cladistics-based paper in *Journal of Human Evolution* (Stringer 1982). This seemed to have little impact on the debate, but at least served to usefully focus my views on the Neanderthals for the next phases of my research.

The impact of changing chronologies

In 1987, one of the first applications of emerging chronometric techniques (thermoluminescence applied to burnt flint) seemed to reinforce the expected pattern in the Levant, dating the recently discovered Neanderthal burial at Kebara in the anticipated time range of about 60 kya (Valladas *et al.* 1987). However, shortly afterwards, the first application was made to the site of the Qafzeh early modern material (Valladas *et al.* 1988), giving an astonishing age estimate of about 90 kya, more than twice the generally expected figure, but in line with earlier suggestions originated by Bar-Yosef & Vandermeersch (1981), and supported from geology (Farrand 1979) and fauna (Tchernov 1981). Further applications of non-radiocarbon-dating methods later amplified the pattern suggested by the age estimates for Qafzeh and Kebara (e.g. see Grün & Stringer 1991). It seemed likely that the early modern burials at both Qafzeh and Skhul dated to between about 90–130 kya, while the Neanderthal burials from Kebara and Amud dated younger than these figures, in the range 50–60 kya. The implications of this newly emerging chronology were already beginning to impact the 1986 and 1987 Sante Fe and Cambridge meetings (see below), and scenarios of unilinear evolution from Neanderthals to moderns in both the Levant and Europe started to be reassessed. Workers like me heightened efforts to try and identify alternative origins for the Skhul and Qafzeh people, whether from local predecessors such as Zuttiyeh, or from further south, in Africa. As the intervening period between the Levantine early moderns and Neanderthals approximated the transition from 'interglacial' Marine Isotope Stage (MIS) 5 to 'glacial' MIS-4, this also led to a proposed scenario where Neanderthals only appeared in the Levant after the onset of glaciation further north (Bar-Yosef 1998). However, more recent direct dating of the Tabun C1 specimen now suggests that this Neanderthal, at least, could also date from MIS-5 (Grün & Stringer 2000).

The European hominin sequence can now be interpreted as showing an accretional appearance of Neanderthal characteristics during the Middle Pleistocene (Stringer 1974; Hublin 1998; 2007). This is perfectly exemplified by the large skeletal sample from the Sima de los Huesos at Atapuerca, now dated *c.* 400–500 kya, which shows an array of Neanderthal characteristics. For the Spanish workers who have described the material (Arsuaga *et al.* 1997), these specimens represent late members of the species *Homo heidelbergensis*, whereas I prefer to classify them as an early form of the probable descendent species *Homo neanderthalensis*. Regardless of these different taxonomic views it is evident that from this time on, Neanderthal features continued to accrue so that by about 125 kya ago (Rink *et al.* 1995), specimens such as those from Krapina (Croatia) and Saccopastore (Italy) were quite comparable to examples from the last glaciation.

New dating using burnt flint (luminescence), mammalian tooth enamel (electron spin resonance), and uranium-series dating, in concert with accelerator radiocarbon dating (requiring much smaller samples of organic material than conventional methods) has generally confirmed the previous picture of the

Middle–Upper Palaeolithic sequence in Europe, but with some additional complexity. While the once-favoured model of a rapid *in situ* evolution of Neanderthals into Cro-Magnons has been resoundingly falsified, confirmation of a rapid replacement of Neanderthals by them is still not possible (*pace* Paul), given remaining uncertainties in dating and in establishing human associations with 'transitional' industries apart from the Châtelperronian. The apparently clear picture of the Middle–Upper Palaeolithic archaeological interface in Europe has continued to become cloudier since the St Césaire discovery. Further Neanderthal skeletal material associated with this industry has been confirmed at Arcy (France) (Hublin *et al.* 1996), and pendants and bone tools have now been identified at both sites (although note the cautions by authors such as Chase 2007, White 2007 and Bar-Yosef 2007). Thus some workers now claim that Neanderthal technological and symbolic capabilities were potentially or actually the same as those of Cro-Magnons (d'Errico & Zilhão 2003), while others (including Paul and I) have used the data to suggest that acculturation of Neanderthal populations may have been occurring. In my opinion it is still difficult from present dating evidence to demonstrate a clear directional wave of advance of the Aurignacian, and its assumed external source remains elusive (e.g. see Davies 2007; Belfer-Cohen & Goring-Morris 2007; Svoboda 2007; Kozłowski 2007; Tostevin 2007; Bar-Yosef 2007). It is even possible that *Homo sapiens* first arrived in Europe with a pre-Aurignacian/Middle Palaeolithic technology, although this is speculating beyond the available evidence (e.g. see Svoboda 2007; Kozłowski 2007; Tostevin 2007; Bar-Yosef 2007). As some of these authors also consider, a precursor industry that might mark the appearance of early modern pioneers, although currently without diagnostic fossil material, is the Bohunician of eastern Europe, dating beyond 40,000 radiocarbon years. In this context it is unfortunate that the enigmatic Oase hominins (see Zilhão *et al.* 2007) at present lack archaeological associations.

The impact of changing chronologies in Africa has been as dramatic as those in the Levant, with the whole time-scale of the African Palaeolithic stretched back (Clark 1975; Klein 1999; Henshilwood 2007; McBrearty 2007; Shea *et al.* 2007; Barham 2007). The Middle Stone Age is now believed to have begun by at least 250 kya and the transition to the Later Stone Age began prior to 40 kya in some regions. Thus the African record can now be seen to be in concert with, or even in advance of, records from Eurasia. The hominin record has been similarly reassessed. Biostratigraphic correlation suggests that the Broken Hill cranium may date from over 300 kya (Klein 1999), although direct dating attempts to test this are currently in progress. A combination of ESR dating on human tooth enamel and luminescence dating of sediments now suggests that the Florisbad cranium dates from about 260 kya (Grün *et al.* 1996), while other fossils that are potentially ancestral to modern humans from Guomde (Kenya), Singa (Sudan) and Djebel Irhoud are now dated by a variety of techniques to more than 130 kya (Bräuer *et al.* 1997; McDermott *et al.* 1996; Smith *et al.* 2007) . The human fossils Omo Kibish 1 and 2 may well date to about 195 kya (McDougall *et al.* 2005), and the Kibish evidence has been supplemented by three other Ethiopian crania from Herto, dated to about 160 kya (White *et al.* 2003). Despite my earlier assertions, it is not yet clear whether the picture of human evolution in Africa over the last 300,000 years does parallel that of Europe. Do both regions show a gradual, accretional transition from *Homo heidelbergensis* to a more derived species — in Europe *Homo neanderthalensis*, and in Africa *Homo sapiens*? Or will patterns turn out to be more complex, with relict populations surviving in isolation alongside more derived contemporaries? Interpretations of the behavioural record are beyond the scope of this review, but there is growing evidence for the early appearance of some aspects of modern human behaviour during the Middle Palaeolithic/ Middle Stone Age, including pigment use and symbolism (Vanhaeren *et al.* 2006; Bouzouggar *et al.* 2007; Henshilwood 2007), but it is still unclear to what extent such patterns were emerging outside of Africa as well (d'Errico & Zilhão 2003), and the extent of genuine continuity both in Africa and elsewhere (e.g. see Shea *et al.* 2007; O'Connell & Allen 2007; Marean 2007; d'Errico & Vanhaeren 2007).

The arrival of *Homo sapiens* in the Far East is still poorly dated and understood, and it is to be hoped that present studies in Arabia and the Indian Subcontinent can help to elucidate patterns of demography and dispersal across southern Asia (Petraglia 2007). The Upper Cave material from Zhoukoudian has recently been supplemented by the Tinyuan material, dated close to 40 kya (Shang *et al.* 2007), while the Laibin partial cranium has been dated at about the same age (Shen *et al.* 2007). However, the much earlier dates claimed for the Liujiang skeleton (Shen *et al.* 2002) remain controversial. Further south, the Niah 'deep' material can be dated in the range 30–40 kya (Barker *et al.* 2007; Rabett & Barker 2007), while in Indonesia, the redating of the Ngandong and Sambungmacan fossils to less than 50 kya by ESR and uranium series (Swisher *et al.* 1996) remains highly controversial, implying as it does a survival of *Homo erectus* as late as Neanderthals survived in Eurasia. The date of arrival of modern humans in Indonesia is still unclear, although an isolated tooth from Punung (possibly *Homo sapiens*: Westaway *et al.* 2007) may date from over 100 kya, and work by the present author and colleagues continues on dating the Wajak crania. Finally, the Indonesian record cannot be discussed without reference to the astonishing discovery of *Homo floresiensis* (Brown *et al.* 2004). Its implications for modern human dispersals in the region, and for human evolution in general, are still being worked through.

Also still unclear is the timing of the first arrival in Australia, and the route taken to get there. The rockshelters of Malakunanja II and Nawalabila in northern Australia may contain artefacts dating from over 50 kya, based on luminescence dates (Roberts *et al.* 1990; 1994), but these age estimates are considered unreliable by other workers (e.g. see O'Connell & Allen 2007). However, these sites do not contain ancient human remains, the oldest of which are still those from the Willandra Lakes region of southeastern Australia. In 1999 the Mungo 3 burial was redated to approximately double the original age estimate from radiocarbon (Bowler & Thorne 1976; Thorne *et al.* 1999), although subsequent work has suggested a younger age of around 42 kya (Bowler *et al.* 2003). But even this younger age (for an apparent red-ochre burial) exceeds any reliable age estimate for modern humans in Europe.

The impact of genetic data

Genetic data now loom large in studies of modern human origins (see Kivisild 2007; Underhill *et al.* 2007), but this only started to be so towards the end of the two decades under discussion here. Earlier studies had to work with population frequencies of genetic markers, the products of the genetic code (e.g. blood groups, proteins). By combining data from populations, attempts were made to reconstruct the genetic history of humans (Cavalli-Sforza & Bodmer 1971; Nei & Roychoudhury 1982). However, the advent of techniques that revealed individual molecular sequence data allowed phylogenetic trees or genealogies of specific genes or DNA segments to be constructed. Two pioneering papers published in *Nature* in 1986 and 1987 heralded the genetic revolution to come. One was population frequency-based, while the other adopted a phylogenetic approach using DNA markers called RFLPs (Restriction Fragmentation Length Polymorphisms). In the 1986 paper, Wainscoat *et al.* (1986) studied polymorphisms close to the beta-globin gene, and showed by genetic distance analyses that African populations were quite distinct from non-African ones, which in turn shared features with each other. The following year Cann *et al.* (1987) published their paper giving a genealogy of 134 mitochondrial DNA 'types' constructed from restriction maps of 148 people from different regions. The genealogy was used to reconstruct increasingly ancient hypothetical ancestors, culminating in one female, most parsimoniously located in Africa. Moreover, using an mtDNA divergence rate calculated from studies of other organisms, it was estimated that this hypothetical female African ancestor lived at about 200 kya, and that younger divergences could be estimated for samples outside of Africa. These conclusions were immediately highly controversial, as was shown by the discussions following Mark Stoneking's and James Wainscoat's co-authored presentations at the 1987 meeting, and Lewin's report on the meeting (Lewin 1987). Of course many developments have followed, not least the recovery of DNA from Neanderthal fossils leading to the recent draft genome sequence (see e.g. Green *et al.* 2010), but the impact of those early studies, including the papers in the *Human Revolution* volume (Mellars & Stringer 1989a) cannot be overestimated.

The impact of conferences and workshops

It is sometimes argued now that conferences make little impact on the progress of science, apart from providing networking opportunities. This is undoubtedly largely due to the long lead-in time for the planning of conferences, compared with the increasingly rapid dissemination of new data now, particularly through electronic media. But in contrast, in the period from 1970–90, conferences were often venues for the first presentation of new data and analyses. Two meetings which I attended certainly influenced the following few years of my scientific research, and I believe they similarly affected other participants: the Santa Fe Advanced Seminar 'The Origins of Modern Human Adaptation' organized by Erik Trinkaus in 1986, and the Cambridge 'Origins and Dispersal of Modern Humans' meeting organized by Paul and me in the following year. The former meeting, with contributed papers published in 1989 (Trinkaus 1989), brought together workers such as Lewis Binford, Milford Wolpoff and Ofer Bar-Yosef in a small closed workshop, with papers presubmitted and circulated, rather than presented. Thus discussion time was maximized, and it was originally intended that the recorded workshop proceedings would be published alongside revised versions of the submitted papers — an idea which I think was abandoned because of the highly discursive nature of some of the discussions! Nevertheless, over three days, I got a much better (and constructive) insight into the thinking of Palaeolithic archaeologists than I had managed over the previous ten years!

The 1987 Cambridge meeting was on a much larger and more public scale, but like the Santa Fe workshop, it witnessed many of the biggest names in palaeoanthropology coming to grips with rapidly emerging data and new ideas. This was set against the developing contest between what I have called 'Classic' Multiregionalism and Recent African Origin ('Out of Africa') that was to focus debate for the next 15 years or so, until Recent African Origin became the dominant paradigm around the beginning of this century (Stringer 2002). Paul and I described the difficult gestation of what became known as the 'Human Revolution' meeting in our introduction to volume 1 of the conference papers (Mellars & Stringer 1989b), although a very different and much more partisan account can be read in the late Peter Ucko's book (Ucko 1987). This was a turbulent time in both palaeoanthropological science and archaeological politics, and the schisms following the World Archaeological Congress of 1986 undoubtedly set back our field for several years. Nevertheless, I think the Human Revolution Conference achieved its academic aims (witness the four-page report in *Science*: Lewin 1987), and set a number of research agendas for the following decade. In particular, the Santa Fe and Cambridge meetings provided a wealth of data and discussion that helped to focus my thinking for the *Science* paper I wrote with Peter Andrews in the following year — a paper which I am proud to say is still one of the most heavily cited in palaeoanthropology (Stringer & Andrews 1988).

Concluding remarks

In my view, the period from 1970–90 advanced palaeoanthropological thinking on modern human origins more than any other two decades in the last century, although it will need historians of science to provide a more objective opinion on that than I can. These years saw substantial advances in our ability to calibrate the most recent stages of human evolution. Some specimens, such as late Neanderthals, had their estimated ages broadly confirmed, while others such as the early modern remains from Skhul and Qafzeh, were shown to be much older than generally believed. In the archaeological arena it was also becoming increasingly evident that at least some of the South African Middle Stone Age sites and their contained human fossils were similarly ancient. A number of important new fossil discoveries were made, such as the partial skeleton from Kebara, and the late Neanderthals from Vindija (in what is now Croatia) and St Césaire. The former two provided important information on Neanderthal variation, while the latter find allowed the enduring mystery of the makers of the Châtelperronian to be addressed with hard data at last. Further dots of new data on spatial maps permitted the first tentative reconstructions of modern human dispersals through Eurasia and Australasia, and such efforts were dramatically catalysed by emerging genetic data near the end of the time period in question.

Although palaeontological and archaeological discoveries had already been pointing in that direction for Paul and me, and a small community of others researchers, there is no doubt that, rightly or wrongly, it took the 1987 *Nature* paper by Cann and colleagues to get the wider world to sit up and take notice of the idea of a Recent African Origin for modern humans. We would have got there without 'Mitochondrial Eve', of course, but it would have taken us a lot longer to achieve general acceptance for an Out of Africa scenario without the complementary genetic data. And as I explained above, I think that the Human Revolution meeting and its ensuing publications played their part too, in focusing discussions about those two decades of palaeoanthropological progress, and summarizing the state of the art in palaeoanthropology as we looked towards the present century. But twenty years on from the Cambridge meeting, we cannot afford to be complacent! General scientific acceptance of a scenario like Out of Africa does not mean it is fully understood, and we and our successors still have much to do to flesh out that basic scenario with hard data from the large number of missing areas that remain.

References

ApSimon, A.M., 1980. The last Neanderthal in France? *Nature* 287, 271–2.

Arsuaga, J.L., J.M. Bermúdez de Castro & E. Carbonell (eds.), 1997. The Sima de los Huesos hominid site. *Journal of Human Evolution* 33 (special issue), 105–421.

Barham, L., 2007. Modern is as modern does? Technological trends and thresholds in the south-central African record, in *Rethinking the Human Revolution: New Behavioural and Biological Perspectives on the Origin and Dispersal of Modern Humans*, eds. P. Mellars, K. Boyle, O. Bar-Yosef & C. Stringer. (McDonald Institute Monographs.) Cambridge: McDonald Institute for Archaeological Research, 165–76.

Bar-Yosef, O., 1998. The chronology of the Middle Paleolithic of the Levant, in *Neandertals and Modern Humans in Western Asia*, eds. T. Akazawa, K. Aoki & O. Bar-Yosef. New York (NY): Plenum Press, 39–56.

Bar-Yosef, O., 2007. The dispersal of modern humans in Eurasia: a cultural interpretation, in *Rethinking the Human Revolution: New Behavioural and Biological Perspectives on the Origin and Dispersal of Modern Humans*, eds. P. Mellars, K. Boyle, O. Bar-Yosef & C. Stringer. (McDonald Institute Monographs.) Cambridge: McDonald Institute for Archaeological Research, 207–17.

Bar-Yosef, O. & B. Vandermeersch, 1981. Notes concerning the possible age of the Mousterian layers in Qafzeh Cave, in *Préhistoire du Levant: chronologie et organisation de l'espace depuis les origines jusqu'au VIe millénaire*, eds. J. Cauvín & P. Sanlaville. Paris: CNRS, 281–5.

Barker G., H. Barton, M. Bird *et al.*, 2007. The 'human revolution' in lowland tropical Southeast Asia: the antiquity and behavior of anatomically modern humans at Niah Cave (Sarawak, Borneo). *Journal of Human Evolution* 52(3), 243–61.

Belfer-Cohen, A. & A.N. Goring-Morris, 2007. From the beginning: Levantine Upper Palaeolithic cultural change and continuity, in *Rethinking the Human Revolution: New Behavioural and Biological Perspectives on the Origin and Dispersal of Modern Humans*, eds. P. Mellars, K. Boyle, O. Bar-Yosef & C. Stringer. (McDonald Institute Monographs.) Cambridge: McDonald Institute for Archaeological Research, 199–205.

Bouzouggar, A., N. Barton, M. Vanhaeren *et al.*, 2007. 82,000-year-old shell beads from North Africa and implications for the origins of modern human behavior. *Proceedings of the National Academy of Sciences of the USA* 104(24), 9964–9.

Bowler, J.M. & A.G. Thorne, 1976. Human remains from Lake Mungo: discovery and excavation of Lake Mungo III, in *The Origin of the Australians*, eds. R. Kirk & A.G. Thorne. Canberra: Australian Institute of Aboriginal Studies, 127–38.

Bowler, J.M., H. Johnston, J.M. Olley *et al.*, 2003. New ages for human occupation and climatic change at Lake Mungo, Australia. *Nature* 421, 837–40.

Bräuer, G., Y. Yokoyama, C. Falguères & E. Mbua, 1997. Modern human origins backdated. *Nature* 386, 337–8.

Brown, P., T. Sutikna, M.J. Morwood *et al.*, 2004. A new small-bodied hominin from the Late Pleistocene of Flores, Indonesia. *Nature* 431, 1055–61.

Cann, R., M. Stoneking & A.C. Wilson, 1987. Mitochondrial DNA and human evolution. *Nature* 325, 31–6.

Cavalli-Sforza, L. & W. Bodmer, 1971. *The Genetics of Human Populations*. San Francisco (CA): W.H. Freeman.

Chase, P.G., 2007. The significance of 'acculturation' depends on the meaning of 'culture', in *Rethinking the Human Revolution: New Behavioural and Biological Perspectives on the Origin and Dispersal of Modern Humans*, eds. P. Mellars, K. Boyle, O. Bar-Yosef & C. Stringer. (McDonald Institute Monographs.) Cambridge: McDonald Institute for Archaeological Research, 55–65.

Clapham, F. (ed.), 1976. *The Rise of Man*. London: Sampson Low.

Clark, J.D., 1975. Africa in prehistory: peripheral or paramount? *Man* 10(2), 175–98.

Davies, W., 2007. Re-evaluating the Aurignacian as an expression of modern human mobility and dispersal, in *Rethinking the Human Revolution: New Behavioural and Biological Perspectives on the Origin and Dispersal of Modern Humans*, eds. P. Mellars, K. Boyle, O. Bar-Yosef & C. Stringer. (McDonald Institute Monographs.) Cambridge: McDonald Institute for Archaeological Research, 263–74.

d'Errico, F. & M. Vanhaeren, 2007. Evolution or revolution? New evidence for the origin of symbolic behaviour in and out of Africa, in *Rethinking the Human Revolution: New Behavioural and Biological Perspectives on the Origin and Dispersal of Modern Humans*, eds. P. Mellars, K. Boyle, O. Bar-Yosef & C. Stringer. (McDonald Institute Monographs.) Cambridge: McDonald Institute for Archaeological Research, 275–86.

d'Errico, F. & J. Zilhão, 2003. A case for Neandertal culture. *Scientific American* 13, 34–5.

Farrand, W.R., 1979. Chronology and palaeoenvironment of Levantine prehistoric sites as seen from sediment studies. *Journal of Archaeological Science* 6(4), 369–92.

Green, R.E., J. Krause, A.W. Briggs *et al.*, 2010. A draft sequence of the Neandertal genome. *Science* 328(5979), 710–22.

Grün, R. & C.B. Stringer, 1991. Electron spin resonance dating and the evolution of modern humans. *Archaeometry* 33(2), 153–99.

Grün, R. & C. Stringer, 2000. Tabun revisited: revised ESR chronology and new ESR and U-series analyses of dental material from Tabun C1. *Journal of Human Evolution* 39(6), 601–12.

Grün, R., J.S. Brink, N.A. Spooner *et al.*, 1996. Direct dating of Florisbad hominid. *Nature* 382, 500–501.

Henshilwood, C.S., 2007. Fully symbolic *sapiens* behaviour: innovations in the Middle Stone Age at Blombos Cave, South Africa, in *Rethinking the Human Revolution: New Behavioural and Biological Perspectives on the Origin and Dispersal of Modern Humans*, eds. P. Mellars, K. Boyle, O. Bar-Yosef & C. Stringer. (McDonald Institute Monographs.) Cambridge: McDonald Institute for Archaeological Research, 123–32.

Howell, F.C., 1957. The evolutionary significance of variation and varieties of 'Neanderthal' man. *The Quarterly Review of Biology* 32(4), 330–47.

Hublin, J.-J., 1998. Climatic changes, paleogeography, and the evolution of the Neandertals, in *Neandertals and Modern Humans in Western Asia*, eds. T. Akazawa, K. Aoki & O. Bar-Yosef. New York (NY): Plenum Press, 295–310.

Hublin, J.-J., 2007. What can Neanderthals tell us about modern origins?, in *Rethinking the Human Revolution: New Behavioural and Biological Perspectives on the Origin and Dispersal of Modern Humans*, eds. P. Mellars, K. Boyle, O. Bar-Yosef & C. Stringer. (McDonald Institute Monographs.) Cambridge: McDonald Institute for Archaeological Research, 235–48.

Hublin, J.-J., F. Spoor, M. Braun, F. Zonneveld & S. Condemi, 1996. A late Neanderthal associated with Upper Palaeolithic artefacts. *Nature* 381, 224–6.

Kivisild, T., 2007. Complete mtDNA sequences — quest on 'Out-of-Africa' route completed?, in *Rethinking the Human Revolution: New Behavioural and Biological Perspectives on the Origin and Dispersal of Modern Humans*, eds. P. Mellars, K. Boyle, O. Bar-Yosef & C. Stringer. (McDonald Institute Monographs.)

Cambridge: McDonald Institute for Archaeological Research, 21–32.
Klein, R.G., 1973. *Ice-Age Hunters of the Ukraine*. Chicago (IL): University of Chicago Press.
Klein, R.G., 1999. *The Human Career: Human Biological and Cultural Origins*. Chicago (IL): University of Chicago Press.
Kozłowski, J.K., 2007. The significance of blade technologies in the period 50–35 kyr BP for the Middle Palaeolithic–Upper Palaeolithic transition in central and eastern Europe, in *Rethinking the Human Revolution: New Behavioural and Biological Perspectives on the Origin and Dispersal of Modern Humans*, eds. P. Mellars, K. Boyle, O. Bar-Yosef & C. Stringer. (McDonald Institute Monographs.) Cambridge: McDonald Institute for Archaeological Research, 317–28.
Lewin, R., 1987. Africa: cradle of modern humans. *Science* 237(4820), 1292–5.
Marean, C.W., 2007. Heading north: an Africanist perspective on the replacement of Neanderthals by modern humans, in *Rethinking the Human Revolution: New Behavioural and Biological Perspectives on the Origin and Dispersal of Modern Humans*, eds. P. Mellars, K. Boyle, O. Bar-Yosef & C. Stringer. (McDonald Institute Monographs.) Cambridge: McDonald Institute for Archaeological Research, 367–79.
McBrearty, S., 2007. Down with the revolution, in *Rethinking the Human Revolution: New Behavioural and Biological Perspectives on the Origin and Dispersal of Modern Humans*, eds. P. Mellars, K. Boyle, O. Bar-Yosef & C. Stringer. (McDonald Institute Monographs.) Cambridge: McDonald Institute for Archaeological Research, 133–51.
McDermott, F., C. Stringer, R. Grün, C.T. Williams, V.K. Din & C. Hawkesworth, 1996. New Late-Pleistocene uranium-thorium and ESR dates for the Singa hominid (Sudan). *Journal of Human Evolution* 31(6), 507–16.
McDougall, I., F.H. Brown & J.G. Fleagle, 2005. Stratigraphic placement and age of modern humans from Kibish, Ethiopia. *Nature* 433, 733–6.
Mellars, P. & C. Stringer (eds.), 1989a. *The Human Revolution: Behavioural and Biological Perspectives on the Origins of Modern Humans*. Edinburgh: Edinburgh University Press.
Mellars, P. & C. Stringer, 1989b. Introduction, in *The Human Revolution: Behavioural and Biological Perspectives on the Origins of Modern Humans*, eds. P. Mellars & C. Stringer. Edinburgh: Edinburgh University Press, 1–14.
Mellars, P., K. Boyle, O. Bar-Yosef & C. Stringer (eds.), 2007. *Rethinking the Human Revolution: New Behavioural and Biological Perspectives on the Origin and Dispersal of Modern Humans*. (McDonald Institute Monographs.) Cambridge: McDonald Institute for Archaeological Research.
Mercier, N., H. Valladas, J.L. Joron, J.L. Reyss, F. Leveque & B. Vandermeersch, 1991. Thermoluminescence dating of the late Neanderthal remains from Saint-Césaire. *Nature* 351, 737–9.
Multiple authors, 2001. Neanderthals meet modern humans. *Athena Review* 2(4).
Nei, M. & A.K. Roychoudhury, 1982. Genetic relationship and evolution of human races. *Evolutionary Biology* 14, 1–59.
O'Connell, J.F. & J. Allen, 2007. Pre-LGM Sahul (Pleistocene Australia-New Guinea) and the archaeology of early modern humans, in *Rethinking the Human Revolution: New Behavioural and Biological Perspectives on the Origin and Dispersal of Modern Humans*, eds. P. Mellars, K. Boyle, O. Bar-Yosef & C. Stringer. (McDonald Institute Monographs.) Cambridge: McDonald Institute for Archaeological Research, 395–410.
Petraglia, M.D., 2007. Mind the gap: factoring the Arabian Peninsula and the Indian Subcontinent into Out of Africa models, in *Rethinking the Human Revolution: New Behavioural and Biological Perspectives on the Origin and Dispersal of Modern Humans*, eds. P. Mellars, K. Boyle, O. Bar-Yosef & C. Stringer. (McDonald Institute Monographs.) Cambridge: McDonald Institute for Archaeological Research, 383–94.
Rabett, R. & G. Barker, 2007. Through the looking glass: new evidence on the presence and behaviour of Late Pleistocene humans at Niah Cave, Sarawak, Borneo, in *Rethinking the Human Revolution: New Behavioural and Biological Perspectives on the Origin and Dispersal of Modern Humans*, eds. P. Mellars, K. Boyle, O. Bar-Yosef & C. Stringer. (McDonald Institute Monographs.) Cambridge: McDonald Institute for Archaeological Research, 411–24.
Rink, W.J., H.P. Schwarcz, F.H. Smith & J. Radovčić, 1995. ESR ages for Krapina hominids. *Nature* 378, 24.
Roberts, R.G., R. Jones & M.A. Smith, 1990. Thermoluminescence dating of a 50,000-year-old human occupation site in northern Australia. *Nature* 345, 153–6.
Roberts, R.G., R. Jones, N.A. Spooner, M.J. Head, A.S. Murray & M.A. Smith, 1994. The human colonization of Australia: optical dates of 53,000 and 60,000 years bracket human arrival at Deaf Adder Gorge, Northern Territory. *Quaternary Science Reviews (Quaternary Geochronology)* 13(5–7), 575–83.
Shang, H., H. Tong, S. Zhang, F. Chen & E. Trinkaus, 2007. An early modern human from Tianyuan Cave, Zhoukoudian, China. *Proceedings of the National Academy of Sciences of the USA* 10(16), 6573–8.
Shea, J.J., J.G. Fleagle & Z. Assefa, 2007. Context and chronology of early *Homo sapiens* fossils from the Omo Kibish formation, Ethiopia, in *Rethinking the Human Revolution: New Behavioural and Biological Perspectives on the Origin and Dispersal of Modern Humans*, eds. P. Mellars, K. Boyle, O. Bar-Yosef & C. Stringer. (McDonald Institute Monographs.) Cambridge: McDonald Institute for Archaeological Research, 153–62.
Shen, G., W. Wang, Q. Wang *et al.*, 2002. U-series dating of Liujiang hominid site in Guangxi, southern China. *Journal of Human Evolution* 43(6), 817–29.
Shen, G., W. Wang, H. Cheng & R.L. Edwards, 2007. Mass spectrometric U-series dating of Laibin hominid site in Guangxi, southern China. *Journal of Archaeological Science* 34(12), 2109–14.
Smith, T.M., P. Tafforeau, D.J. Reid *et al.*, 2007. Earliest evidence of modern human life history in North African

early *Homo sapiens*. *Proceedings of the National Academy of Sciences of the USA* 104(15), 6128–33.
Stringer, C.B., 1974. Population relationships of later Pleistocene hominids: a multivariate study of available crania. *Journal of Archaeological Sciences* 1(4), 317–42.
Stringer, C.B., 1982. Towards a solution to the Neanderthal problem. *Journal of Human Evolution* 11(5), 431–8.
Stringer, C., 2002. Modern human origins: progress and prospects. *Philosophical Transactions of the Royal Society, London Series B* 357, 563–79.
Stringer, C.B., 2007. The origin and dispersal of *Homo sapiens*: our current state of knowledge, in *Rethinking the Human Revolution: New Behavioural and Biological Perspectives on the Origin and Dispersal of Modern Humans*, eds. P. Mellars, K. Boyle, O. Bar-Yosef & C. Stringer. (McDonald Institute Monographs.) Cambridge: McDonald Institute for Archaeological Research, 15–20.
Stringer, C.B. & P. Andrews, 1988. Genetic and fossil evidence for the origin of modern humans. *Science* 239(4845), 1263–8.
Stringer, C.B. & E. Trinkaus, 1981. The Shanidar Neanderthal crania, in *Aspects of Human Evolution*, ed. C. Stringer. London: Taylor & Francis, 129–65.
Svoboda, J.A., 2007. On modern human penetration to northern Eurasia: the multiple advances hypothesis, in *Rethinking the Human Revolution: New Behavioural and Biological Perspectives on the Origin and Dispersal of Modern Humans*, eds. P. Mellars, K. Boyle, O. Bar-Yosef & C. Stringer. (McDonald Institute Monographs.) Cambridge: McDonald Institute for Archaeological Research, 329–39.
Swisher, C.C., W.J. Rink, S.C. Antón *et al.*, 1996. Latest *Homo erectus* of Java: potential contemporaneity with *Homo sapiens* in Southeast Asia. *Science* 274(5294), 1870–74.
Tchernov, E., 1981. The biostratigraphy of the Levant, in *Préhistoire du Levant: chronologie et organisation de l'espace depuis les origines jusqu'au VIe millénaire*, eds. J. Cauvín & P. Sanlaville. Paris: CNRS, 67–97.
Thorne, A., R. Grün, G. Mortimer *et al.*, 1999. Australia's oldest human remains: age of the Lake Mungo 3 skeleton. *Journal of Human Evolution* 36(6), 591–612.
Tostevin, G.B., 2007. Social intimacy, artefact visibility and acculturation models of Neanderthal–modern human interaction, in *Rethinking the Human Revolution: New Behavioural and Biological Perspectives on the Origin and Dispersal of Modern Humans*, eds. P. Mellars, K. Boyle, O. Bar-Yosef & C. Stringer. (McDonald Institute Monographs.) Cambridge: McDonald Institute for Archaeological Research, 341–57.
Trinkaus, E., 1984. Western Asia, in *The Origins of Modern Humans*, eds. F. Smith & F. Spencer. New York (NY): Alan Liss, 251–93.
Trinkaus, E. (ed.), 1989. *The Emergence of Modern Humans.* Cambridge: Cambridge University Press.
Ucko, P., 1987. *Academic Freedom and Apartheid: the Story of the World Archaeological Congress*. London: Duckworth.
Underhill, P.A., N.M. Myres, S. Rootsi *et al.*, 2007. New phylogenetic relationships for Y-chromosome haplogroup I: reappraising its phylogeography and prehistory, in *Rethinking the Human Revolution: New Behavioural and Biological Perspectives on the Origin and Dispersal of Modern Humans*, eds. P. Mellars, K. Boyle, O. Bar-Yosef & C. Stringer. (McDonald Institute Monographs.) Cambridge: McDonald Institute for Archaeological Research, 33–42.
Valladas, H., J.L. Joron, G. Valladas *et al.*, 1987. Thermoluminescence dates for the Neanderthal burial site at Kebara in Israel. *Nature* 330, 159–60.
Valladas, H., J.L. Reyss, J.L. Joron, G. Valladas, O. Bar-Yosef & B. Vandermeersch, 1988. Thermoluminescence dating of Mousterian Proto-Cro-Magnon remains from Israel and the origin of modern man. *Nature* 331, 614–16.
Vandermeersch, B., 1989. The evolution of modern humans: recent evidence from Southwest Asia, in *The Human Revolution: Behavioural and Biological Perspectives in the Origins of Modern Humans*, eds. P. Mellars & C. Stringer. Edinburgh: Edinburgh University Press, 155–64.
Vanhaeren, M., F. d'Errico, C. Stringer, S.L. James, J.A. Todd & H.K. Mienis, 2006. Middle Paleolithic shell beads in Israel and Algeria. *Science* 312(5781), 1785–8.
Wainscoat, J.S., A.V.S. Hill, A.L. Boyce *et al.*, 1986. Evolutionary relationships of human populations from an analysis of nuclear DNA polymorphisms. *Nature* 319, 491–3.
Westaway, K.E., M.J. Morwood, R.G. Roberts *et al.*, 2007. Age and biostratigraphic significance of the Punung rainforest fauna, East Java, Indonesia, and implications for *Pongo* and *Homo*. *Journal of Human Evolution* 53(6), 709–17.
White, R., 1982. Rethinking the Middle/Upper Paleolithic transition. *Current Anthropology* 23(2), 169–92.
White, R., 2007. Systems of personal ornamentation in the early Upper Palaeolithic: methodological challenges and new observations, in *Rethinking the Human Revolution: New Behavioural and Biological Perspectives on the Origin and Dispersal of Modern Humans*, eds. P. Mellars, K. Boyle, O. Bar-Yosef & C. Stringer. (McDonald Institute Monographs.) Cambridge: McDonald Institute for Archaeological Research, 287–302.
White, T.D., B. Asfaw, D. DeGusta *et al.*, 2003. Pleistocene *Homo sapiens* from Middle Awash, Ethiopia. *Nature* 423, 742–7.
Wolpoff, M.H., A. ApSimon, C.B. Stringer, R.G. Kruszynski & R.M. Jacobi, 1981. Allez Neanderthal. *Nature* 289, 823–4.
Zilhão, J., E. Trinkaus, S. Constantin *et al.*, 2007. The Peştera cu Oase people, Europe's earliest modern humans, in *Rethinking the Human Revolution: New Behavioural and Biological Perspectives on the Origin and Dispersal of Modern Humans*, eds. P. Mellars, K. Boyle, O. Bar-Yosef & C. Stringer. (McDonald Institute Monographs.) Cambridge: McDonald Institute for Archaeological Research, 249–62.

Chapter 4

Thinking Through the Upper Palaeolithic Revolution

Clive Gamble

It is thirty-five years since Paul Mellars, using Palaeolithic data from Europe, defined the Upper Palaeolithic transition (Table 4.1). Since then there has been great interest in a much broader, and older, human revolution, marked by his three edited volumes that dared to think (Mellars & Stringer 1989; Mellars 1990), and then rethink (Mellars *et al.* 2007), this major change in human evolution. Along the way there have been doubters, at first physical anthropologists in the multiregional school of modern human origins, and latterly some archaeologists whose placards bear the slogans 'Down with the revolution' (McBrearty 2007), 'Sex happens: hybrids are modern too' (Darte *et al.* 1999) and 'Neolithic symbols rule, OK!' (Renfrew 2007).

This is the stuff of debate fuelled by new evidence most importantly since 1973 from genetics and science-based dating. And Paul has never shirked from robustly rebutting the claims that have called into question the reality and significance of this transition (Mellars 2005; 2006). My purpose in this paper is not however to write a history of the debate since that is patently premature as the papers in this volume testify.

Table 4.1. *The major changes between the Middle and Upper Palaeolithic in southwest France (Mellars 1973).*

Material technology
A greater range and complexity of tool forms and a replacement of *stability* in Middle Palaeolithic tool forms with rapid change during the Upper Palaeolithic; a development in bone, ivory and antler working; the appearance of personal ornaments.
Subsistence activities
A greater emphasis on a single species (often reindeer); a broadening of the subsistence base to include small game; the possible development of large-scale co-operative hunting and a greater efficiency in hunting due to the invention of the bow and arrow; very possibly these changes were accompanied by improvements in food storage and preservation techniques.
Demography and social organization
A substantial increase in population density and the maximum size of the co-residential group as inferred from the number of sites and the dimensions of settlements; group aggregation occurs to participate in co-operative hunting of migratory herd animals such as reindeer; increase in corporate awareness.

Instead I want to use my scepticism of the usefulness of the human revolution model (Gamble 2007) to comment more generally on transitions during the Palaeolithic. I want to examine how we carve history at its joints, those points of transition, and suggest that we need to take a much longer view if we are to understand change using archaeological data. What is significant is the way we conceptualize the mind engaged in the carving. Too little attention has been paid to this issue besides asserting the importance of symbols and language. I want to suggest there are alternatives to this Cartesian view of cognition and where the model of an extended mind (Clark & Chalmers 1998) changes the timing of the transition and forces us to reconsider our data. Think in metaphors rather than symbols is my suggestion, and if I am right, one of the first casualties will be that archaeological invention: the modern human.

Three revolutions

The use of the term revolution to explain change in prehistory was justified by Gordon Childe (1935, 7) as a way of making historical sense of the three-age system. He regarded the Stone (Palaeolithic and Neolithic), Bronze and Iron Ages as indicative of significant stages in human progress. This fourfold division encapsulated for him 'real revolutions that affected all departments of human life' (Childe 1935, 7). He had an extremely low opinion of hunters and gatherers and so his two key events were the Neolithic followed by the Urban Revolution: the transition to village agriculture followed by the rise of civilization.

The human revolution which Mellars has been so closely involved with postdates Childe's death by a decade. One of its unacknowledged influences was the maverick claim that domestication was as old as the Neanderthals (Higgs & Jarman 1969). The claim by Higgs that there never was a Neolithic revolution, just a continuum of ever closer man–animal, man–plant relationships, was one element in a broader re-evaluation of hunters and gatherers. Most important was the global comparison of extant hunter-gatherers (Lee & DeVore 1968) and the highly influential studies of their decision-making that provided the study of the Palaeolithic with hitherto unsuspected evidence for rational behaviour (Clark 1954; Binford & Binford 1966). These approaches overturned the conception, which began with Lubbock (1865) and continued unchallenged through to Childe, that hunters, both ancient and modern, lacked the sensibilities and cognitive qualities of the civilized person, the modern human. For example, the influential developmental psychologist Lev Vygotsky, writing in the 1930s, placed Primitive Man (by which he meant hunters and gatherers and any society without literacy) as a second stage in the development from apes to civilized people. The following gives a flavour of the dismissive world view which archaeologists shared: 'Primitive man has no concepts, and finds abstract generic names completely alien. Primitive and civilised man use words in quite different ways' (Luria & Vygotsky 1992,69). The familiar contrast is drawn between magic and reason, symbols and formal systems such as algebra. And if Primitive Man required different methods of analysis then so too did their ancient counterparts.

So, freed from the shackles of this Neolithic supremacy and supported by a new synthesis of variably mobile, numerically small populations, the way was opened for sketching a much older human revolution. This process began with the Middle–Upper Palaeolithic transition in Europe (Klein 1973; Mellars 1973), as shown in Table 4.1, that has since been developed into a check-list for modern behaviour using Palaeolithic evidence (Table 4.2). And during this development the human revolution was increasingly associated with the notion that it represented the origins of the modern mind as defined by the use of material symbols and language (Noble & Davidson 1996; Mithen 1996; Lewis-Williams 2003; Mellars 1996).

Since then the model has been criticized for its Eurocentric focus and much older data have been found in Africa for the traits in Table 4.2 (McBrearty & Brooks 2000; Henshilwood & Marean 2003; d'Errico *et al.* 2003). The current position is summarized in Table 4.3, where the appearance of modern humans, ourselves, equipped with symbolic behaviour, is disputed by at least four chronological camps.

To complicate matters further this appearance is not always clear cut. As a result the distinction between Anatomically (AMH) and Fully Modern Human (FMH) has been made to square the circle of looks and ability: anatomy and culture. However, to my mind AMH seems a very similar concept to

Table 4.2. *Klein's (1995) traits to recognize, worldwide, fully modern behaviour from 50–40,000 years ago.*

1	Substantial growth in the diversity and standardization of artefact types
2	Rapid increase in the rate of artefactual change through time and in the degree of artefact diversity through space
3	First shaping of bone, ivory, shell and related materials into formal artefacts, e.g. points, awls, needles, pins etc.
4	Earliest appearance of incontrovertible art
5	Oldest undeniable evidence for spatial organization of camp floors, including elaborate hearths and the oldest indisputable structural 'ruins'
6	Oldest evidence for the transport of large quantities of highly desirable stone raw material over scores or even hundreds of kilometres
7	Earliest secure evidence for ceremony or ritual, expressed both in art and in relatively elaborate graves
8	First evidence for human ability to live in the coldest, most continental parts of Eurasia
9	First evidence for human population densities approaching those of historic hunter-gatherers in similar environments
10	First evidence for fishing and for other significant advances in human ability to acquire energy from nature

Vygostky's use of Primitive Man as a developmental stage en route to Civilized Man. They were biologically similar to us. They had symbols but they were not our symbols. They used words differently. Change AMH to PMH rather than FMH and you will see instantly how untenable the concept really is.

Elsewhere, I have argued (Gamble 2007, 36) that AMH is a hybrid classification, introduced *c.* 1970, whose purpose is to reconcile contradictory data from biology–genetics–archaeology and so preserve the model of orderly transitions that is our main description of change (Bar-Yosef 1998). What is striking about AMH and FMH is their appeal to recency (Proctor 2003); we became who we are late in human evolution. The result supports a core Western belief, that our modernity is always deferred. This belief underpinned Childe's Neolithic revolution and more recently Renfrew's (2003) reworking of it as a sapient paradox. The human revolution claims that modernity started much earlier and among hunters rather than farmers (Table 4.3). But the important point is that modernity at 300, 200, 100 or 60 kyr ago is still *recent* when judged against six million years of hominin evolution. Furthermore, like the claims of the Neolithic revolution, modernity could only occur *within* the species *Homo sapiens*. Hence the importance of classifying other fossils as either *Homo neanderthalensis* or *Homo sapiens neanderthalensis*, or more recently *Homo floresiensis* and *Homo sapiens idaltu*. Neanderthals and the hobbit are not AMH, the traits are stacked against them. But then neither is *idaltu*. These 160-kyr-old Herto hominins apparently represent, with commendable imprecision,

Table 4.3. *Carving prehistory at its joints when it comes to the origins of modern humans.*

Age (approximate)		Modern human variant	Focus	References
>15 kyr ago	No human revolution	AMH, FMH	Africa	(McBrearty 2007)
>60 kyr ago	Human revolution	AMH, FMH	Africa, western Asia	(Henshilwood *et al.* 2002)
<60 kyr ago	Upper Palaeolithic revolution	FMH	Global diaspora (see this volume)	(Bar-Yosef 2002)
<15 kyr ago	Neolithic revolution	FMH	Independent centres	(Hodder 1990)

'the probable immediate ancestors of anatomically modern humans' (White *et al.* 2003, 742), with their odd mixture of skulls that were highly polished after death, Acheulean bifaces and a transitional cranial morphology. This is what it means to be a classificatory hybrid, invented to accommodate a model of change that in fact has changed little from juxtaposing the Primitive with the Civilized, the hunter with the farmer, the innate with the symbolic.

The mind in the skull

Archaeologists have got themselves into terminological difficulties with AMH and FMH because they have adhered to a single view of cognition: one that stops at the skin and where cognition is contained within the skull. The archaeological use of symbols, so central to supporters of both the human and Neolithic revolutions as the moment when we became human, depends upon this conceptualization and where cognition is an internal process, acting on the environment that surrounds it but separated from it. This is the rational mind that has dominated the study of cognition since the seventeenth century and where Descartes is a foundation figure (Coward & Gamble 2008). For archaeologists, the evolution of this contained mind is measured by its ability to manipulate the environment first through innate behaviour and then through the imposition of symbolically mediated behaviour. As Hutchins (1995, 355) explains, when technology and material culture provide the proxies for an internal theory of the mind, the following conclusion is inescapable:

> If one believes that technology is the consequence of cognitive capabilities, and if one further believes that the only place to look for the sources of cognitive capabilities is inside individual minds, then observed differences in level of technology between a 'technologically advanced' and a 'technologically primitive' culture will inevitably be seen as evidence of advanced and primitive minds.

His comment was directed at psychologists like Luria and Vygotsky rather than archaeologists, but it equally applies to explanations of the difference between AMH and FMH in terms of 'substantial amounts of brain power' (Henshilwood & Marean 2003). And the reason is that the internal mind model subscribes to a late modernity. Whether a human, Upper or Neolithic revolution (Table 4.3) is favoured depends on the depth and quality of the symbolic evidence available. For this reason two important archaeological studies by Noble & Davidson (1996) and Mithen (1996) come to broadly similar conclusions about a recent modernity but emphasize different mechanisms. In Noble & Davidson's study language emerges as a more effective way to impose symbolic order on the outside world, and this explains the explosion in symbolic proxies in the Upper Palaeolithic revolution, while for Mithen, cognitive fluidity linked the mental modules together to create similar gains for the mind as a symbolic processor that began during the older human revolution.

Thinking outside the box

But there is an alternative. The extended-mind model argues that cognition occurs inside and outside the body (Clark & Chalmers 1998; Hutchins 1995; Clark 1997). It does not stop at the skin. Neither does cognition simply manipulate symbols to interpret the external environment, with the implication that human evolution is about becoming better manipulators of symbols since this is what marks our distinctiveness as a species.

Rowlands (2003, 166) provides an example with his Barking Dog principle. He argues that evolutionary fitness is greater when individuals get the environment to do some of their cognitive tasks. In other words why have a dog and bark? Imagine a situation where it is necessary to perform an adaptive task such as acquiring food. It is clearly disadvantageous to evolve an internal mechanism alone if an alternative exists, whereby the same result can be achieved by combining internal mechanisms with the manipulation of the external environment. This is where technology is so adaptive. Not because it solves problems by applying rational thought to them, but rather because it is part of our architecture of cognition and available for exaptation when required. According to the external mind model we think *through* things (Knappett 2005) rather than conceiving *of* objects as the internal model of mind proposes.

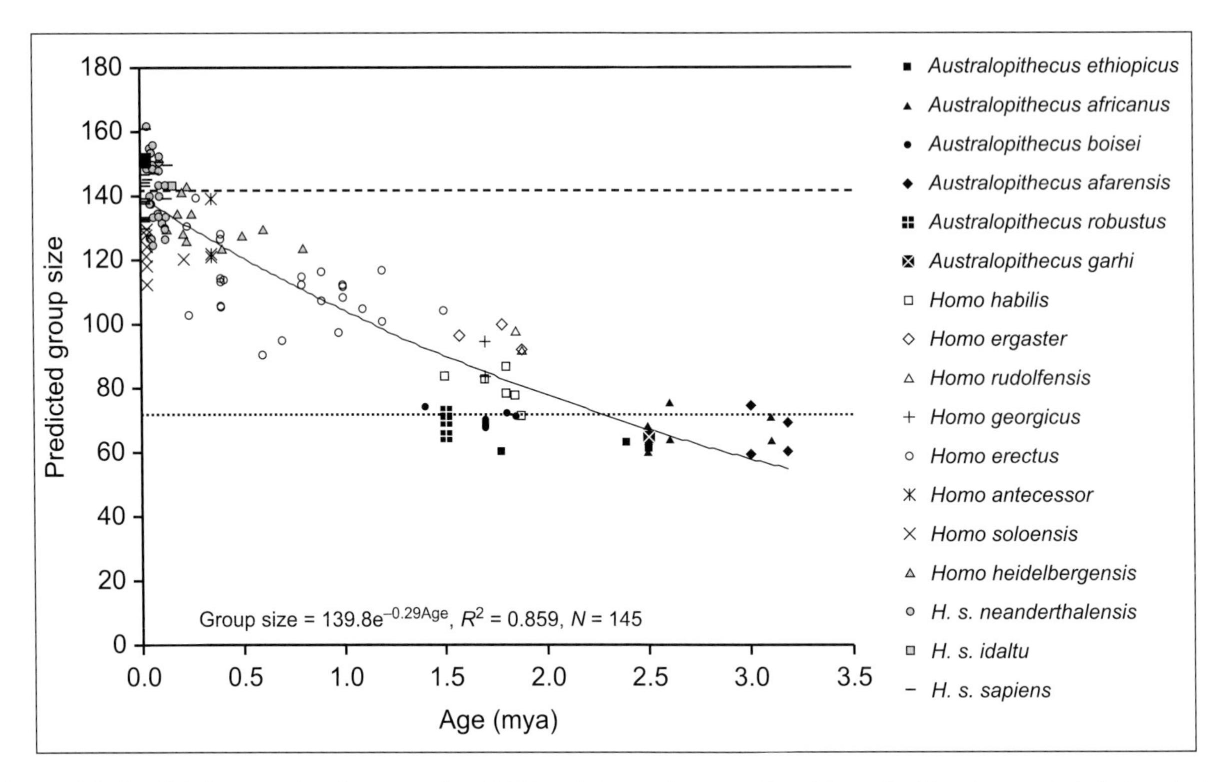

Figure 4.1. *Predicted group sizes for a sample of 145 hominin specimens gathered from the literature and plotted against geological age. The procedure outlined by Aiello & Dunbar (1993) is followed and where cranial capacity is used to plot the neo-cortex:cortex ratio observed in extant primates. The horizontal lines indicate threshold group sizes that require 20 per cent (lower line) and 40 per cent (upper line) of the time budget to be devoted to grooming as estimated by Dunbar. The fitted line is an exponential relating group size to age. (Graph courtesy of Matt Grove and reproduced with permission.)*

One further example shows the utility of the external model of the mind. Cooking is now regarded as an important adaptation since encephalization was achieved at the cost of a smaller stomach (Aiello & Wheeler 1995; Wrangham *et al.* 1999) and this had dietary implications. In particular, cooking meat in a fire acts as an external stomach since it aids digestion and increases the extractive efficiency of a smaller gut. What was once entirely an internal process has now become an adaptation linking the internal to the external environment. Therefore, fire management and cooking were part of the cognitive architecture of hominins from an early date (Gowlett *et al.* 1981), because they coupled an internal (brain/gut ratio) development with the external manipulation of the environment (cooking). A barking dog indeed!

Social brains beyond *Homo sapiens*

These couplings of biology and culture long before a human revolution raise more problems for a recent modernity (Table 4.3). In particular the notion that modern humans can only be found within *Homo sapiens* is called into question. When our cognitive architecture is no longer seen as defined by symbolic proxies alone, but by its extension into the surrounding world of objects and things, then our supposed uniqueness begins to look rather normal. In other words we share the characteristic of an external cognition with animals, some of whom are also thinking through things such as nests, termite probes and sensory cues to mark territories.

But for the moment let us stay with the hominins and examine a world that does not require modern humans and their counterparts AMH and FMH. The most decisive evidence that none of the revolutions in Table 4.3 was significant comes from the study of encephalization. As shown in Figure 4.1 the trend in brain size shows a marked increase after 600 kyr ago when values for the Encephalization Quotient (EQ) approach to within 10 per cent of contemporary populations (Ruff *et al.* 1997; Rightmire 2004). This trend has been compared by Aiello & Dunbar (1993) with the brain sizes of extant apes and monkeys. Here they drew attention to the correlation between brain size, largely a function of the neo-cortex:cortex ratio, and the size of the group for that same species. Group in this context is not necessarily a collection of permanent residents moving together as a unit. Primates are highly prone to fission–fusion and hence a dispersed form of sociality. Group size is therefore measured by the patterns of interaction during their annual movements. What Aiello & Dunbar discovered was that as brain size increased so too did group size. Returning to Figure 4.1, the predicted group sizes for hominins are shown, based on their brain size and calculated from the primate graph. These rise from the chimpanzee levels of 70 achieved by many of the Australopithecines, to 120 for those rapidly encephalizing hominins 600 kyr ago. From there it is a comparatively small rise to the predicted 150 for *Homo sapiens* and *Homo neanderthalensis* in the last 100 kyr. It was also Aiello & Dunbar's proposal that language would be selected for to provide the means of integrating these larger groups in a fission–fusion society.

The social-brain hypothesis, derived from these studies, proposes that hominin social life drove the enlargement of the brain during hominin evolution. The benefits of association provided the strong selection needed to re-order the ratio between gut and brain. But what is fascinating for archaeologists is that the timing of the encephalization event 600 kyr ago produces as great a disconnect between biology and culture as Renfrew (2007) claims in his sapient paradox between the modern humans of the human revolution and the much later explosion of symbolism in the villages and towns of the earliest Neolithic. Only this time we have to wait hundreds of thousands of years before finding, in credible numbers, convincing proxies for symbolic behaviour in the form of art, engraving, burial ritual etc.

This chronological disconnect is a particular puzzle for the internal model of the mind. If the brain is so large and symbols are what characterizes large human brains then why no obvious proxies for the archaeologist to discover? The answer, enshrined in the Upper Palaeolithic revolution, is to find a mind within a mind; Mithen's cognitive fluidity, Noble & Davidson's late appearance of language, or Renfrew's sapient paradox. In this way modernity is deferred within the brain, like a homunculus sitting inside a homunculus.

But the chronological disconnect between biology and culture, while of great interest, is less of a surprise to the extended model of the mind. On the principle of the Barking Dog, distributed cognition has been a feature of all hominins since at least the oldest stone tools, currently 2.5 Myr ago (Semaw *et al.* 1997). The tool-using abilities of New and Old World monkeys (Moura & Lee 2004) point to a much longer ancestry yet to be discovered. Rather than cite symbols as exclusive to FMH minds it is instead the case that through the extension of cognition out into the world we share much more with all our hominin ancestors. Our insistence on the uniqueness of the modern human (AMH or FMH) is clouding our judgement when it comes to the process of human evolution. Revolution becomes a necessary concept to accommodate such contradictory evidence.

Metaphors rather than symbols

However, it is the case that Australopithecine minds were different to ours. For example, Figure 4.1 shows that they could still conduct their social lives using primate solutions such as fingertip-grooming rather than develop ancillary methods such as language to integrate larger groups. Deacon (1997, 347) goes further, arguing that as Australopithecines were tool-using then the later encephalization that characterizes *Homo*, accompanied by language, was a consequence rather than a cause of this behaviour. For Deacon (1997, 43), language is about symbolic reference and only hominins achieve it. Therefore the main task is to figure out how combinations of words refer to things. It is in this context that technology, a prior adaptation, acted as an external support that led to language (Deacon 1997, 347). Deacon concludes that the key to the development of symbolic reference,

> is the co-evolutionary perspective which recognizes that the evolution of language took place neither inside nor outside brains, but at the interface where cultural evolutionary processes affect biological evolutionary processes. (Deacon 1997, 409)

Deacon's demonstration of how an extended mind model can address a key target of evolutionary study such as language is informative for this discussion. But how does it help archaeologists? Although an alternative to the internal model of mind as a processor of symbols is welcome, we are still left with the problem of identifying symbols in the absence of language.

Archaeologists understand symbols as standing *for* something else (Noble & Davidson 1996, 216); the crown for the king, the spear for the hunter. But without directly observing behaviour, and listening to its soundtrack, the ascription of meaning to these symbolic references, becomes impossible. What ensues are contested interpretations of objects such as the Bilzingsleben incised bones (Mania & Mania 2005) or the Berekhat Ram 'figurine' (d'Errico & Nowell 2000) that predate the Upper Palaeolithic revolution with its explosion of symbolic proxies, but not the encephalization event at 600 kyr ago. I doubt if we will ever agree if these were symbols, or not.

But there is an alternative that builds on Deacon's notion of technology as an external support and from which language and symbolic reference is later exapted. Here, things and objects act as solid metaphors (Tilley 1999). Instead of a symbol standing *for* something else, a metaphor depends on our *experiencing* one kind of thing in terms of another (Lakoff & Johnson 1980). However, if archaeologists are to meet this definition they must defy tradition and imbue hominins with senses and emotions that are all too often denied as either unrecorded or off-limits to rational investigation.

I have discussed the range of Palaeolithic metaphors at length elsewhere (Gamble 2007). Since experience is grounded in our bodies, we can establish links between our understanding of our bodies as instruments (kicking, cutting, inscribing, pointing etc.) and containers (digesting, gestating, excreting and yes, thinking). A social technology is informed by these experiences and the emotions they provoke. Instruments such as spears, pens and chopsticks are external supports that afford different sensations to boxes, houses and clothes that wrap the body in a series of highly varied skins. And it is through these material metaphors that we extend out into the world. They are integral to our cognitive architecture and have been for several million years.

These metaphors also allow us to describe change. Much of the prehistory of hominin technology has been dominated by instruments. Preservation plays its part but even at a well-preserved Lower Pleistocene site such as Gesher-Benot-Ya'aqov (Goren-Inbar *et al.* 2002) and the Middle Pleistocene locale of Schöningen (Thieme 2005) the evidence for instruments is overwhelming.

Conclusion

What the human and Upper Palaeolithic revolution record at different times and places is the ascendancy of material metaphors based on the container (Gamble 2007, ch. 7). These take many forms familiar to us in our own container-obsessed world; jewellery that encircled the body, graves in which corpses were laid, coverings of red ochre, tents and huts, musical instruments (sic.) such as flutes, paintings that covered rock walls and many more. Add the many examples from the papers in this volume and you will see how their forms and frequency change by latitude and longitude across great arcs of the Old World; from Europe to Russia and Korea, and from South Africa to India and Australia. Our complex cultural worlds are worlds of containers. Contrast this with a similar Old World distribution for Acheulean bifaces (Gamble & Marshall 2001), a social technology of the hand-extending instrument that lasted for upwards of a million years. Such contrasts show that our amazing cultural diversity lies not only in adapting to local ecologies (Binford 2001) but, and as hominins have always done, in the material construction of our social lives by thinking outside our bodies. Material metaphors grounded in bodily experiences existed long before words. They are preserved by the million in the archaeological record and they should be the focus of our studies of difference and change in the past.

One final word. None of this should imply for a moment that there was a Container revolution. Such a false title could only come by continuing the pursuit of change in the Palaeolithic with an internal model of the mind. By employing a different concept of the mind, and relating it to the encephalization of hominins, I hope to have shown that modern thought (whatever that is) has a much greater antiquity and extends far back into the hominin lineage. This journey started with Paul Mellars's insights into the Upper Palaeolithic transition and is set to run for a lot longer.

Acknowledgements

My first introduction to the Upper Palaeolithic revolution came when Paul, before he moved to Sheffield, was brought in to lecture to a small class of Cambridge undergraduates. This was welcome relief from the flint typology of the innumerable layers in La Ferrassie, which came from Charles McBurney, and the prehistory of humans-as-red-deer that we learned from Eric Higgs. The Palaeolithic, it seemed, involved a choice between extremes and Paul's all too short guest appearance showed there was a middle way with interesting questions. There was also the great benefit that Anny was on hand in the Haddon library to explain to a befuddled first year how to find a book. Advice I have never forgotten. A few years later Paul was the external examiner for my PhD thesis, very Higgsian in tone, which was redeemed in his eyes by my brief excursion into the German evidence for that same momentous change. The Palaeolithic has been such a delight to research and debate for the last thirty years due to Paul's clarity concerning what was important and what needed to be sorted out. I don't expect him to agree with this paper for a moment; but it certainly couldn't have been written without either him or Anny.

I would also like to thank the British Academy Centenary Project: *From Lucy to Language: the Archaeology of the Social Brain*, for financial support and two of its Fellows Matt Grove, for Figure 4.1, and Fiona Coward for challenging every sentence.

References

Aiello, L.C. & R.I.M. Dunbar, 1993. Neocortex size, group size, and the evolution of language. *Current Anthropology* 34(2), 184–93.

Aiello, L.C. & P. Wheeler, 1995. The expensive-tissue hypothesis: the brain and the digestive system in

human and primate evolution. *Current Anthropology* 36(2), 199–221.
Bar-Yosef, O., 1998. On the nature of transitions: the Middle to Upper Palaeolithic and the Neolithic Revolution. *Cambridge Archaeological Journal* 8(2), 141–63.
Bar-Yosef, O., 2002. The Upper Paleolithic revolution. *Annual Review of Anthropology* 31, 363–93.
Binford, L.R., 2001. *Constructing Frames of Reference: an Analytical Method for Archeological Theory Building using Ethnographic and Environmental Data Sets.* Berkeley (CA): University of California Press.
Binford, L.R. & S.R. Binford, 1966. A preliminary analysis of functional variability in the Mousterian of Levallois facies. *American Anthropologist* 68(2), 238–95.
Childe, V.G., 1935. Changing methods and aims in prehistory. *Proceedings of the Prehistoric Society* 1, 1–15.
Clark, A., 1997. *Being There: Putting Brain, Body, and World Together Again.* Cambridge (MA): MIT Press.
Clark, A. & D. Chalmers, 1998. The extended mind. *Analysis* 58(1), 7–19.
Clark, J.G.D., 1954. *Excavations at Star Carr: an Early Mesolithic Site at Seamer near Scarborough, Yorkshire.* Cambridge: Cambridge University Press.
Coward, F. & C. Gamble, 2008. Big brains, small worlds: material culture and the evolution of the mind. *Philosophical Transactions of the Royal Society* Series B 363, 1969–79.
Darte, C., J. Maurício, P.B. Pettitt *et al.*, 1999. The early Upper Palaeolithic human skeleton from the Abrigo do Lagar Velho (Portugal) and modern human emergence in Iberia. *Proceedings of the National Academy of Sciences of the USA* 96(13), 7604–9.
Deacon, T.W., 1997. *The Symbolic Species: the Co-evolution of Language and the Brain.* Harmondsworth: Allen Lane.
d'Errico, F. & A. Nowell, 2000. A new look at the Berekhat Ram figurine: implications for the origins of symbolism. *Cambridge Archaeological Journal* 10(1), 123–67.
d'Errico, F., C. Henshilwood, G. Lawson *et al.*, 2003. Archaeological evidence for the emergence of language, symbolism, and music — an alternative multidisciplinary perspective. *Journal of World Prehistory* 17(1), 1–70.
Gamble, C., 2007. *Origins and Revolutions: Human Identity in Earliest Prehistory.* Cambridge: Cambridge University Press.
Gamble, C.S. & G. Marshall, 2001. The shape of handaxes, the structure of the Acheulean world, in *A Very Remote Period Indeed: Papers on the Palaeolithic Presented to Derek Roe*, eds. S. Milliken & J. Cook. Oxford: Oxbow Books, 19–27.
Goren-Inbar, N., E. Werker & C.S. Feibel, 2002. *The Acheulian Site of Gesher Benot Ya'aqov, Israel: the Wood Assemblage.* Oxford: Oxbow Books.
Gowlett, J.A.J., J.W.K. Harris, D. Walton & B.A. Wood, 1981. Early archaeological sites, hominid remains and traces of fire from Chesowanja, Kenya. *Nature* 294, 125–9.
Henshilwood, C.S. & C.W. Marean, 2003. The origin of modern human behaviour: critique of the models and their test implications. *Current Anthropology* 44(5), 627–51.
Henshilwood, C.S., F. d'Errico, R. Yates *et al.*, 2002. Emergence of modern human behaviour: Middle Stone Age engravings from South Africa. *Science* 295(5558), 1278–80.
Higgs, E.S. & M.R. Jarman, 1969. The origins of agriculture: a reconsideration. *Antiquity* 43, 31–41.
Hodder, I., 1990. *The Domestication of Europe: Structure and Contingency in Neolithic Societies.* Oxford: Blackwell.
Hutchins, E., 1995. *Cognition in the Wild.* Cambridge (MA): MIT Press.
Klein, R.G., 1973. *Ice-Age Hunters of the Ukraine.* Chicago (IL): University of Chicago Press.
Klein, R.G., 1995. Anatomy, behavior, and modern human origins. *Journal of World Prehistory* 9(2), 167–98.
Knappett, C., 2005. *Thinking Through Material Culture: an Interdisciplinary Perspective.* Pittsburgh (PA): University of Pennsylvania Press.
Lakoff, G. & M. Johnson, 1980. *Metaphors We Live By.* Chicago (IL): University of Chicago Press.
Lee, R.B. & I. DeVore (eds.), 1968. *Man the Hunter.* Chicago (IL): Aldine.
Lewis-Williams, D., 2003. *The Mind in the Cave.* London: Thames and Hudson.
Lubbock, J., 1865. *Pre-Historic Times, as illustrated by Ancient Remains and the Manners and Customs of Modern Savages.* London: Williams and Norgate.
Luria, A.R. & L.S. Vygotsky, 1992. *Ape, Primitive Man, and Child: Essays in the History of Behavior.* New York (NY): Harvester Wheatsheaf.
Mania, D. & U. Mania, 2005. The natural and socio-cultural environment of *Homo erectus* at Bilzingsleben, Germany, in *The Individual Hominid in Context: Archaeological Investigations of Lower and Middle Palaeolithic Landscapes, Locales and Artefacts*, eds. C. Gamble & M. Porr. London: Routledge, 98–114.
McBrearty, S., 2007. Down with the revolution, in *Rethinking the Human Revolution: New Behavioural and Biological Perspectives on the Origin and Dispersal of Modern Humans*, eds. P. Mellars, K. Boyle, O. Bar-Yosef & C. Stringer. (McDonald Institute Monographs.) Cambridge: McDonald Institute for Archaeological Research, 133–51.
McBrearty, S. & A.S. Brooks, 2000. The revolution that wasn't: a new interpretation of the origin of modern human behavior. *Journal of Human Evolution* 39(5), 453–563.
Mellars, P., 1973. The character of the Middle–Upper Palaeolithic transition in southwest France, in *The Explanation of Culture Change: Models in Prehistory*, ed. C. Renfrew. London: Duckworth, 255–76.
Mellars, P. (ed.), 1990. *The Emergence of Modern Humans: an Archaeological Perspective.* Edinburgh: Edinburgh University Press.
Mellars, P., 1996. *The Neanderthal Legacy: an Archaeological Perspective from Western Europe.* Princeton (NJ): Princeton University Press.
Mellars, P., 2005. The impossible coincidence: a single-species model for the origins of modern human behavior in Europe. *Evolutionary Anthropology* 14(1), 12–27.
Mellars, P., 2006. Why did modern human populations disperse from Africa *ca.* 60,000 years ago? A new model. *Proceedings of the National Academy of Sciences of the USA* 103, 9381–6.

Mellars, P. & C. Stringer (eds.), 1989. *The Human Revolution: Behavioural and Biological Perspectives on the Origins of Modern Humans*. Edinburgh: Edinburgh University Press.

Mellars, P., K. Boyle, O. Bar-Yosef & C. Stringer (eds.), 2007. *Rethinking the Human Revolution: New Behavioural and Biological Perspectives on the Origin and Dispersal of Modern Humans*. (McDonald Institute Monographs.) Cambridge: McDonald Institute for Archaeological Research.

Mithen, S., 1996. *The Prehistory of the Mind*. London: Thames and Hudson.

Moura, A.C. de A. & P.C. Lee, 2004. Caphuchin stone tool use in Caatinga dry forest. *Science* 306(5703), 1909.

Noble, W. & I. Davidson, 1996. *Human Evolution, Language and Mind: a Psychological and Archaeological Inquiry*. Cambridge: Cambridge University Press.

Proctor, R.N., 2003. Three roots of human recency: molecular anthropology, the refigured Acheulean, and the UNESCO response to Auschwitz. *Current Anthropology* 44(2), 213–39.

Renfrew, C., 2003. *Figuring it Out: What are We? Where do We Come From? The Parallel Visions of Artists and Archaeologists*. London: Thames and Hudson.

Renfrew, C., 2007. *Prehistory: Making of the Human Mind*. London: Weidenfeld and Nicolson.

Rightmire, G.P., 2004. Brain size and encephalization in early to mid-Pleistocene *Homo*. *American Journal of Physical Anthropology* 124(2), 109–23.

Rowlands, M., 2003. *Externalism: Putting Mind and World Back Together Again*. Chesham: Acumen.

Ruff, C.B., E. Trinkaus & T.W. Holliday, 1997. Body mass and encephalization in Pleistocene *Homo*. *Nature* 387, 173–6.

Semaw, S., P. Renne, J.W.K. Harris *et al.*, 1997. 2.5 million-year-old stone tools from Gona, Ethiopia. *Nature* 385, 333–6.

Thieme, H., 2005. The Lower Palaeolithic art of hunting: the case of Schöningen 13 II-4, Lower Saxony, Germany, in *The Individual Hominid in Context: Archaeological Investigations of Lower and Middle Palaeolithic Landscapes, Locales and Artefacts*, eds. C. Gamble & M. Porr. London: Routledge, 115–32.

Tilley, C., 1999. *Metaphor and Material Culture*. Oxford: Blackwell.

White, T.D., B. Asfaw, D. Degusta *et al.*, 2003. Pleistocene *Homo sapiens* from Middle Awash, Ethiopia. *Nature* 423, 742–7.

Wrangham, R.W., J.H. Jones, G. Laden, D. Pilbeam & N. Conklin-Brittain, 1999. The raw and the stolen: cooking and the ecology of human origins. *Current Anthropology* 40(5), 567–94.

Chapter 5

The Effect of Organic Preservation on Behavioural Interpretations at the South African Middle Stone Age Sites of Rose Cottage and Sibudu

Lyn Wadley

The point at which symbolic behaviour began is likely to be the point at which people in the past developed thought patterns recognizable by people today. Elsewhere I suggested that the most securely recognized types of symbolic behaviour are 'externally stored' in art, personal ornaments, changing lithic style and spatial patterns. At Rose Cottage Cave the Middle Stone Age (MSA) layers have no organic preservation except for charcoal and ash. Thus lithics and spatial patterns of lithics in association with hearths could be examined, but not organic material culture items. In contrast, Sibudu MSA organic preservation is good and worked bone, perforated marine shells, carbonized seeds and burnt bedding are preserved in MSA layers between 70,000 and 60,000 years ago. The Sibudu evidence supports the early appearance of shell ornaments and worked bone at Blombos Cave. However, if organic preservation is a requirement for secure recognition of symbolic behaviour, many sites can automatically be eliminated as useless, even though their occupants were as modern as those who lived in better-preserved sites. The problem can be circumvented by looking broadly at complex cognition, rather than simply at symbolism. There are some steps contained within technological processes that can only be performed by people with complex cognition. Looking for these steps is one way around the dilemma caused by the presence/absence of organic artefacts.

Although symbolic thoughts may have existed in people's heads in the deep past, we cannot, as archaeologists, hope to recognize such symbolism until it was expressed in an 'out-of-brain' form (Donald 1991; 1998). I have previously argued (Wadley 2001) that modern behaviour is about social relationships that are expressed, negotiated, legitimized and transmitted through symbolism that can be recognized in the manipulation of space and material culture. Artwork, personal ornamentation, rapidly changing lithic style and the social use of space were suggested as means by which symbolism can be recognized archaeologically and Henshilwood & Marean (2003) have made use of this same trait list. I reasoned that the four types of symbolic storage need not be linked as a package of behaviours, an idea that supported the McBrearty & Brooks (2000) suggestion that modern traits did not arrive as a 'revolution'. I suggested further that any one of the behavioural traits is adequate for the recognition of early symbolic behaviour. This model was effectively applied by Brumm & Moore (2005) to demonstrate the early presence of symbolism in Australia.

In 2001, when I proposed the model, there was no evidence for personal ornamentation earlier than about 40,000 years ago in Africa, but this situation changed with the discovery at Blombos Cave of perforated shells, *Nassarius kraussianus,* with an age in excess of 70,000 years ago (Henshilwood *et al.* 2004; d'Errico *et al.* 2005) and the discovery of even older shell beads (>80,000 years ago) in Morocco (Bouzouggar *et al.* 2007; d'Errico *et al.* 2009). These discoveries have, for me (several other archaeologists accepted early symbolism in Africa years ago), unequivocally shifted the chronological boundaries for the origin of symbolism in Africa. The nature of archaeological research is such that these early shell beads may be only the beginning of such discoveries. While definitions of art, lithic style and a modern use of space are open to debate, the marine-shell beads are unambiguous ornaments, which must be symbolic because they act as indices of social identity (Wadley 2006a). Ornaments can be part of embodied experience that constructs identities in social contexts (Fisher & Loren 2003) and social and physical aspects, such as gender, sexuality, age, race or religion, are key to the constructions of identity (Joyce 2003) which are con-

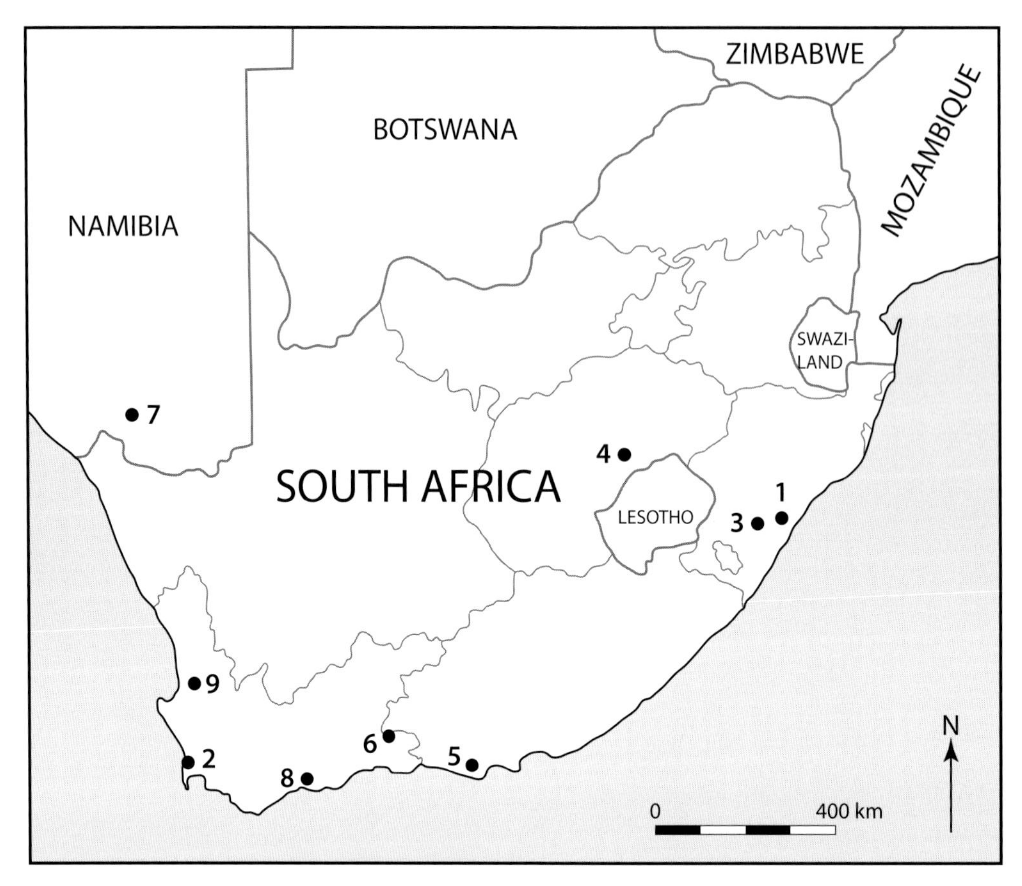

Figure 5.1. *Middle Stone Age sites mentioned in the text: 1) Sibudu; 2) Peers Cave; 3) Umhlatuzana; 4) Rose Cottage; 5) Klasies; 6) Boomplaas; 7) Apollo 11; 8) Blombos; 9) Diepkloof.*

stituted with material culture. What one puts on one's body or on other possessions need not only represent the identity one wishes to convey at the time; it can also be about one's position in the landscape and in a social context (Fisher & Loren 2003). Thus, artefacts such as ornaments can have multiple meanings that may be embedded in a number of different cultural experiences (Wilkie & Bartoy 2000).

While a single example of early symbolic behaviour from a single site, such as Blombos, is sufficient to confirm the use of symbolism before 70,000 years ago, it would be good to corroborate this evidence with data from other sites because multiple cases of symbolic behaviour are more convincing than isolated examples. I explore the evidence from two other South African sites: Rose Cottage and Sibudu.

Background to Rose Cottage

Rose Cottage is a landlocked site more than 300 km from the coast. It is situated in grassland vegetation just five kilometres from Ladybrand in the eastern Free State (Fig. 5.1). The cave is about 20 m long and 10 m wide with an altitude of approximately 1400 m above mean sea level. It was first excavated in the 1940s by Malan and in the early 1960s by Beaumont; the more recent excavations by Wadley began in 1987 and ended in 1998 (Wadley 1997). Excavations into the Later Stone Age (LSA) deposits were conducted in a 32 m^2 grid, but the grid was reduced to 23 m^2 for the Middle Stone Age (MSA) layers that are discussed here. An additional small excavation into the MSA deposits abutting the Malan excavation was made by Harper (1997).

Deposits are more than 6 m deep. The youngest LSA occupation has an age of approximately 500 radiocarbon years and the earliest occupation during the MSA has an age estimate, based on optically stimulated luminescence (OSL), of 95.9±6.6 kya (Pienaar *et al.* 2008) (Table 5.1). The oldest pre-Howiesons Poort layers of the site have a MSA stone industry containing points, side-scrapers and blades (Wadley & Harper 1989; Harper 1997; Soriano *et al.* 2007). A rich Howiesons Poort Industry with backed tools made on small blades replaced the earliest MSA occupations. The same opaline and tuffaceous rocks were used

Table 5.1. *Rose Cottage and Sibudu ages (Rose Cottage ages: Valladas* et al. *2005; Pienaar* et al. *2008; Sibudu ages: Jacobs* et al. *2008a,b). * = single aliquot OSL age estimates; the remaining OSL ages were calculated from single-grain analysis.*

Broad industrial affiliation	**Rose Cottage**			**Sibudu**
	^{14}C calibrated (BC)	**TL (kya)**	**OSL (kya)**	**OSL (kya)**
All MSA assemblages more recent than the Howiesons Poort	26,315 (Pta-6303)			
	27,309 (Pta-6202)		27.0±2 1 *	
	27,177 (Pta-7126)			
	30,298 (Pta-7184)			
	26,610 (Pta-5596)		36.8±2.1 *	
	34,375 (Pta-7805)			38.0±2.6
	36,609 (Pta-7796)			39.1±2.5
	34,925 (Pta-5592)			49.9±2.5
	34,375 (Pta-7763)			49.1±2.1
				46.6±2.3
				47.6±1.9
				46.0±1.9
				49.4±2.3
				57.6±2.1
				59.6±2.3
		47.1±10.2	34.8±2.3 *	59.0±2.2
		49.4±10.1	56.9±5.3 *	58.3±2.0
		50.5±4.6	61.8±3.0 *	58.6±2.1
				58.2±2.4
Howiesons Poort		58.6±6.6		61.7±1.5
		41.7±3.7	62.5±3.1 *	63.8±2.5
		56.3±4.5	54.4±5.0 *	64.7±1.9
		60.0±4.6		
		48.9±5.3	67.7±2.9 *	
Pre-Howiesons Poort			61.6±6.2 * 74.1±4.0 *	70.5±2.0
		64.5±6.6	89.4±6.0 *	72.5±2.0
		68.4±8.3		73.2±2.3
		72.5±6.8		77.2±2.1
		76.3±14.8		

throughout the MSA sequence even in the Howiesons Poort Industry (Harper 1997). The Howiesons Poort Industry at Rose Cottage is innovative because of its technique of direct marginal percussion using a soft stone hammer (Soriano *et al.* 2007). The technique did not disappear rapidly, but was progressively abandoned during the course of the Howiesons Poort Industry itself (Soriano *et al.* 2007). Thus Howiesons Poort technology is not a single, unchanging entity. Segments and other backed tools in the Rose Cottage Howiesons Poort were examined microscopically and were found to have traces of ochre and plant material along their backed edges where they would have been hafted to handles or shafts (Gibson *et al.* 2004; Wadley *et al.* 2004). This suggests that compound adhesives were being manufactured at the site and this issue will be returned to later.

Rose Cottage also has a suite of final MSA and transitional MSA/LSA assemblages (Clark 1997a,b) with ages between about 30,000 and 20,600 radiocarbon years (Table 5.1), and a long LSA sequence (Wadley 1997).

Rose Cottage has no organic preservation other than charcoal in layers older than about 20,000 years ago. We therefore do not know whether worked bone or shell artefacts were originally present in MSA layers at Rose Cottage. Based on the assumption that spatial patterns, as well as the presence of symbolic material culture, can inform us of behaviours implying modern mental abilities, the Rose Cottage lithic distributions were examined in association with features such as hearths (Wadley 2001; 2004a; 2006b). Circular to sub-circular hearths are present throughout the sequence; these are always flat on the floor of the cave and are never built with stone foundations or surrounds. The analyses showed that even the occupations between about 30,000 and 26,000 radiocarbon years ago have unstructured camp organization with artefacts strewn in close association with many overlapping hearths. In contrast, LSA spatial patterns, from layers with ages more recent than 13,000 radiocarbon years ago, include a few widely-spaced hearths and discrete groupings of material culture items and food waste. This suggests that segregated activities took place in the LSA, whereas this was not the case earlier, where spatial patterns appear less complex than in the LSA. I have elsewhere (Wadley 2001; 2004a; 2006b) suggested that people living in this cave were not using space symbolically during the MSA and that there is no early evidence for symbolic behaviour at this site.

Background to Sibudu

Sibudu is located approximately 40 km north of Durban, about 15 km inland of the Indian Ocean (Fig. 5.1), in coastal forest with adjacent savanna, on a steep cliff overlooking the Tongati River. The shelter is 55 m long and about 18 m in breadth. The excavation grid is in the northern part of the shelter at an altitude of approximately 100 m above mean sea level.

The present excavations, which are ongoing, began in 1998 and 21 m^2 of MSA deposit have been excavated by the Wadley team. The excavation grid includes a 6 m^2 sounding that has been deepened each season. A few years back, a rock base at a depth of about 3 m was reached in four of the squares. The pre-60,000 year old deposits have been reached only in the deep sounding. Optically stimulated luminescence (OSL) dating through single-grain analyses of sedimentary quartz has resulted in ages with good precision (Jacobs *et al.* 2008a,b) (Table 5.1). Four samples pre-date 70,000 years ago, three pre-date 60,000 years ago, six samples have a weighted mean age of 58.5±1.4 kya, six samples have a weighted mean age of 47.7±1.4 kya and two samples have a weighted mean age of 38.6±1.9 kya. The three young phases were separated by occupational hiatuses of 10.8±1.3 kya and 9.1±3.6 kya. The depth of sediments and volumes of cultural material were substantial at *c.* 58,000 and *c.* 48,000 years ago, suggesting intense, although relatively short, pulses of occupation.

MSA occupations occur immediately below Iron Age occupations; no LSA occupations are present in the site. The MSA cultural sequence is long, with pre-Still Bay, Still Bay, Howiesons Poort, post-Howiesons Poort, late and final MSA assemblages (Villa *et al.* 2005; Wadley 2005; 2007; Cochrane 2006; 2008; Delagnes *et al.* 2006; Villa & Lenoir 2006; Wadley & Jacobs 2006; Wadley & Mohapi 2008). The Sibudu Still Bay has bifacial points and bifacial tools, although most are broken. The Howiesons Poort Industry above the Still Bay is rich in backed tools, especially segments, but it lacks points. By about 58,000 years ago, backed tools were replaced by points and scrapers and between this period and 38,000 years ago, unifacially or bifacially retouched lithic points seem to have been consistently used as parts of weapons for hunting (Villa *et al.* 2005; Wadley 2005; Villa & Lenoir 2006). Use-trace analysis supports this interpretation; residues are extraordinarily well-preserved on stone tools from Sibudu and use-trace analyses have provided detailed information about aspects of MSA technology (Lombard 2004; 2005; 2006a,b; Williamson 2004; 2005). Quartz and quartzite predominated amongst the rock types used for stone tools in the earliest of the *c.* 58 kya layers (Cochrane 2006). Quartz and quartzite are not an important part of the lithic industries anywhere else in the sequence, where hornfels and dolerite are principally knapped, even in the Howiesons Poort. The Sibudu cultural sequence can be compared to that from Umhlatuzana Rockshelter (Kaplan 1990), approximately 90 km southwest of Sibudu (Fig. 5.1).

The combination at Sibudu of good organic preservation, deep deposits and many clearly defined stratigraphic layers provides fine resolution of data. Sibudu has detailed evidence for environmental conditions in OIS-4 and OIS-3 (Allott 2006; Plug 2004; 2006; Schiegl *et al.* 2004; Wadley 2004b; 2006c; Cain 2006; Glenny 2006; Herries 2006; Renaut & Bamford 2006; Schiegl & Conard 2006; Sievers 2006; Wells 2006; Clark & Plug 2008). The Howiesons Poort, pre-dating 60 kya, seems to have been created in an environment that included moist, evergreen forest (Allott 2006) with giant rat, blue duiker and other forest animals (Glenny 2006; Clark & Plug 2008). The Howiesons Poort hunting emphasis on small browsers switched in the post-Howiesons Poort to large and medium-sized animals. The people who lived at Sibudu were competent encounter-hunters who were able to kill a wide range of animals, including zebra, buffalo and bushpig (Plug 2004; Cain 2006; Wells 2006; Clark & Plug 2008).

The increased woodland at *c.* 48 kya seems to have contained more deciduous taxa than previously. Warmer temperatures and increased evapotranspiration at *c.* 48 kya may have resulted in an expansion of deciduous woodland and an increase in associated animals, but hunting with points continued.

Much worked bone (only some of it is published) is present in Howiesons Poort layers, including a bone point that may have been an arrowhead (Backwell *et al.* 2008). Little worked bone was present in the post-60 kya layers, however, a fragment of caudal rib with ten regular parallel notches made with a retouched stone tool (Cain 2004) was found in a layer with an age of *c.* 58 kya. The notches lack residues (B. Williamson pers. comm.) and there does not seem to be an obvious function for the piece. A thin bone pin that may have been ground and polished (Cain 2004) was found in a layer with an age of *c.* 38 kya.

Marine shell is present in Still Bay and Howiesons Poort layers, but not abundantly enough to represent food remains, so it seems likely that shell was deliberately transported from the coast as a raw material (Plug 2006). Amongst the shells are tiny perforated *Afrolittorina africana* that were possibly beads (d'Errico *et al.* 2008). The early worked bone and shell occurring at Sibudu shows that Blombos, about 1200 km southwest of Sibudu, is not an isolated case of the early use of such material culture in southern Africa.

The presence of hundreds of whole, sometimes immature, carbonized sedge nutlets (*Schoenoplectus* sp.) in most of the Sibudu MSA deposits (Sievers 2006) suggests that people were bringing sedges from the river into the shelter. Pollen from sedges, recovered from several layers, supports the suggestion that people rather than animals brought whole sedges to the shelter because sedge pollen is not windborne (Renaut & Bamford 2006). Today these spongy-textured sedges are particularly popular for traditional Zulu sleeping mats (Van Wyk & Gericke 2000). No matting has yet been found at Sibudu, but people may have used bundles of sedges for bedding; this is the first evidence in the MSA for such use of these plants. Micromorphological studies of Sibudu sediments have revealed parallel layering of sedge phytoliths that support the interpretation of sedge bedding (Goldberg *et al.* 2009).

Individual hearths at Sibudu look like the ones from Rose Cottage. However, there are many more hearths in the post-60 kya than the earlier layers. These often comprise overlapping burning events that spread horizontally across the excavation grid. A prominent feature of the *c.* 48 kya occupations is a palimpsest of burnt deposits that occurs towards the back of the shelter. Repeated burning episodes appear to be represented here; some of them may be accidental and post-depositional, whilst others may represent deliberate burning of old bedding and site maintenance (Goldberg *et al.* 2009). As in other sites (Speth 2006), these Sibudu palimpsests of hearths are set on the floor of the cave well away from the walls.

The contribution of Rose Cottage and Sibudu to knowledge about the MSA

Rose Cottage and Sibudu occur far from the Cape where most South African MSA research has been conducted. The two sites have settings different from each other: Rose Cottage is located in grassland far from the coast, while Sibudu is in forest/savanna near, but not on the coast. Both sites were occupied in Marine Isotope Stages (MIS) 4 and 3, but, unlike Sibudu, Rose Cottage was subsequently occupied in MIS-2 and it contains LSA occupation. The Sibudu Still Bay age of 70,500 years ago overlaps with ages for the Still Bay at Blombos, Diepkloof and Apollo 11 (Jacobs *et al.* 2008b), implying that makers of this industry, like the makers of the Howiesons Poort, coexisted in widespread parts of southern Africa. Age estimates for the Howiesons Poort Industry fall neatly between about 65,000 and 62,000 years ago at both sites, which also have long, well-dated post-Howiesons Poort sequences. Both sites have evidence for continued use of the same rock types throughout their occupations. This is in contrast to sites like Klasies River, where locally available rock types used for pre-Howiesons Poort tools were abandoned in favour of rocks brought from longer distances for use in the Howiesons Poort (Deacon 1995; Wurz 2002).

Neither Rose Cottage nor Sibudu display MSA spatial patterning as complex as that evidenced in LSA occupations; people living in these sites in the MSA seem to have been unconcerned about emphasizing group or individual identity through their manipulation of space. Similar spatial configurations to those documented in the Rose Cottage final MSA layers occur in the Middle Palaeolithic of Europe where, for example, jumbled hearths are a feature of occupation floors (Stringer & Gamble 1993; Mellars 1996) and refuse areas are sometimes apparent, but no specialized task groups are recognized (Vaquero *et al.* 2001, 591). The use of space at Sibudu is even more difficult to unravel than that at Rose Cottage because there are many overlapping hearths and burning events. However, because of Sibudu's organic preservation, we know that its occupants used sedge bedding and that they sometimes burnt this bedding (Sievers 2006; Goldberg *et al.* 2009) and cleaned up the site by burning bone (Cain 2005) and scraping out hearth ashes (Schiegl *et al.* 2004; Goldberg *et al.* 2009).

Material culture items recovered from Sibudu demonstrate that sites far from the Cape can yield a similar range of cultural items if preservation conditions are suitable. Sibudu worked bone lends support to the evidence for early worked bone at Blombos and Peers Cave (Henshilwood *et al.* 2001; d'Errico & Henshilwood 2007). Furthermore, the perforated *Afrolittorina africana* shells from Sibudu are reminiscent of the perforated *Nassarius kraussianus* shells at Blombos (d'Errico *et al.* 2005) and at sites much farther north in Morocco (d'Errico *et al.* 2009). Ostrich eggshell is absent from Sibudu MSA layers; ostriches do not occur in the area today and may not have done in the past either. Consequently, Sibudu lacks engraved eggshell like that in the Howiesons Poort Industry at Diepkloof (Parkington *et al.* 2005; Rigaud *et al.* 2006) and it also lacks eggshell beads of the kind found in MSA sites elsewhere in South Africa (Deacon 1995; Mason 1962; Plug 1982; Vanhaeren 2005). Ostriches do thrive in the Rose Cottage area, but the lack of organic preservation at the site precludes the occurrence of eggshell in the MSA deposits.

Repeated comments that Rose Cottage lacks the potential for yielding organic artefacts give the impression that sites lacking good organic preservation are not worth excavating. I suggest that this is too negative an approach and that we need to think creatively about how to deal with sites that share Rose Cottage Cave's challenges. It must be possible to look beyond organic artefacts that are supposedly mediated by symbolism. One potential way may be to examine ancient technical strategies to find within them evidence for mental steps that can only be taken by minds like our own. What is the complex cognition that we possess? It must include amongst its attributes 'cognitive fluidity' (Mithen 1996), the ability to exploit innovative thoughts, a capacity for novel and sustained multilevel operations (Amati & Shallice 2007), multi-tasking, abstract thought and the use of recursion and concepts of past and future (Barnard *et al.* 2007).

I have chosen to examine compound adhesive that occurs as microscopic residues at Sibudu and Rose Cottage on Howiesons Poort backed tools as well as on retouched points from pre-Howiesons Poort and post-Howiesons Poort contexts. The residues led me to explore the method of manufacture of compound adhesives.

Replications to create compound glues from plant gum and ochre powder demonstrate the complexity of the process (Wadley *et al.* 2009). The early artisans seem deliberately to have effected physical transformations to their ingredients; these included chemical changes from acidic to less acidic pH, dehydration of the adhesive near wood fires, and changes to mechanical workability and electrostatic forces. They did not simply create red glue; they irreversibly transformed gum and ochre to a concrete-like product. Experiments make it plain that some of the steps required for making compound adhesive are impossible without multi-tasking because the variability of natural ingredients means that no set recipe can be used and that the artisan has to make ongoing adjustments to ingredients while creating the glue (Wadley *et al.* 2009). The artisan needs to hold in mind what was previously done in order to carry out what is still needed for mixing glue, maintaining fire temperature and mentally rotating stone tools. Abstract thought is required in order to explain to others properties such as viscosity and workability. The mental attributes required for compound glue manufacture imply that there is overlap between the cognitive abilities of modern people and people in the MSA. The study provides a novel way of recognizing complex cognition in the MSA without necessarily invoking the concept of symbolism. This type of approach, which should be extended to technologies other than glue-making, is particularly important for sites like Rose Cottage, where the lack of organic preservation rules out the possibility of finding worked bone or symbolically-mediated items such as shell beads that may once have existed.

Fully modern, symbolling humans did not necessarily produce a repetitive package of symbolic traces (Brumm & Moore 2005) and we should not expect that the same symbolically-mediated cultural items occurred everywhere even when lithic sequences of sites overlap. However, deciding whether symbolic traces are absent because of cultural differences or issues of preservation is testing, as is demonstrated by the case studies of Rose Cottage and Sibudu. Looking beyond symbolism for recognizable attributes of complex cognition in technical processes seems to hold some promise for resolving the issue.

Acknowledgements

I thank Wendy Voorvelt for drawing Figure 5.1. Funding for the excavation of Rose Cottage and Sibudu has been received from the Centre for Scientific Research and the National Research Foundation, South Africa. The ideas incorporated in this paper are not necessarily held by either funding body. I thank the Institute for Human Evolution and the School of Geography, Archaeology and Environmental Studies for their continued support.

References

Allott, L.F., 2006. Archaeological charcoal as a window on palaeovegetation and wood use during the Middle Stone Age at Sibudu Cave. *Southern African Humanities* 18(1), 173–201.

Amati, D. & T. Shallice, 2007. On the emergence of modern humans. *Cognition* 103(3), 358–85.

Backwell, L., F. d'Errico & L. Wadley, 2008. Middle Stone Age bone tools from the Howiesons Poort layers, Sibudu Cave, South Africa. *Journal of Archaeological Science* 35, 1566–80.

Barnard, P.J., D.J Duke, R.W. Byrne & I. Davidson, 2007. Differentiation in cognitive and emotional meanings: an evolutionary analysis. *Cognition and Emotion* 21, 1155–83.

Bouzouggar A, N. Barton, M. Vanhaeren *et al.*, 2007. 82,000-year-old shell beads from North Africa and implications for the origins of modern human behavior. *Proceedings of the National Academy of Sciences of the USA* 104, 9964–9.

Brumm, A. & M.W. Moore, 2005. Symbolic revolutions and the Australian archaeological record. *Cambridge Archaeological Journal* 15(2), 157–75.

Cain, C.R., 2004. Notched, flaked and ground bone artefacts from Middle Stone Age and Iron Age layers of Sibudu Cave, KwaZulu-Natal, South Africa. *South African Journal of Science* 100, 195–7.

Cain, C.R., 2005. Using burned bone to look at Middle Stone Age occupation and behavior. *Journal of Archaeological Science* 32, 873–84.

Cain, C.R., 2006. Human activity suggested by the taphonomy of 60 ka and 50 ka faunal remains from Sibudu Cave. *Southern African Humanities* 18(1), 241–60.

Clark, A.M.B., 1997a. The MSA/LSA transition in southern Africa: new technological evidence from Rose Cottage Cave. *South African Archaeological Bulletin* 52, 113–21.

Clark, A.M.B., 1997b. The final Middle Stone Age at Rose Cottage Cave: a distinct industry in the Basutolian ecozone. *South African Journal of Science* 93, 449–58.

Clark, J.L. & I. Plug, 2008. Animal exploitation strategies during the South African Middle Stone Age: Howiesons Poort and post-Howiesons Poort fauna from Sibudu Cave. *Journal of Human Evolution* 54, 886–98.

Cochrane, G.W.G., 2006. An analysis of lithic artefacts from the ~60 ka layers of Sibudu Cave. *Southern African Humanities* 18(1), 69–88.

Cochrane, G.W.G., 2008. The transition from Howieson's Poort to post-Howieson's Poort Industries in southern Africa. *South African Archaeological Society Goodwin Series* 10, 157–67.

Deacon, H.J., 1995. Two late Pleistocene–Holocene archaeological depositories from the Southern Cape, South Africa. *South African Archaeological Bulletin* 50, 121–31.

Delagnes, A., L. Wadley, P. Villa & M. Lombard, 2006. Crystal quartz backed tools from the Howiesons Poort at Sibudu Cave. *Southern African Humanities* 18(1), 43–56.

Donald, M., 1991. *Origins of the Modern Mind: Three Stages in the Evolution of Culture and Cognition*. Cambridge (MA): Harvard University Press.

Donald, M., 1998. Hominid enculturation and cognitive evolution, in *Cognition and Material Culture: the Archaeology of Symbolic Storage*, eds. C. Renfrew & C. Scarre. (McDonald Institute Monographs.) Cambridge: McDonald Institute for Archaeological Research, 7–17.

d'Errico, F. & C.S. Henshilwood, 2007. Additional evidence for bone technology in the southern African Middle Stone Age. *Journal of Human Evolution* 52(2), 142–63.

d'Errico, F., C. Henshilwood, M. Vanhaeren & K. Van Niekerk, 2005. *Nassarius kraussianus* shell beads from Blombos Cave: evidence for symbolic behaviour in the Middle Stone Age. *Journal of Human Evolution* 48, 2–14.

d'Errico, F., M.Vanhaeren & L.Wadley, 2008. Possible shell beads from the Middle Stone Age of Sibudu Cave. *Journal of Archaeological Science* 35, 2675–85.

d'Errico, F., M. Vanhaeren, N. Barton *et al.*, 2009. Additional evidence on the use of personal ornaments in the Middle Paleolithic of North Africa. *Proceedings of the National Academy of Sciences of the USA*, doi_10.1073_pnas.0903532106.

Fisher, G. & D.D. Loren, 2003. Embodying identity in archaeology. *Cambridge Archaeological Journal* 13(2), 225–30.

Gibson, N.E., L.Wadley & B.S. Williamson, 2004. Microscopic residues as evidence of hafting on backed tools from the 60 000 to 68 000 year-old Howiesons Poort layers of Rose Cottage Cave, South Africa. *Southern African Humanities* 16, 1–11.

Glenny, W., 2006. Report on the micromammal assemblage analysis from Sibudu Cave. *Southern African Humanities* 18 (1), 270–88.

Goldberg, P., C.E. Miller, S. Schiegl *et al.*, 2009. Bedding, hearths, and site maintenance in the Middle Stone Age of Sibudu Cave, KwaZulu-Natal, South Africa. *Archaeological and Anthropological Sciences* 1(2), 95–122.

Harper, P.T.N., 1997. The Middle Stone Age sequences at Rose Cottage Cave: a search for continuity and discontinuity. *South African Journal of Science* 93, 470–75.

Henshilwood, C.S. & C.W. Marean, 2003. The origin of modern human behaviour. *Current Anthropology* 44(5), 627–51.

Henshilwood, C., F. d'Errico, C.W. Marean, R.G. Milo & R. Yates, 2001. An early bone tool industry from the Middle Stone Age at Blombos Cave, South Africa: implications for the origins of modern human behaviour, symbolism and language. *Journal of Human Evolution* 41, 631–78.

Henshilwood, C., F. d'Errico, M. Vanhaeren, K. Van Niekerk & Z. Jacobs, 2004. Middle Stone Age shell beads from South Africa. *Science* 304, 404.

Herries, A.I. R., 2006. Archaeomagnetic evidence for climate change at Sibudu Cave. *Southern African Humanities* 18(1), 131–47.

Jacobs, Z., A.G. Wintle, G.A.T. Duller, R.G. Roberts & L. Wadley, 2008a. New ages for the post-Howiesons Poort, late and final Middle Stone Age at Sibudu Cave, South Africa. *Journal of Archaeological Science* 35, 1790–807.

Jacobs, Z., R.G. Roberts, R.F. Galbraith *et al.*, 2008b. Ages for the Middle Stone Age of Southern Africa: implications for human behavior and dispersal. *Science* 322, 733–5.

Joyce, R.A., 2003. Making something of herself: embodiment in life and death at Playa de los Muertos, Honduras. *Cambridge Archaeological Journal* 13(2), 248–61.

Kaplan, J., 1990. The Umhlatuzana rock shelter sequence: 100 000 years of Stone Age history. *Natal Museum Journal of Humanities* 2, 1–94.

Lombard, M., 2004. Distribution patterns of organic residues on Middle Stone Age points from Sibudu Cave, KwaZulu-Natal, South Africa. *South African Archaeological Bulletin* 59, 37–44.

Lombard, M., 2005. Evidence of hunting and hafting during the Middle Stone Age at Sibudu Cave, KwaZulu-Natal, South Africa: a multianalytical approach. *Journal of Human Evolution* 48, 279–300.

Lombard, M., 2006a. First impressions on the functions and hafting technology of Still Bay pointed artefacts from Sibudu Cave. *Southern African Humanities* 18(1), 27–41.

Lombard, M., 2006b. Direct evidence for the use of ochre in the hafting technology of Middle Stone Age tools from Sibudu Cave, KwaZulu-Natal. *Southern African Humanities* 18(1), 57–67.

Mason, R.J., 1962. *Prehistory of the Transvaal*. Johannesburg: Witwatersrand University Press.

McBrearty, S. & A.S. Brooks, 2000. The revolution that wasn't: a new interpretation of the origin of modern human behaviour. *Journal of Human Evolution* 39, 453–563.

Mellars, P., 1996. *The Neanderthal Legacy: an Archaeological Perspective from Western Europe*. Princeton (NJ): Princeton University Press.

Mithen, S., 1996. *The Prehistory of the Mind*. London: Thames and Hudson.

Parkington, J., C. Poggenpoel, J.-P. Rigaud & P.-J. Texier, 2005. From tool to symbol: the behavioural context of intentionally marked ostrich eggshell from Diepkloof, Western Cape, in *From Tools to Symbols: From Early Hominids to Modern Humans*, eds. F. d'Errico & L. Backwell. Johannesburg: Witwatersrand University Press, 475–92.

Pienaar, M., S. Woodborne & L. Wadley, 2008. Optically stimulated luminescence dating at Rose Cottage Cave. *South African Journal of Science* 104, 65–70.

Plug, I., 1982. Bone tools and shell, bone and ostrich eggshell beads from Bushman Rock Shelter (BRS), eastern Transvaal. *South African Archaeological Bulletin* 37, 57–62.

Plug, I., 2004. Resource exploitation: animal use during the Middle Stone Age at Sibudu Cave, KwaZulu-Natal. *South African Journal of Science* 100, 151–8.

Plug, I., 2006. Aquatic animals and their associates from the Middle Stone Age levels at Sibudu Cave. *Southern African Humanities* 18 (1), 289–99.

Renaut, R. & M. Bamford, 2006. Results of preliminary palynological analysis of samples from Sibudu Cave. *Southern African Humanities* 18(1), 235–40.

Rigaud, J.-P., P.-J. Texier, J. Parkington & C. Poggenpoel, 2006. Le mobilier Stillbay et Howiesons Poort de l'abri Diepkloof. La chronologie du Middle Stone Age sud-africain et ses implications. *Palévol* 5(6), 839–49.

Schiegl, S. & N.J. Conard, 2006. The Middle Stone Age sediments at Sibudu: results from FTIR spectroscopy and microscopic analyses. *Southern African Humanities* 18 (1), 149–72.

Schiegl, S., P. Stockhammer, C. Scott & L. Wadley, 2004. A mineralogical and phytolith study of the Middle Stone Age hearths in Sibudu Cave, KwaZulu-Natal, South Africa. *South African Journal of Science* 100, 185–94.

Sievers, C., 2006. Seeds from the Middle Stone Age layers at Sibudu Cave. *Southern African Humanities* 18(1), 203–22.

Soriano, S., P. Villa & L. Wadley, 2007. Blade technology and tool forms in the Middle Stone Age of South Africa: the Howiesons Poort and post-Howiesons Poort at Rose Cottage Cave. *Journal of Archaeological Science* 34(5), 681–703.

Speth, J. D., 2006. Housekeeping, Neandertal style: hearth placement and midden formation in Kebara Cave (Israel), in *Transitions before the Transition: Evolution and Stability in the Middle Palaeolithic and Middle Stone Age*, eds. E. Hovers & S.L. Kuhn. New York (NY): Springer, 171–88.

Stringer, C. & C. Gamble, 1993. *In Search of the Neanderthals*. London: Thames & Hudson.

Valladas, H., L. Wadley, N. Mercier *et al.*, 2005. Thermoluminescence dating on burnt lithics from Middle Stone Age layers at Rose Cottage Cave. *South African Journal of Science* 101, 169–74.

Vanhaeren, M., 2005. Speaking with beads: the evolutionary significance of personal ornaments, in *From Tools to Symbols: From Early Hominids to Modern Humans*, eds. F. d'Errico & L. Backwell. Johannesburg: Witwatersrand University Press, 525–53.

Vaquero, M., G. Chacón, C. Fernández, K. Martinez & J.M. Rando, 2001. Intrasite spatial patterning and transport in the Abric Romaní Middle Paleolithic site (Capellades, Barcellona, Spain), in *Settlement Dynamics of the Middle Palaeolithic and Middle Stone Age*, vol. I, ed. N.J. Conard. Tübingen: Kerns Verlag, 573–95.

Villa, P. & M. Lenoir, 2006. Hunting weapons of the Middle Stone Age and the Middle Palaeolithic: spear points from Sibudu, Rose Cottage and Bouheben. *Southern African Humanities* 18(1), 89–122.

Villa, P., A. Delagnes & L. Wadley, 2005. A late Middle Stone Age artefact assemblage from Sibudu (KwaZulu-Natal): comparisons with the European Middle Palaeolithic. *Journal of Archaeological Science* 32, 399–422.

Wadley, L., 1997. Rose Cottage Cave: archaeological work 1987 to 1997. *South African Journal of Science* 93, 439–44.

Wadley, L., 2001. What is cultural modernity? A general view and a South African perspective from Rose Cottage Cave. *Cambridge Archaeological Journal* 11(2), 201–21.

Wadley, L., 2004a. Late Middle Stone Age spatial patterns in Rose Cottage Cave, South Africa, in *Settlement Dynamics of the Middle Palaeolithic and Middle Stone Age*, vol. II, ed. N.J. Conard. Tubingen: Kerns Verlag, 23–36.

Wadley, L., 2004b. Vegetation changes between 61 500 and 26 000 years ago: the evidence from seeds in Sibudu Cave, KwaZulu-Natal. *South African Journal of Science* 100, 167–73.

Wadley, L., 2005. A typological study of the final Middle Stone Age stone tools from Sibudu Cave, KwaZulu-Natal. *South African Archaeological Bulletin* 60, 51–63.

Wadley, L., 2006a. Revisiting cultural modernity and the role of ochre in the Middle Stone Age, in *The Prehistory of Africa*, ed. H. Soodyall. Johannesburg: Jonathan Ball Publishers, 49–63.

Wadley, L., 2006b. The use of space in the late Middle Stone Age of Rose Cottage Cave, South Africa, in *Transitions before the Transition: Evolution and Stability in the Middle Palaeolithic and Middle Stone Age*, eds. E. Hovers & S.L. Kuhn. New York (NY): Springer, 279–94.

Wadley, L., 2006c. Partners in grime: results of multi-disciplinary archaeology at Sibudu Cave. *Southern African Humanities* 18(1), 315–41.

Wadley, L., 2007. Announcing a Still Bay Industry at Sibudu Cave. *Journal of Human Evolution* 52, 681–9.

Wadley, L. & P.T. Harper, 1989. Rose Cottage Cave revisited: Malan's Middle Stone Age collection. *South African Archaeological Bulletin* 44, 23–32.

Wadley, L. & Z. Jacobs, 2006. Sibudu Cave: background to the excavations, stratigraphy and dating. *Southern African Humanities* 18 (1), 1–26.

Wadley, L. & M. Mohapi, 2008. A segment is not a monolith: evidence from the Howiesons Poort of Sibudu, South Africa. *Journal of Archaeological Science* 35, 2594–605.

Wadley, L., B.S. Williamson & M. Lombard, 2004. Ochre in hafting in Middle Stone Age southern Africa: a practical role. *Antiquity* 78, 661–75.

Wadley, L., T. Hodgskiss & M. Grant, 2009. Implications for complex cognition from the hafting of tools with compound adhesives in the Middle Stone Age, South Africa. *Proceedings of the National Academy of Sciences of the USA* 106(24), 9590–94.

Wells, C.R., 2006. A sample integrity analysis of faunal remains from the RSp layer at Sibudu Cave. *Southern African Humanities* 18(1), 261–77.

Wilkie, L.A. & K.M. Bartoy, 2000. A critical archaeology revisited. *Current Anthropology* 41(5), 747–61.

Williamson, B.S., 2004. Middle Stone Age tool function from residue analysis at Sibudu Cave. *South African Journal of Science* 100, 174–8.

Williamson, B.S., 2005. Subsistence strategies in the Middle Stone Age at Sibudu Cave: the microscopic evidence from stone tool residues, in *From Tools to Symbols: From Early Hominids to Modern Humans,* eds. F. d'Errico & L. Backwell. Johannesburg: Witwatersrand University Press, 493–511.

Wurz, S., 2002. Variability in the Middle Stone Age lithic sequence, 115,000–60,000 years ago at Klasies River, South Africa. *Journal of Archaeological Science* 29, 1001–15.

Van Wyk, B.-E. & N. Gericke, 2000. *People's Plants: a Guide to Useful Plants of Southern Africa.* Pretoria: Briza.

Chapter 6

The Dispersal of Modern Humans into Australia

Peter Veth

Preamble

On current evidence it appears that modern humans dispersed from Africa, across southern Asia and into Greater Australia (or Sahul) some time between 100,000 to 50,000 years ago (cf. Balme *et al.* 2009; O'Connell & Allen 2007; Lahr & Foley 2004). This spread was likely rapid and I believe these peoples were 'infinitely adaptable' (Veth 2005a). In this paper I will argue that the movement of anatomically modern humans through the increasingly depauperate islands and biogeographic boundaries of the Indonesian Archipelago towards Greater Australia was a unique event — especially given that Australia was the most arid continent colonized by modern humans (Fig. 6.1). A combination of factors including island biogeography and increasing aridity served in concert, I believe, to create uniquely modern cultural configurations. Some of these include the earliest global evidence for long-distance sea voyaging, cremation of the dead, production of 'art', personal ornamentation, long-distance exchange of goods and graphic systems and, not the least, entry into and persistence within a major desert continent.

Pleistocene cycles of both cold and hot deserts — such as the ice sheets of Europe and the expanding deserts of Australia (cf. Smith & Hesse 2005) provide a canvas for population dispersal and isolation. Refuges — whether located in oceans, ice sheets or deserts (after Veth 1993) are arguably the powerhouses for biological evolution, cultural innovation and ultimately the emergence of differences in social dynamics.

A Eurocentric perspective of deserts, as captured in the imagination of novelists, film makers, theologians and even earlier anthropologists has seen them as harsh and uncompromising landscapes. They are equated with extreme experiences including revelations, religious epiphanies, treks through the 'wilderness' and privations of the most horrible kind.

And yet archaeological evidence from around the world suggests that people actually chose to enter desert environments willingly or persisted in their occupation of landscapes that subsequently became arid (cf. chapters in Veth *et al.* 2005b). This occurred as early as 40,000 to 60,000 years ago — and possibly earlier. Persistence of modern humans in the glacial landscapes of Europe and Eurasia also occurred during this time, likely underpinned by complex information-sharing behaviours — reflected in widespread sharing of iconographic objects such as Venus figurines. Indeed Gamble *et al.* (2004) have noted parallels in regional occupation patterns from the northern hemisphere in response to glacial cycles associated with the Last Glacial Maximum (LGM) to archaeological trends from the LGM of the southern hemisphere (after Veth 1993; 2005a). This creates the fascinating scenario in which modern humans appear to have explored and settled the worlds' deserts at the same time they developed other 'modern' attributes underpinning their successful global diaspera; such as complex information exchange and long-distance maritime voyaging (Balme *et al.* 2009; Veth 2005b).

How old are the Australians?

Despite extravagant claims that Australians may be up 1,000,000 years old and that *Homo floresiensis* (the Hobbit) may have reached Australian shores (R. Roberts, ABC TV 2004) our collaborative research over the last 15 years from East Timor and the Aru Islands, combined with what we know from the rest of Greater Australia (Sahul), demonstrates the presence of anatomically modern humans whose *behaviours* can only be described as 'modern'. All of these sites date to within the last 60,000 BP years — even adopting the most 'optimistic' dating scenarios (cf. O'Connor & Chappell 2003).

A great deal has been written recently about the date people first arrived in Australia and the reliability of both the different radiometric methods used to achieve this end and the measurements obtained (cf. chapters in O'Connor & Veth 2000). There is essentially

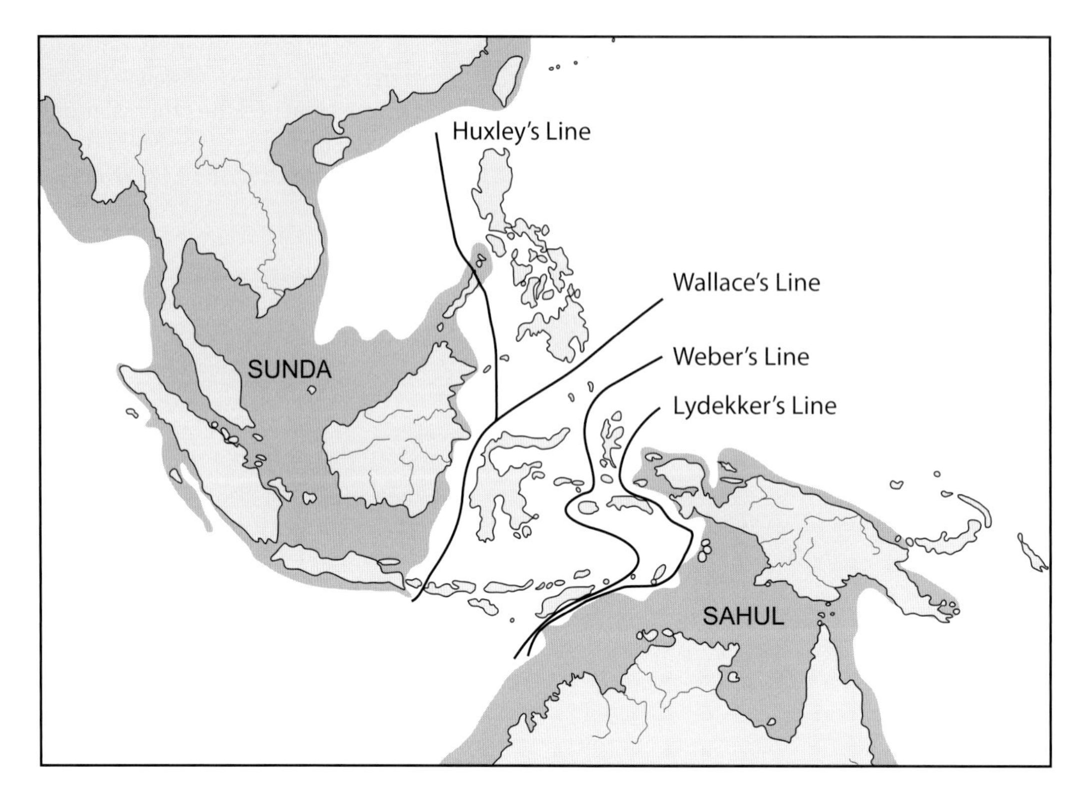

Figure 6.1. *Location of landmasses of Sunda and Sahul (as configured during lower sea stands of the Late Pleistocene), islands of Wallacea which were not connected during the time of the diaspera of hominins and major biogeographic boundaries of relevance to faunal distributions. Until the recent reporting of* Homo floresiensis *(cf. Morwood* et al. *2004) it was held that only anatomically modern humans had crossed Wallace's Line (i.e. had made any kind of water crossing; here between Bali and Lombok).*

a 'long chronology' of about 60,000 years based on luminescence dating techniques, and a 'short chronology' of no more than 45,000 years based on traditional radiocarbon dating approaches (O'Connell & Allen 2004; 2007).

In addition to concerns about luminescence techniques generally, many proponents of the short chronology question the archaeological association of cultural remains with the non-cultural sediments these techniques actually date. Others argue that occupation of Australia around 60,000 years ago ignores evidence concerning the chronology of dispersal of modern humans elsewhere in the world, something the 'long chronologists' argue is being, or soon will be undermined, by new discoveries in East Asia.

Sue O'Connor, Peter Hiscock and I (amongst others) have been active in this debate and have drawn attention to the fact that not enough effort has been expended on discussing the nature of human activity following colonization and the changes that occurred during the Pleistocene as colonists began to explore the enormous and diverse landmass of Sahul (i.e. the expanded land mass of Australia, Papua New Guinea, Tasmania and the Aru Islands which were joined during low sea stands for much of the Pleistocene). For example, to what degree were colonists coastally-tethered (if at all) and what was their capacity to make the arid landscapes of the interior their territory (see Hiscock 2008; O'Connor & Chappell 2003; O'Connor & Veth 2000)? There are three principal reasons for the relative dearth of Pleistocene site reports. Firstly, the issue of chronology has taken on a life of its own and might even be seen now to lead the investigation of Pleistocene archaeology in Australia (Allen 2003, 38). Secondly, there is still a paucity of published analysis of material finds from, or following, the period of initial occupation and then on through the major environmental fluctuations of the Pleistocene — and specifically the long, cold and cyclically arid LGM lasting from *c.* 28,000–18,000 BP (see Lilley 2006; and comments below). Thirdly, many early sites only have stone artefacts preserved, and very small samples at that, upon which it is difficult to build comprehensive interpretations of past lifeways. For example, at Devil's Lair in the southwest of the continent, currently the oldest Australian site based on the radiocarbon chronology, a mere 15 stone artefacts occur in the earliest claimed human occupation deposits dated between 44,500 BP and 47,000 BP (Turney *et al.* 2001).

In recent times Australian archaeologists have increasingly eschewed earlier stereotypes that contrasted the Pleistocene as a time of relative stasis in which hunter-gatherers were few, sparsely spread over the continent and culturally 'simple'. The traditional view saw the Holocene as the time of rapid socio-economic and technological change. In a recent paper O'Connor & Veth (2006) have characterized archaeological evidence for different regions of Australia from the Pleistocene and into the Holocene of Australia.

We suggested that marked changes occurred in settlement dynamics and demography in the Pleistocene from the time of first occupation and through the arid phases associated with the LGM. We argued that these changes were reflected in archaeological sequences as changes in intensity of occupation intensity and frequency of site visits which could be measured through variations in quantities of cultural materials discarded against radiocarbon ages and other proxy data. In a nutshell — the chapter makes an extended case for early complex behaviours and the existence of different regional signatures *before* the Last Glacial Maximum.

These behaviours included different technological organizational strategies (specifically with reference to lithics), varying configurations in group mobility and contrasting dietary suites. In fact we mischievously labelled some of these early complex cultural complexes as evidence for 'Pleistocene intensification' — mainly as a counterpoint to the tediously unidirectional, cultural evolutionist models (recently critiqued in Hiscock 2008, 218) which only see efflorescence in social and technological complexity during the last 4000 to 3000 years (as summarized in Mulvaney & Kamminga 1999). Having said that, the development of a Neo-Marxist social archaeology, as seeded by Harry Lourandos for the late Holocene of Australia (cf. Veth 2006), did fuel ideational and symbolic explanations for real changes in the archaeological record. These included new art traditions — such as the totemic *Wandjina* paintings of the Kimberley, increases in use of ceremonial aggregation sites and more bounded territories for mobile hunter-gatherers in fertile areas of the continent.

Across Wallace's Line

In looking at a 'transect of evidence' for humans from Sunda, across Wallacea, to Sahul we are faced with a new dispersal line for *Homo erectus* (and possibly its descendants) which currently ends at Flores (see Fig. 6.1). On current evidence there is no basis to believe that *Homo erectus* or *Homo floresiensis* (the now famous Hobbit of Flores) reached the other islands of Wallacea (such as Sulawesi, islands of the Maluku region, Timor and Roti) — let alone Greater Australia. Following a brief overview of some current criteria for modern human behaviours, and their material manifestations in Timor and Sahul, we can then take a fresh look westwards towards the Wallace Line.

There is now unequivocal evidence for human occupation from at least 30,000 to 42,000 solar years from East Timor, Roti, Aru, Halmahera, Ambon, Gebe and Sulawesi (Fig. 6.2). There are no earlier dates or skeletal evidence for the genus *Homo* from these myriad islands of Wallacea. There has been exciting new evidence for the dispersal of *Homo erectus*, and its possible descendant *Homo floresiensis*, as far east as the island of Flores (cf. Brown *et al.* 2004; Morwood *et al.* 2004; Westaway *et al.* 2007). This would assume the dispersal of pre-modern hominins from Java across Wallace's Line — bridging the Bali–Lombok divide — which only involves a short sea crossing or drifting on prevailing currents southward from Sulawesi down to islands of east Indonesia.

Early sites from East Timor and the Aru Islands

Collaborative survey and excavation programs by Sue O'Connor, Matthew Spriggs and myself (with a range of Australian and Indonesian colleagues) were carried out over ten years in the Aru Islands (Maluku Province) of eastern Indonesia and in East Timor (key East Timor sites and the Aru Islands shown in Fig. 6.2). The impetus for our research was the presence of unequivocal 45,000-year-old dates for occupation of Australia and the lack of equivalent-aged sites on the islands of Wallacea (O'Connor *et al.* 2005; O'Connor & Veth 2006).

The team worked in the Aru Islands from 1995–99, deciding to cease work then owing to escalating religious violence. On invitation of the provisional Timor Leste government it then moved its research focus to East Timor in 2000. That work was conducted collaboratively for the next five years and then by O'Connor until the present.

Research in the Aru Islands (once part of Greater Australia) provided the first Pleistocene-aged cultural assemblages dating back to over 30,000 years from large limestone caves, with high levels of chronometric hygiene including cross-checking of dates from charcoal features, *Celtis* seeds, Uranium-Thorium dates from flowstones capping and underlying deposits, AMS dates on shell artefacts (XRD-tested) and a variety of bone dates (O'Connor *et al.* 2005). The faunal

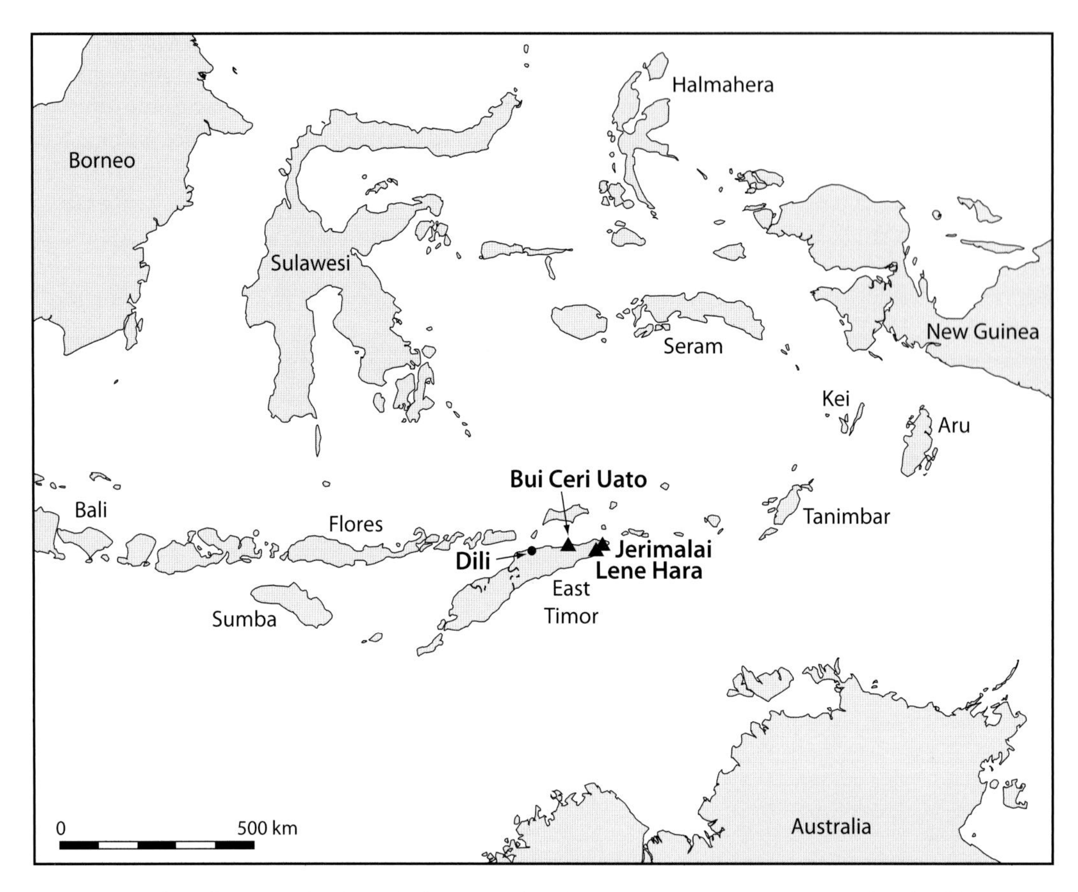

Figure 6.2. *Location of Wallacean islands including Sumba, Flores, Timor, Sulawesi and islands of the Maluku region such as Seram, Halmahera, Kei and Tanimbar. These have not been joined during the time of likely dispersal of* Homo erectus *or any other ancestral hominin. The Aru Islands, Papua New Guinea and Australia were joined as one landmass for much of this period, however, as Sahul (see Fig. 6.1). Two key Pleistocene sites of East Timor shown including Lene Hara and Ian Glover's original site of Bui Ceri Uato (see discussion of Pleistocene sequences in O'Connor* et al. *2002a; map adapted from O'Connor 2010, fig. 4.1).*

sequence suggests that from before 30,000 years ago the site was located near pockets of riparian forest along watercourses surrounded by open grassed savannah. Predation focused on the large Agile wallaby, *Macropus agilis,* and several smaller wallabies (*Thylogale* spp.) with lesser proportions of phalangerid, bandicoot and python. Game was so abundant (apparently 'naïve' in the earliest levels) that carcasses appear to have only been partially processed for their available protein. Following insulation of Aru, wetter forests spread, macropod populations declined and there was a shift towards smaller forest game such as phalangers and fruitbats.

A significant discovery from Aru, which has been reported on elsewhere (chapters in O'Connor *et al.* 2005), is the earliest skeleton from northern Sahul; that of a woman buried at around 20,500 BP. She was a tall, large-brained woman, and apparently in a secondary burial context.

Systematic surveys and excavation of sites by researchers from Australia, Hawaii and Indonesia in southeast, central and northern Maluku (the Aru Islands, Buru, Halmahera, Gebe and Morotai) have all now provided sites of a similar Pleistocene age.

The Pleistocene/Holocene units from the Aru Islands reveal the presence of complex implements, such as edge-ground points made on macropod bone, shell scrapers on *Geloina* and most importantly changing dietary and technological suites in concert with changes in climate, vegetation and landmass.

Five years of survey, recording and excavation of open and rockshelter sites along the northern and eastern portion of East Timor have produced an array of sites spanning some 42,000 years (O'Connor 2007; 2010; O'Connor *et al.* 2002a,b; O'Connor & Veth 2005; Spriggs *et al.* 2003; Veth *et al.* 2004).

In 2002 we carried out further excavations at Lene Hara cave — which we had previously dated to *c.* 35,000 BP (Fig. 6.2). At another part of this large cave a very rich sequence was revealed which has been dated to *c.* 10,000 BP containing a considerable ceramic assemblage in the upper unit. Notably the excavations produced a range of shell beads, flaked shell tools and polished shell fish hooks. Direct AMS dates on the lowest hook (and beads) have returned figures of approximately 10,000 BP — clearly pre-dating any Austronesian techno-complex by at least 5000 years (O'Connor & Veth 2005). The East Timor fish hook dated to 9741±60 BP (R-28145/3) represents the first firm evidence for the manufacture of fish hooks at this early period in Island Southeast Asia, Melanesia or the Pacific and suggests continuity in fishing practices between pre-Neolithic and Neolithic levels in East Timor. Its similarity to small baited jabbing hooks from Lapita contexts in the Pacific suggests that this technology may have its origins in early Holocene or even Pleistocene contexts in the rich marine environs of the Indonesian Archipelago, rather than from Austronesian dispersal as part of a Lapita package.

Remains of large pelagic off-shore fish are found in the same units as these early fish hooks and have now been recovered from the nearby site of Jerimalai remarkably back to *c.* 42 kya BP (O'Connor 2010). Competent off-shore marine resource exploitation is evident from these sites — providing further evidence for the maritime capabilities of pre-Austronesian peoples. The 200 km sea journey to Manus at 20,000 BP and the earlier colonization(s) of Sahul increasingly speak to maritime skills normally assigned to the founders of Near and Remote Oceania.

Other major cave sites excavated in East Timor include Matja Kuru 1 and 2. These sites are located near the largest freshwater lake in East Timor and lie 10 km from the coast. They date back to 32,000 BP and contain significant maritime fauna — as well as prey from the lake such as freshwater turtle.

A notable feature of Matja Kuru 2 is a dog burial directly dated by AMS to 3150–3100 cal. BP — the oldest from Island Southeast Asia. The practice of burying magical dogs continues up until this day.

A major point of comparison emerges between the Aru Islands and East Timor. Prior to the introduction of domestic animals the endemic (non-marine) fauna of East Timor was extremely depauperate compared to Aru; the latter with its Sahul connections. Whilst Timor had only rats, bats and reptiles; Aru had these as well as a range of small-, medium- and large-bodied marsupials and the ground-dwelling cassowary. Although at roughly similar latitudes the onset of aridity appears to have advantaged hunter-gather groups in Aru as the closed rainforest communities gave way to the grassland habitats required by the larger macropods such as *Macropus agilis.*

Conversely in East Timor, evidence of occupation during this period is sparse. Even prior to the onset of aridity interior assemblages suggest reliance on coastal resources. Drier conditions in Wallacea would presumably have reduced forest cover and fruit- and seed-bearing trees. Although the habitat requirements of the giant rat are unknown, more open vegetation is unlikely to have advantaged them or the large bats. Perhaps the focus of occupation shifted closer to the coast or into wooded areas at higher altitudes.

The point here is that groups appear to be demonstrating high degrees of adaptability by moving through the Wallacean Islands (East Timor was not connected to Australia) and then making landfall on Sahul (the Aru islands were connected to Australia) in an archaeologically instantaneous 'moment', insofar as available dates allow us to infer.

Desert sites from Australia

From Australia we now have a range of interior 'desert-edge' sites from the Pleistocene — where groups appear either to have persisted, or changed their territory, in response to increasing aridity associated with the longer, colder LGM lasting from 28,000 to 18,000 BP with perhaps the driest episode occurring towards the end of this phase (after Lambeck & Chappell 2001; Chappell 2004; Veth *et al.* 2009).

Richard Gould's pioneering work in the Western Desert between 1965 and 1980 was strongly influenced by his ethnoarchaeological experience. Following his excavations at Puntutjarpa (Figs. 6.3 & 6.4) and James Range rockshelters, he portrayed a uniform and conservative 'Desert Culture' similar to that of the North American Great Basin (Gould 1969; 1980). The culture was Holocene in age, assumed the ethnographically-described configuration of Australian desert culture and was argued to display great uniformity in material culture, economy and settlement patterns through time.

Work in the adjacent Hammersley Ranges (Fig. 6.4) of the Pilbara uplands by others at the same time suggested instead a deeper antiquity of occupation and greater dynamism in settlement patterns and other aspects of culture (Veth 1995; 2005a).

During the mid-1980s there was a convergence in models by three researchers working in different areas of the arid zone: Smith (1988) focusing on the ranges of central Australia, Hiscock studying gorge refugia in northeastern Australia, and Veth (1989) working in the Western Desert and Pilbara. In a nutshell, these studies suggested three things. First, occupation of the arid zone was Pleistocene in age. Second, there had been changes in the size of group

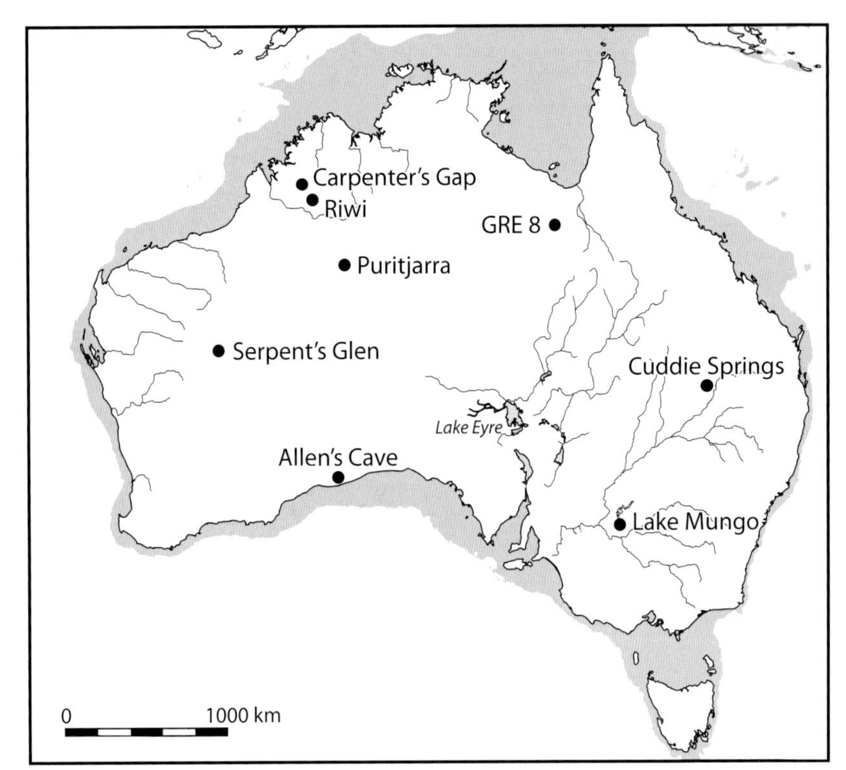

Figure 6.3. *Contemporary Australian landmass and shaded extension of continental shelf reflecting –130 m bathymetric contour associated with the peak of the Last Glacial Maximum (after Lambeck & Chappell 2001). Key desert-edge sites shown here include Carpenter's Gap and Gregory Gorge 8 which have persistence of occupation and significant shifts in their lithic and dietary assemblages during the LGM. Riwi and Serpent's Glen have a hiatus in occupation coincident with the LGM. Puritjarra has sparse occupation during the LGM. Allen's Cave, Cuddie Springs and the Lake Mungo sites reveal occupation during the LGM with the latter demonstrating major shifts in dietary breadth, prey choice and inferred group mobility in response to the onset of aridity; see chapters in Veth* et al. *(2005). (Map adapted from O'Connor 2010, fig. 4.1.)*

territories in response to environmental shifts attending changes in aridity throughout the Pleistocene. People withdrew into biogeographic refuges such as permanently-watered gorges in drier periods and expanded out again with climatic amelioration (see Fig. 6.3). Finally, social and economic changes were apparent in the late Holocene.

These patterns constitute a very different scenario from the conservative, Holocene-aged 'Desert Culture' that had been the dominant paradigm up until that time.

On the strength of these new understandings, I began to look at diversity in the Australian deserts. I divided them using a biogeographic model based on the distribution of ecological 'refuges' from aridity, 'corridors' between refuges and 'barriers' to movement (Veth 1993). The model made a case for a delay in the occupation of linear dunefields until the mid-Holocene. Critique and empirical testing of the model's assumptions over the last 20 years has seen its core themes generally reinforced and supported (cf. review chapters in Smith & Hesse 2005; Veth *et al.* 2005b). The proposal that early occupation of the sandy deserts was delayed, however, has now been overturned by the discovery of Pleistocene sites such as Serpent's Glen rockshelter, Riwi and the Parnkupirti site in these areas (cf. O'Connor *et al.* 1998; Veth *et al.* 2009; Fig. 6.4).

Island biogeography, deserts and human adaptability

The *infinite adaptability* of early groups and their apparent ability to utilize most landscapes (and resource suites) is evident from Sulawesi, Muluku Province, West Papua and northern Australia.

Selective pressures — from islandization and aridity — appear to have been successfully mediated by *cultural* responses such as changes in diet, territoriality and settlement patterns (cf. Veth 2003).

Most of the recent discussion about what traits or sets of traits can be said to constitute modern human behaviour derive from the African literature where debate has centred on the Middle Stone Age and whether the archaeological evidence for the period from 300,000–100,000 years ago can be said to represent the actions of behaviourally modern *Homo sapiens*. The archaeological signatures of such modern behaviours are typically argued (see Brumm & Moore 2005; Davidson & Noble 1992) to be:

- manipulation and storage of symbolic information, represented archaeologically by art and ornamentation;
- campsite structures that reveal complex group interactions;
- ritualized burial practices;
- stylistic expression in artefacts (argued to define

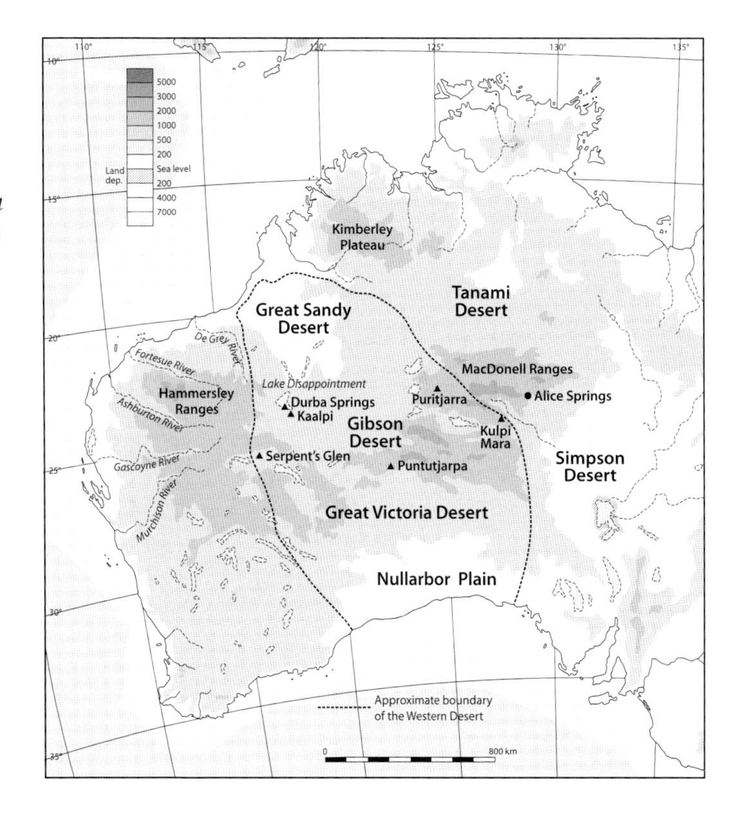

Figure 6.4. *Location of the major contemporary desert regions of Australia, including the Western Desert and the Hammersley Range Uplands, as mentioned in the text. Also shown are the key Pleistocene desert-edge sites of Serpent's Glen, Puritjarra and Kulpi Mara (the latter on the western edge of the Central Australian ranges). The Western Desert is characterized by uncoordinated and internal drainage and highly variable precipitation which is quite low on average at 300 mm p.a. or less (after Veth 2005a; base map after Collins atlas 1:20,000,000).*

group identity); and

- evidence of long-distance procurement of raw materials (argued to display planning depth and execution or exchange).

An approach that requires the presence of a 'suite' of archaeological evidence as evidence for modern behaviours is favoured here.

We can now have a summary look at some of the evidence that exists for modern behaviours from Wallacea and Australia.

The first evidence for long-distance sea voyaging outside of Africa comes from the settlement of Sulawesi, the islands of Maluku and Sahul and movements onwards and east to Island Melanesia. *Long-distance voyaging* occurs in the last 45–60,000 years (depending on whether you accept the short or long chronology).

On the question of information exchange witnessed in art production and ochre distribution networks — these actions are dated from approximately 25,000 to 40,000 BP from Wallacea and Sahul. In Australia lumps of hematite with ground facets are found in the earliest assemblages. Mike Smith has shown that early ochres from Puritjarra are sourced from distant quarries reflecting 'open' and shifting social interaction systems over 30,000 BP (chapters in Smith & Hesse 2005). An ochre-covered slab is dated from the site of Carpenter's Gap in the Kimberley to 42,000 BP (O'Connor 1999; O'Connor & Fankhauser 2001) with some Bradshaw paintings dated to before 20,000 and heavily altered 'archaic faces' of the arid interior hypothesized to date to before the LGM (cf. McDonald & Veth 2006) — the later occurring over two thirds of the continent. Art or the marking of bedrock, parietal media and bodily ornamentation was apparently part of the cultural repertoire of the first Australians.

The ability for populations to re-structure geographically during major climate change — such as the LGM — is suggested from sequences from East Timor, the Aru Islands and northern Australia. Just as Clive Gamble and colleagues (2004) have argued that major shifts in population occurred during the LGM between northern Europe, France and Iberia (using radiocarbon determinations as proxies for occupation), so I and colleagues believe that major changes in settlement occurred in the Southern Hemisphere.

The presence and antiquity of ritual burial practice and grave goods in early sites is ubiquitous. From the secondary burial of Lemdubu Woman in the Aru Islands to the cremations of Willandra Lakes — these practices are well in place early on. Ochre-covered burials at Lake Mungo (Fig. 6.3) combine evidence of several modern traits (cf. Hiscock 2008; Mulvaney & Kamminga 1999).

The antiquity and nature of personal ornamentation from Mandu Mandu rockshelter, northwest

Australia in *Conus* shell beads dated to greater than 32,000 years ago (Morse 1993) and tusk-shell fragments of the order Dentaliidae from Riwi rockshelter in the Kimberley (Balme 2000) at 30,000 years ago (Fig. 6.3) all argue for early elaborations in symbolic and personal identifying behaviour. The Riwi shells are 300–500 km from their source.

As Balme & Morse have recently argued (2006, 1) personal ornaments such as beads and pendants are one of the cornerstones of modern human behaviour (cf. d'Errico *et al.* 2005; Mellars 2005). They (Balme & Morse 2006, 9) emphasize that this is a global phenomena, arguing that:

> Shell beads are one part of a suite of evidence for symbolic behaviour found in the Australian Pleistocene record ... that together illustrate the consistency of behaviours and the importance of bodily ornaments associated with anatomically modern humans throughout the world ...

Australia also has evidence for the widespread use of composite and complex tools such as edge-ground and waisted stone axes and shell adzes. There are flakes from stone axes dated to before 35,000 BP from northern Australia and shell adzes from the Pleistocene from Manus and Gebe (O'Connor 1999; O'Connor & Veth 2000). In the African context researchers have argued that composite tools are by definition indicators of specifically designed and style-laden behaviours indicative of cultural modernity.

While it has been suggested that the efflorescence of symbolic activity witnessed in the Upper Palaeolithic of Europe only occurs in the mid- to late Holocene in Australia (after Brumm & Moore 2005, 168; cf. Balme & Morse 2006, 10) and that this reflects decreasing social isolation (and presumably emerging territoriality), I prefer an alternative explanation. Hard evidence for long-distance movement of individuals or 'exotic' materials such as shellfish, ochres and lithics before 30,000 years ago in Australia and the case for regional cultural configurations (*sensu* O'Connor & Veth 2006) argue against social isolation. Indeed it is precisely the maintenance of long-distance social connections between highly mobile and low-density groups (Veth 2005c) that provides the fuel for social cohesion. This is witnessed in cycles of aggregation followed by dispersal of groups due to ecological and social tensions. The 'broad-grained' landscapes of glacial Europe and arid Australia would both predict for early and dual symbolic activities: for maintaining the inclusiveness of disparate groups and also signalling difference during the inevitable tensions arising from the 'hot-tubs' of intergroup gatherings.

One of the conclusions of the volume *Desert Peoples: Archaeological Perspectives* (Veth *et al.* 2005b) is that the *timing* for colonization of, or persistence within, global deserts is only reliably dated to the last 40–60,000 years. In our view complex information exchange is implicated here as a risk-minimizing strategy in new landscapes.

Finally, as Holdaway, Hiscock and others have argued (cf. Lilley 2006) there are regionally distinct strategies for organizing lithic production by the terminal Pleistocene for many parts of Australia.

Coda

One of the highest impact finds of the last decade has been that of *Homo floresiensis* (The Hobbit) on the island of Flores (Fig. 6.2) — in the middle of an island chain populated by modern humans dating back to the Pleistocene (cf. Balme *et al.* 2009). At Liang Bua lithic assemblages and *Homo floresiensis* remains have been argued to be in association. There are, however, no ochres, polished bone or shell, crafted objects nor indeed direct evidence for fire that may be reliably associated with *Homo floresiensis* — these being common markers of modern humans (Balme *et al.* 2009; Morwood *et al.* 2004; 2005; Niven 2007; Westaway *et al.* 2007).

Given that Flores was surrounded by islands with modern humans, to the west, east and north, by at least 30,000 to 40,000 years ago — it is almost certain that modern humans did settle Flores by this early stage. It will be simply a matter of time with the targeting of appropriate cave and shelter sites and deployment of appropriate dating methodologies that this apparent 'evolutionary gap' in the southern dispersal of modern humans from Africa to Australia will be filled in.

References

Allen, J., 2003. Discovering the Pleistocene in Island Melanesia, in *Pacific Archaeology: Assessments and Prospects. Proceedings of the International Conference for the 50th anniversary of the First Lapita Excavation (July 1952)*, ed. C. Sand. (Les Cahiers de l'Archéologie en Nouvelle-Calédonie 15.) Nouméa: Département Archéologie, Service des Musées et du Patrimoine de Nouvelle-Calédonie, 33–42.

Balme, J., 2000. Excavation revealing 40,000 years of occupation at Mimbi Caves, south central Kimberley, Western Australia. *Australian Archaeology* 51, 1–5.

Balme, J. & K. Morse, 2006. Shell beads and social behaviour in Pleistocene Australia. *Antiquity* 80(310), 799–811.

Balme, J., I. Davidson, J. McDonald, N. Stern & P. Veth, 2009. Symbolic behaviour and the peopling of the southern arc route to Australia. *Quaternary International* 202(1–2), 59–68.

Brown, P., T. Sutikna, M.J. Morwood *et al.*, 2004. A new small-bodied hominin from the Late Pleistocene of Flores, Indonesia. *Nature* 431, 1055–61.

Brumm, A. & M.W. Moore, 2005. Symbolic revolutions and the Australian archaeological record. *Cambridge Archaeological Journal* 15(2), 157–75.

Chappell, J., 2004. Cold, Dry and Large? Greater Australia Through the Later Stages of the last Ice Age. Unpublished Plenary Address of the Australian Archaeological Association Annual General Meeting, Jindabyne.

Davidson, I. & W. Noble, 1992. Why the first colonisation of the Australian region is the earliest evidence of modern human behaviour. *Archaeology in Oceania* 27, 135–42.

d'Errico, F., C. Henshilwood, M. Vanhaeren & K. van Niekerk, 2005. *Nassarius karassianus* shell beads from Blombos Cave: evidence for symbolic evidence in the Middle Stone Age. *Journal of Human Evolution* 48, 3–24.

Gamble, C., W. Davies, P. Pettitt & M. Richards, 2004. Climate change and evolving human diversity in Europe during the last glacial. *Philosophical Transactions of the Royal Society of London* Series B 359(1442), 243–54.

Gould, R.A., 1969. Subsistence behaviour among the Western Desert Aborigines of Australia. *Oceania* 39(4), 253–74.

Gould, R.A., 1980. *Living Archaeology*. (New Studies in Archaeology.) Cambridge: Cambridge University Press.

Hiscock, P., 2008. *Archaeology of Ancient Australia.* New York (NY): Routledge.

Lahr, M.M. & R. Foley, 2004. Palaeoanthropology: human evolution writ small. *Nature* 431, 1043–4.

Lambeck, K. & J. Chappell, 2001. Sea level change through the last glacial cycle. *Science* 292(5517), 679–86.

Lilley, I., 2006. Archaeology in Oceania: themes and issues, in *Archaeology of Oceania: Australia and the Pacific Islands,* ed. I. Lilley. Oxford: Blackwell, 1–28.

McDonald, J. & P. Veth, 2006. Rock art and social identity: a comparison of graphic systems in arid and fertile environments, in *Archaeology of Oceania: Australia and the Pacific Islands,* ed. I. Lilley. Oxford: Blackwell, 96–115.

Mellars, P., 2005. The impossible coincidence.: a single-species model for the origins of modern human behavior in Europe. *Evolutionary Anthropology* 14, 12–27.

Morse, K., 1993. Shell beads from Mandu Mandu Creek rockshelter, Cape Range peninsula, Western Australia, dated to before 30,000 BP. *Antiquity* 67, 877–83.

Morwood, M.J., R.P. Soejono, R.G. Roberts *et al.*, 2004. Archaeology and age of a new hominin from Flores in eastern Indonesia. *Nature* 431, 1087–91.

Morwood, M.J., P. Brown, Jatmiko *et al.*, 2005. Further evidence for small-bodied hominins from the Late Pleistocene of Flores, Indonesia. *Nature* 437, 1012–17.

Mulvaney, D.J. & J. Kamminga, 1999. *Prehistory of Australia.* Sydney: Allen and Unwin.

Niven, J.E., 2007. Brains, islands and evolution: breaking all the rules. *Trends in Ecology & Evolution* 22(2), 57–9.

O'Connell, J.F. & J. Allen, 2004. Dating the colonization of Sahul (Pleistocene Australia–New Guinea): a review of recent research. *Journal of Archaeological Science* 31(6), 835–53.

O'Connell, J.F. & J. Allen, 2007. Pre-LGM Sahul (Pleistocene Australia–New Guinea) and the archaeology of early modern humans, in *Rethinking the Human Revolution: New Behavioural and Biological Perspectives on the Origins and Dispersal of Modern Humans*, eds. P. Mellars, K. Boyle, O. Bar-Yosef & C. Stringer. (McDonald Institute Monographs.) Cambridge: McDonald Institute for Archaeological Research, 395–410.

O'Connor, S., 1999. *30,000 Years of Aboriginal Occupation: Kimberley, Northwest Australia.* (Terra Australis 14.) Canberra: ANH Publications; Centre for Archaeological Research, Australian National University.

O'Connor, S., 2007. New evidence from East Timor contributes to our understanding of earliest modern human colonisation east of the Sunda Shelf. *Antiquity* 81, 523–35.

O'Connor, S., 2010. Pleistocene migration and colonization in the Indo-Pacific region, in *The Global Origins and Development of Seafaring*, eds. A. Anderson, J.H. Barrett & K.V. Boyle. (McDonald Institute Monographs.) Cambridge: McDonald Institute for Archaeological Research, 41–55.

O'Connor, S. & J. Chappell, 2003. Colonisation and coastal subsistence in Australia and Papua New Guinea: different timing, different modes, in *Pacific Archaeology: Assessments and Prospects. Proceedings of the International Conference for the 50th anniversary of the First Lapita Excavation (July 1952)*, ed. C. Sand. (Les Cahiers de l'Archéologie en Nouvelle-Calédonie 15.) Nouméa: Département Archéologie, Service des Musées et du Patrimoine de Nouvelle-Calédonie, 17–32.

O'Connor, S. & B. Fankhauser, 2001. One step closer: an ochre covered rock from Carpenters Gap Shelter 1, Kimberley region, Western Australia, in *Histories of Old Ages: Essays in Honour of Rhys Jones*, eds. A. Anderson, I. Lilley & S. O'Connor. Canberra: Pandanus Books; Centre for Archaeological Research, Australian National University, 287–300.

O'Connor, S. & P. Veth (eds.), 2000. *East of Wallace's Line: Studies of Past and Present Maritime Cultures in the Indo-Pacific Region.* (Modern Quaternary Research in South East Asia 16: V–VI.) Rotterdam: A.A. Balkema.

O'Connor, S. & P. Veth, 2005. Early Holocene shell fish hooks from Lene Hara Cave, East Timor establish complex fishing technology was in use in Island Southeast Asia five thousand years before Austronesian settlement. *Antiquity* 79, 249–56.

O'Connor, S. & P. Veth, 2006. Revisiting the past: changing interpretations of Pleistocene settlement subsistence and demography in northern Australia, in *Archaeology of Oceania: Australia and the Pacific Islands*, ed. I. Lilley. Oxford: Blackwell, 29–47.

O'Connor, S., P. Veth & C. Campbell, 1998. Serpent's Glen rockshelter: report of the first Pleistocene-aged occupation from the Western Desert. *Australian Archaeology* 46, 12–22.

O'Connor, S., M. Spriggs & P. Veth, 2002a. Excavation at Lene Hara Cave establishes occupation in East Timor at least 30,000–35,000 years ago. *Antiquity* 76, 45–50.

O'Connor, S., M. Spriggs & P. Veth, 2002b. Direct dating of

shell beads from Lene Hara Cave, East Timor. *Australian Archaeology* 55, 18–21.
O'Connor, S., M. Spriggs & P. Veth (eds.), 2005. *The Archaeology of the Aru Islands, Eastern Indonesia.* (Terra Australis 22.) Canberra: Pandanus Books; Centre for Archaeological Research, Australian National University.
Smith, M.A., 1988. The Pattern and Timing of Prehistoric Settlement in Central Australia. Unpublished PhD thesis, University of New England.
Smith, M. & P. Hesse (eds.), 2005. *23°S: Archaeology and Environmental History of the Southern Deserts.* Canberra: National Museum of Australia.
Spriggs, M., S. O'Connor & P. Veth, 2003. Vestiges of early pre-agricultural economy in the landscape of East Timor: recent research, in *Fishbones and Glittering Emblems: Proceedings of the 9th International Conference of the EurASEAA in Sigtuna Sweden*, eds. A. Karlström & A. Källén. Stockholm: Museum of Far Eastern Antiquities, 49–58.
Turney, C.S.M., M.I. Bird, L.K. Fifield *et al.*, 2001. Early human occupation at Devil's Lair, southwestern Australia 50,000 years ago. *Quaternary Research* 55(1), 3–13.
Veth, P.M., 1989. Islands in the interior: a model for the colonization of Australia's arid zone. *Archaeology in Oceania* 24(3), 81–92.
Veth, P.M., 1993. *Islands in the Interior: the Dynamics of Prehistoric Adaptations within the Arid Zone of Australia.* (Archaeological Series 3.) Ann Arbor (MI): International Monographs in Prehistory.
Veth, P., 1995. Aridity and settlement in northwest Australia. *Antiquity* 69, 733–46.
Veth, P., 2003. *'Abandonment' or Maintenance of Country?: a Critical Examination of Mobility Patterns and Implications for Native Title.* (Land, Rights, Laws: Issues of Native Title vol. 2: Issues, paper no. 2.) Canberra: Native Title Research Unit, Australian Institute of Aboriginal and Torres Strait Islander Studies. Ebook.
Veth, P., 2005a. Between the desert and the sea: archaeologies of the Western Desert and Pilbara regions, Australia, in *23°S: Archaeology and Environmental History of the Southern Deserts*, eds. M.A. Smith & P. Hesse. Canberra: National Museum of Australia, 132–41.
Veth, P., 2005b. Conclusion — major themes and future research directions, in *Desert Peoples: Archaeological Perspectives*, eds. P. Veth, M. Smith & P. Hiscock. Oxford: Blackwell Publishing, 293–300.
Veth, P., 2005c. Cycles of aridity and human mobility: risk-minimization amongst late Pleistocene foragers of the Western Desert, Australia, in *Desert Peoples: Archaeological Perspectives*, eds. P. Veth, M. Smith & P. Hiscock. Oxford: Blackwell Publishing, 100–115.
Veth, P., 2006. Social dynamism in the archaeology of the Western Desert, in *The Social Archaeology of Australian Indigenous Societies: Essays on Aboriginal History in Honour of Harry Lourandos*, eds. B. David, B. Barker & I.J. McNiven. Canberra: Australian Aboriginal Studies Press, 242–53.
Veth, P., M. Spriggs & S. O'Connor, 2004. Looking Back West: Towards the Wallace Line. Unpublished paper delivered to the Australian Archaeological Association Annual General Meeting, Armidale.
Veth, P., M. Spriggs & S. O'Connor, 2005a. The continuity of cave use in the tropics: examples from East Timor and the Aru Islands, Maluku. *Asian Perspectives* 44(1), 180–92.
Veth, P., M. Smith & P. Hiscock, 2005b. *Desert Peoples: Archaeological Perspectives.* Oxford: Blackwell Publishing.
Veth, P., M. Smith, J. Bowler, K. Fitzsimmons, A. Williams & P. Hiscock, 2009. Excavations at Parnkupirti, Lake Gregory, Great Sandy Desert: OSL ages for occupation before the Last Glacial Maximum. *Australian Archaeology* 69, 1–10.
Westaway, K.E., M.J. Morwood, R.G. Roberts *et al.*, 2007. Establishing the time of initial human occupation of Liang Bua, western Flores, Indonesia. *Quaternary Geochronology* 2(1–4), 337–43.

Chapter 7

Indian Lithic Technology Prior to the 74,000 BP Toba Super-eruption: Searching for an Early Modern Human Signature

Michael Haslam, Chris Clarkson, Michael Petraglia, Ravi Korisettar, Janardhana B., Nicole Boivin, Peter Ditchfield, Sacha Jones & Alex Mackay

The South Asian Middle Palaeolithic, or Middle Stone Age (Misra 1989; Allchin & Allchin 1997), is a largely techno-typological construct poorly constrained by chronometric data. As recently reviewed by James & Petraglia (2005; see also James 2007), this phase is characterized by a geographically variable flake-based industry employing both prepared and unprepared cores, with limited use of retouch and widespread but non-ubiquitous diminutive bifaces. The prepared cores do not adhere to any one particular technique (although discoidal and Levallois reduction are the most common), and show a continuation from the previous Acheulean traditions (Petraglia *et al.* 2003). Evidence for the production of blades in low numbers, and in the absence of standardized unidirectional blade core technology, comes from sites such as Patpara (Blumenschine *et al.* 1983), Bhimbetka (Misra 1985), Patne (James 2007), the Thar desert (Allchin *et al.* 1978) and the Kortallayar Basin (Pappu 2001). The few available dates (most importantly from the 16R dune sequence in Rajasthan: Misra 1995) indicate that Middle Palaeolithic radial core and scraper technologies extend back to perhaps 150,000 years ago, and continue in part until at least 38 kyr BP (Petraglia *et al.* 2009a; see also references in James & Petraglia 2005). Subsequently we see the chronologically variable introduction of technologies that depart from the Middle Palaeolithic, such as microliths and systematic production of blades with increasing use of cryptocrystalline materials, accompanied in some cases by ornaments (Deraniyagala 1992; Sali 1989) and initial adoption of systematic backing technologies currently dated to *c.* 35–30 kyr BP (Clarkson *et al.* 2009; Petraglia *et al.* 2009a).

South Asian genetic and geographic data have recently received increasing attention (Kivisild *et al.* 1999; Field & Lahr 2005; Sun *et al.* 2006; Endicott *et al.* 2007; Field *et al.* 2007; Atkinson *et al.* 2008; Oppenheimer 2009), and a dearth of hominin fossils prior to *c.* 30 kyr BP precludes direct morphological analysis (Stock *et al.* 2007). However, the known Late Pleistocene archaeological record of this region has yet to be comprehensively integrated into contemporary evolutionary debates. To a certain extent, progress in this regard has been hindered by a lack of well-dated stratigraphically secure sites, along with under-development of South Asia-centred frameworks into which findings may be incorporated (Petraglia 2007). One potentially critical palaeoenvironmental event concentrated on South (and parts of Southeast) Asia is the Sumatran Toba volcanic super-eruption of *c.* 74,000 years ago (e.g. Ninkovich *et al.* 1978; Westgate *et al.* 1998; Jones 2007). With ejecta known as the Youngest Toba Tuff (YTT), this significant natural event has acted as a focal point for ongoing debates over population bottlenecks and the potential influence of a volcanic winter on the transition from Oxygen Isotope Stage (OIS) 5 to OIS-4, as well as discussions of the direct impact of Toba ash fall on resident hominin populations (e.g. Rampino & Self 1993; Ambrose 1998; 2003; Oppenheimer 2002; Gathorne-Hardy & Harcourt-Smith 2003). The occurrence of the ash in terrestrial contexts also provides a convenient chronostratigraphic marker (Acharyya & Basu 1993), and consequently an opportunity for assessing if modern human populations were present in the Indian subcontinent at the time of the eruption.

Here we present data on lithic technology from two sites in the Jurreru River Valley of Andhra

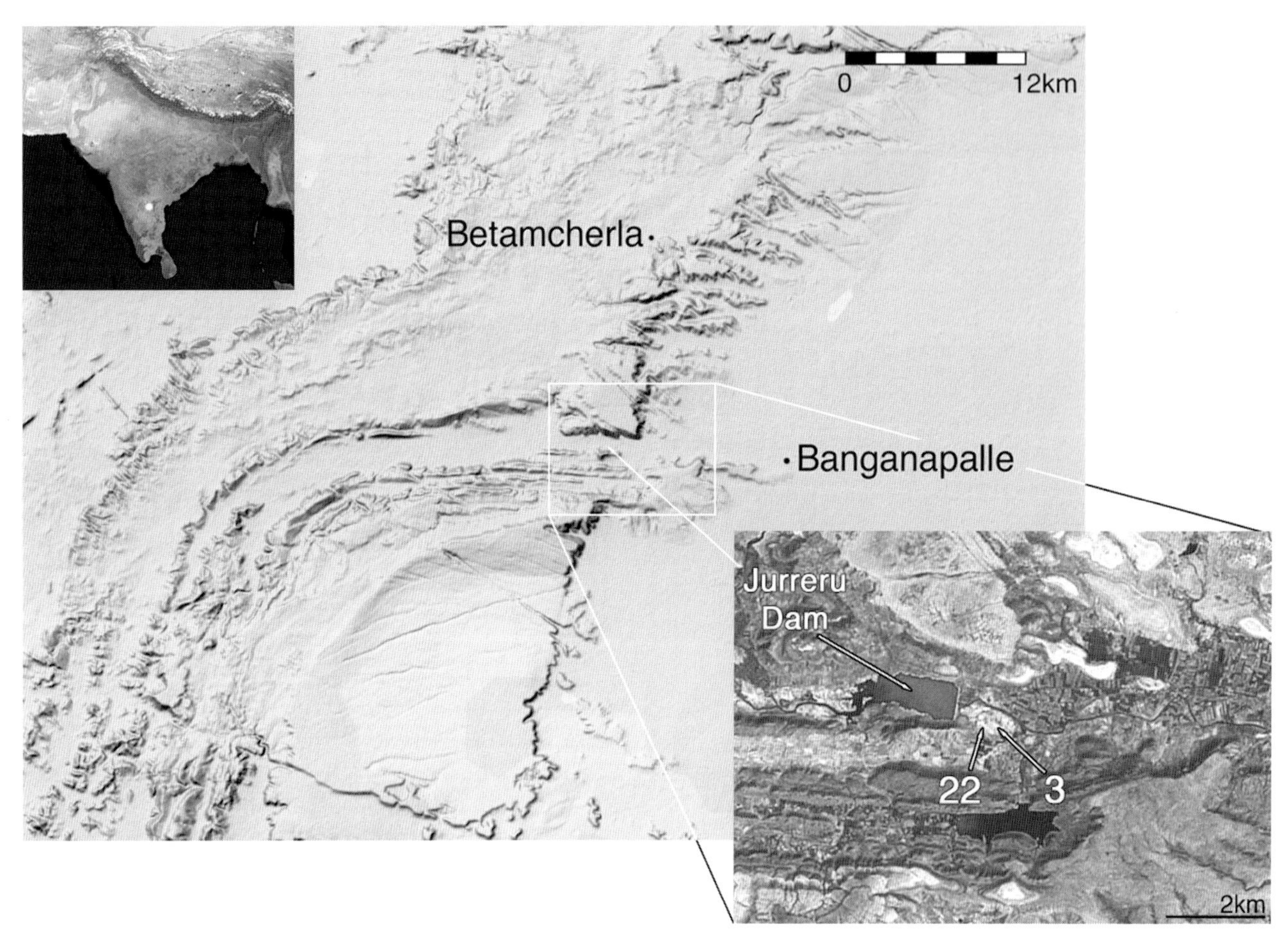

Figure 7.1. *Jwalapuram Localities 3 and 22 in relation to (top left) South Asia, (centre) the Erramala Hills, Andhra Pradesh, and (bottom right) the Jurreru valley and dam. The light-coloured area surrounding the sites, visible immediately to the east of the dam, is exposed Toba ash.*

Pradesh, south-central India, both of which occur stratigraphically immediately below and are sealed by extensive deposits of ash geochemically identified as YTT (Petraglia *et al.* 2007). Analyses of these assemblages provide an initial insight into hominin technological and social conditions at the time of the super-eruption, and supply much-needed data for attempts to discern a signature of modern human arrival in South Asia and the nature of the shift from Middle Palaeolithic to Later Palaeolithic industries.

YTT and the Jurreru River Valley sites

The Jurreru River cuts an eastwards course through the Erramala Hills in the Kurnool District of Andhra Pradesh (Fig. 7.1). Jwalapuram Localities 3 and 22 are located to the south of the river where the valley widens, between the village of Jwalapuram to the east and the present-day Jurreru Dam to the west, and approximately 11 km west of the town of Banganapalle. Archaeological survey of the valley began in 2003, when a published account of Quaternary ash accumulations in the valley (Rao & Rao 1992) led to the discovery of extensive tephra deposits exposed by local mining practices. Subsequent open-site and rockshelter excavations in the valley revealed evidence of hominin occupation from the Acheulean through Indian Middle Palaeolithic and microlithic phases up to the present day (Petraglia *et al.* 2009b).

Electron probe microanalysis (EPMA) characterization of the Jurreru valley volcanic glass sherds demonstrate that they represent a distal deposit of YTT (for methods see Petraglia *et al.* 2007, supporting online material), a finding which is consistent with all previous tephra identifications on the Indian subcontinent (Acharyya & Basu 1993; Shane *et al.* 1995; Westgate *et al.* 1998; although see Lee *et al.* 2004). The sites are approximately 2700 km from the Toba caldera in Sumatra. A total of 225 thickness measurements on ash exposures provide a volume estimate

Figure 7.2. *Jwalapuram Locality 3 (below-ash) completed excavation, facing south. Scale is 50 cm.*

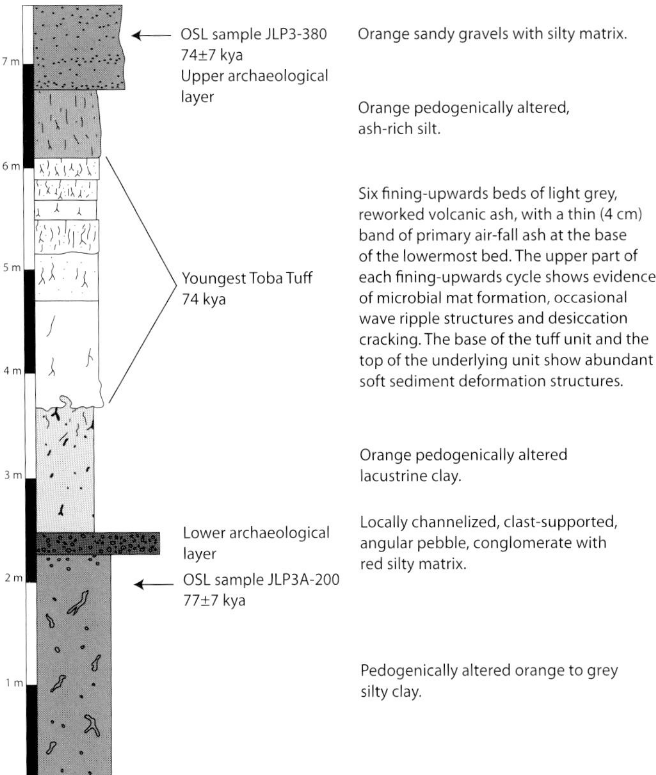

Figure 7.3. *Composite stratigraphy for Jwalapuram Locality 3. The below-ash component comprises sediments from the base of the diagram to the base of the YTT.*

for the remnant Jwalapuram tephra of $7\pm0.7 \times 10^5$ m^3. The majority of the ash beds are composed of material that originally fell on the surrounding valley sides and upstream, but was rapidly mobilized and redeposited on the valley bottom before significant entrainment within other sediments could occur. The thickest known tephra accumulation (>2.5 m) occurs in what would have been a local topographical low, and preserves at its base a 4–5 cm-thick layer of primary ash fall. This primary layer rests on clays displaying soft sediment deformation structures, indicating that initial deposition at this location was most likely in a relatively still-water environment. The sites described here targeted strata both directly beneath and at some distance from this thickest tephra deposit.

Locality 3, below YTT

To obtain a full sedimentary sequence from above and below the thickest exposed tephra bed, as well as from the tephra itself, two locations approximately 20 m apart in the same ash quarry were explored (Petraglia *et al.* 2007). These sites are located less than a kilometre west of Jwalapuram village and were collectively designated Jwalapuram Locality 3. The starting surface of the below-ash excavation was the exposed quarry floor, where local mining activity had removed all tephra down to the clay substrate interface (Fig. 7.2). This excavation surface directly underlay the remnant primary ash layer, and is hypothesized to have been submerged at the time of ash fall.

In total, a 3 m by 3 m area was excavated below the tephra in 5–10 cm levels. Culturally sterile and pedogenically altered orange-brown clay sediments were encountered for the first *c.* 110 cm below the tephra (Fig. 7.3). This level contained possible manganese granules and calcium carbonate nodules; the latter are common *in situ* post-depositional occurrences in the valley. Between 110 and 145 cm depth the clays gave way to a compact clast-supported angular conglomerate with a red silty matrix, containing a total of 214 lithic artefacts. Beneath this level the sediments were once again sterile silty clays with calcium carbonate concretions. Excavation ceased at a depth of 385 cm below the tephra. An optically stimulated luminescence sample (OSL; JLP3A-200) was taken from within the silty clay at a depth of *c.* 175 cm, and returned an age of 77±6 kyr BP based on the weighted mean palaeodose for the single-grain data set (see Petraglia *et al.* 2007, supporting online material). This brackets the artefact sample to a period within a few thousand years before the Toba eruption, with the formation of more than a metre of clays subsequent to artefact deposition and prior to the ash fall likely associated with floodplain or wetland development.

Figure 7.4. *Jwalapuram Locality 22 excavation in progress, facing west.*

Locality 22

Subsequent excavation aimed to identify and excavate a locality on the margins of the suspected paludal area, to characterize occupation deposits at the time of the Toba eruption and to supplement the findings from Jwalapuram Locality 3. It was hypothesized that the marginal sedimentary sequence would be typified by distinct palaeosol development capped by thin, possibly discontinuous, reworked YTT mixed with terrigenous sediment and deposited sub-aerially (i.e. without evidence of water-settling). A section cutting through a disused well west of Locality 3 indicated these characteristics, and a 5 m by 5 m trench designated Jwalapuram Locality 22 was placed *c.* 10 m to the west of the well site (Fig. 7.4). The surrounding landscape is predominantly flat scrub and acacia, with agricultural fields to the south, ash mining to the east, the Jurreru River to the north, and an east–west trending ridge of hills to the west that forms the southern margin of the Jurreru Dam. Jwalapuram Locality 22 is located approximately 420 m west of Locality 3.

The surface of the site and the immediate surrounding area was liberally covered in fresh limestone flakes, resulting from modern flaking activity and chiefly derived from construction of limestone shoring for the nearby well. These flakes were found only within the upper zone of modern disturbance, in otherwise culturally sterile deposits. Excavation at Locality 22 was conducted in 5–10 cm levels, and reached a maximum depth of 375 cm, at which point the local water table was encountered. YTT deposits at the site comprised isolated ash lenses and reasonably pure ash pods within an ashy silt matrix (Fig. 7.5). The tephra content of the silt increased from approximately 170 cm depth to an abrupt interface with a thick pedogenically altered clayey silt palaeosol at 190 to 200 cm. The latter stratum, with occasional sand inclusions, extended to the base of the excavation. Few artefacts were encountered below the top 10 cm of the buried palaeosol. A clay-rich band, with iron and manganese granules concentrated into possible scour fills, was located some 50 cm above the ash/paleosol interface. This layer contained a small number of lithic artefacts including a microblade core, and two pieces of bone. Compact red silt sediments overlying this layer were culturally sterile, with bands of large calcareous concretions resulting from *in situ* diagenesis.

The boundary between the ashy silt and underlying palaeosol is sharp but undulating, providing a further point of difference with the typically flat (water-derived) lower tephra-clay interface seen at Jwalapuram Locality 3. The palaeosol surface and the 5–10 cm below this revealed a total of 796 flaked stone artefacts (Fig. 7.6), with the large majority of these lying horizontally and displaying few signs of extended transport (i.e. size sorting or channel accumulation). Nevertheless, indicators of trampling and abrasion do occur (see below). Distributed

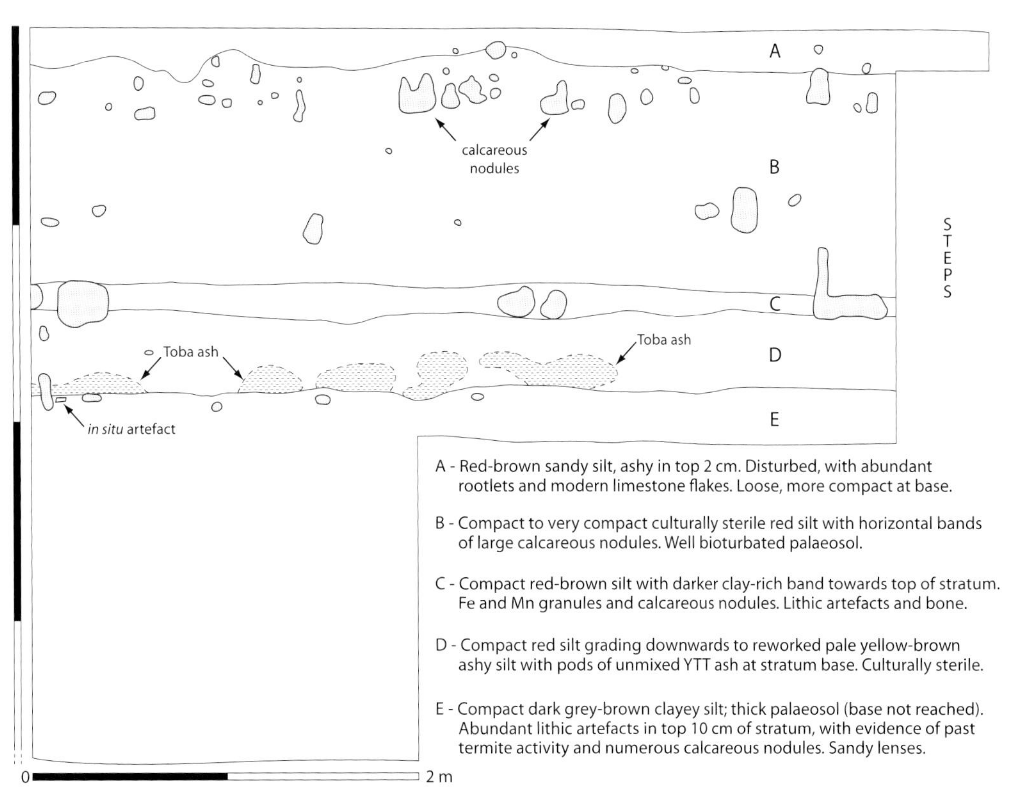

Figure 7.5. *Jwalapuram Locality 22 north profile.*

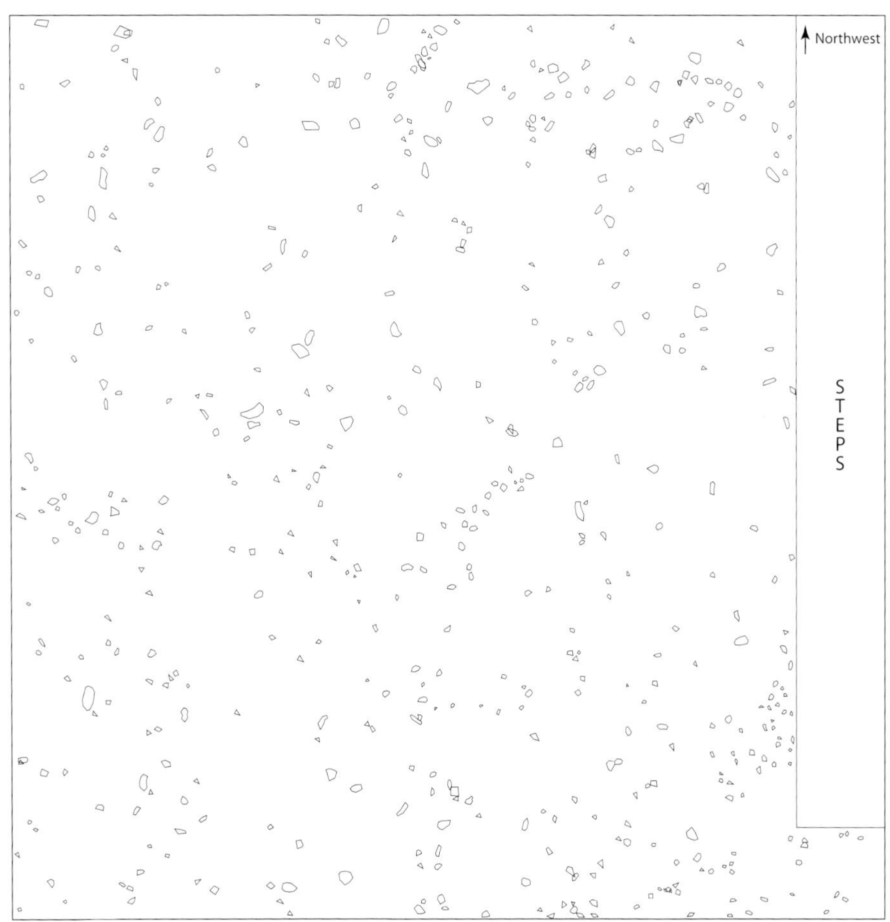

Figure 7.6. *Spatial distribution of artefacts recovered from the palaeosol immediately underlying the c. 74 kyr* BP *YTT ash layer, Jwalapuram Locality 22.*

through the first few centimetres of the palaeosol were a number of horizontal, concreted and elongate micritic infilled structures, which are identified as termite passage-ways. These were interspersed with diagenetic calcium carbonate concretions, as seen at other stratigraphic boundaries at the site. Collectively, these findings are currently interpreted as the upper layer of a natural soil that was exposed (or lightly vegetated) and occupied immediately prior to the Toba super-eruption.

Lithic technology and its implications

We present here an initial examination of the lithic component of these secure contexts. Note also that although artefacts were recovered from both above and below the Toba ash at both localities, only data from beneath the YTT deposits are discussed here.

Techno-typological characteristics of the Locality 3 and Locality 22 assemblages

The Jwalapuram Locality 22 assemblage consists of 796 stone artefacts and the Locality 3 assemblage includes 214 artefacts. Detailed attribute analysis (see Petraglia *et al.* 2007, supporting online material) has to date been conducted on all the Locality 3 material and 344 of the Locality 22 artefacts, and it is these data that are discussed here. The composition of both assemblages is shown in Table 7.1, presenting a breakdown into standard technological classes as well as more formal typological elements where appropriate. The Jwalapuram Locality 22 assemblage is larger and hence a broader range of technological and typological elements is present, consistent with the sample-size effects seen in most lithic assemblages (Grayson & Cole 1998; Hiscock 2001). Both assemblages are dominated by unretouched flakes, comprising 72 to 77 per cent of the assemblages. Flaked pieces, which clearly derive from conchoidal manufacture but lack the diagnostic features required to determine technological category, also make up a large proportion of the assemblages at 7 to 11 per cent.

The varied core assemblage from Jwalapuram Locality 22 indicates that multiple core-reduction strategies were in use at the time of the Toba eruption, with bidirectional, Levallois, discoidal and single and multiplatform core reduction present. Only a single multiplatform core was recovered from the smaller Locality 3 assemblage. Multiplatform cores are easily the most common form at Locality 22, and it is therefore not surprising that the sole core from Locality 3 is of this type. A small weathered biface, manufactured from the same locally available limestone as the majority of other artefacts in the assemblage, was also found below the ash at Jwalapuram Locality 22.

Table 7.1. *Techno-typological composition of the below-ash Jwalapuram Localities 22 and 3 assemblages.*

Type	22	3	% of 22	% of 3	Total
Core types					
Bidirectional core	3	0	0.9	0.0	**3**
Discoidal core	1	0	0.3	0.0	**1**
Levallois core fragment	1	0	0.3	0.0	**1**
Single platform core	2	0	0.6	0.0	**2**
Biface	1	0	0.3	0	**1**
Core fragment	0	1	0.0	0.5	**1**
Multiplatform core	7	1	2.0	0.5	**8**
Flakes and hammerstones					
Hammerstone	0	2	0.0	0.9	**2**
Flake	265	156	77.3	72.9	**421**
Flaked piece	24	24	7.0	11.2	**48**
Flake types					
Blade	2	7	0.6	3.3	**9**
Éclat debordant	2	0	0.6	0.0	**2**
Levallois flake	2	0	0.6	0.0	**2**
Microblade segment	2	0	0.6	0.0	**2**
Retouched					
Unifacial point	1	0	0.3	0.0	**1**
Bifacial point	1	0	0.3	0.0	**1**
Burin	1	0	0.3	0.0	**1**
Burinated scraper	0	1	0.0	0.5	**1**
Tanged point	1	0	0.3	0.0	**1**
Scrapers and retouched fragments	28	22	8.2	10.3	**50**
Totals	**344**	**214**			**558**

The larger range of flake types found at Locality 22 (Figs. 7.7 & 7.8) is consistent with the range of core forms found at the site: Levallois flakes and *éclat debordants* (flakes that remove the lateral edge of radial cores) both indicate that radial flaking from facetted platforms was taking place at the site. Blades and microblades are also present in small numbers, though blades are more common at Jwalapuram Locality 3.

Retouched implements and fragments mostly consist of various kinds of scrapers (including notches), making up 8–10 per cent of the assemblages (for illustrations of retouched Locality 3 artefacts: see Petraglia *et al.* 2007). However, burins and points (tanged and untanged, unifacial and bifacial) are also present. A tanged point from Locality 22 has ventral trimming of the bulb to create two pronounced concavities on either side of the facetted platform, and distal unifacial retouch that forms a stout tip. No comparative study of tanged points from Middle Palaeolithic contexts in India has been attempted, and

Table 7.2. *Comparison of size and shape for complete flakes from below-ash Jwalapuram Localities 22 and 3.*

Attribute	Site	N	Mean	Standard deviation	*t*	df	Mean difference	*p*
Weight (g)	22	43	26.38	59.96	–0.02	122	–0.30	0.98
	3	81	26.68	72.27				
Length (mm)	22	41	30.13	22.71	–1.04	70	–4.30	0.30
	3	78	34.44	18.99				
Proximal width (mm)	22	40	19.79	14.45	0.44	59	1.12	0.66
	3	75	18.67	9.79				
Medial width (mm)	22	43	23.29	17.09	–1.23	73	–3.74	0.22
	3	76	27.03	13.82				
Distal width (mm)	22	39	17.71	13.55	–0.08	64	–0.20	0.94
	3	74	17.91	10.81				
Maximum dimension (mm)	22	43	37.80	24.30	–0.36	68	–1.54	0.72
	3	81	39.33	18.41				
Thickness (mm)	22	43	11.09	8.74	0.12	85	0.20	0.91
	3	78	10.90	8.57				
Platform width (mm)	22	40	18.61	14.03	1.17	53	2.81	0.25
	3	69	15.80	7.83				
Platform thickness (mm)	22	40	8.28	7.52	1.00	58	1.32	0.32
	3	71	6.96	4.90				
Platform angle	22	36	69.94	13.64	–1.11	72	–3.11	0.27
	3	70	73.06	13.91				
Length:Width	22	43	1.28	0.75	–0.68	76	–0.09	0.50
	3	76	1.37	0.64				
Mean GIUR	22	43	0.15	0.39	1.24	51	0.08	0.22
	3	81	0.07	0.18				
Angle of margin contraction or expansion	22	43	7.70	24.46	1.30	77	5.75	0.20
	3	81	1.95	21.43				
Mean retouched edge angle	22	43	12.57	25.66	0.67	77	3.14	0.50
	3	81	9.43	22.72				
Length:Thickness	22	43	3.17	1.75	–1.34	64	–0.40	0.19
	3	78	3.57	1.20				

none has been precisely dated, so we are as yet unable to make statements about the similarity of this point to other examples from South Asia.

Shape and size comparisons for the Locality 3 and Locality 22 flake assemblages

In virtually all respects, the complete flakes from Jwalapuram Localities 22 and 3 are identical. Student *t*-tests performed on a comparison of the means for a wide range of dimensional attributes, as well as ratios of two dimensions expressing flake shape and a measure of retouch intensity (Kuhn 1990), all show non-significant results (see Table 7.2). Likewise, the proportions of different scar orientations (unidirectional from the proximal end versus radial,

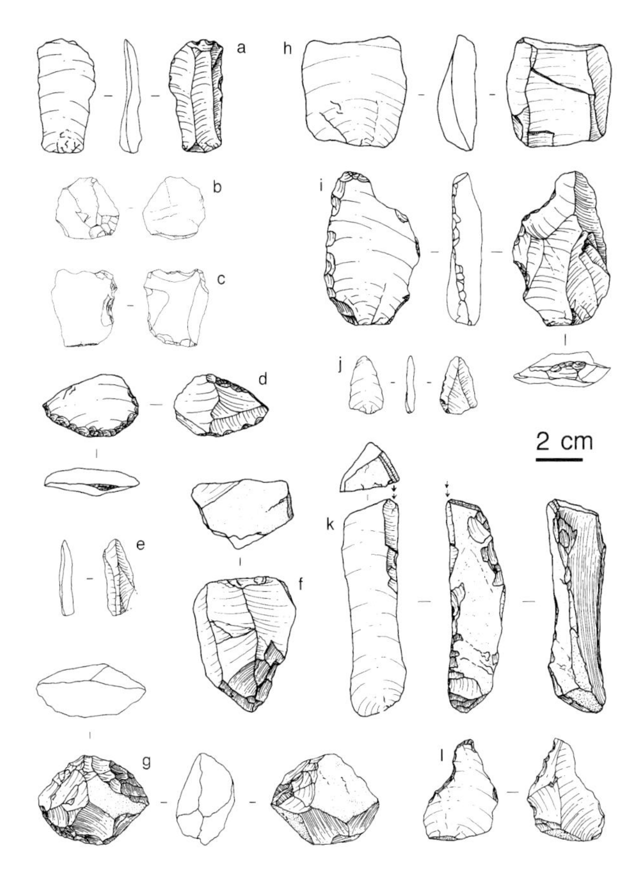

Figure 7.7. *Jwalapuram Locality 22 lithic artefacts: (a) blade; (b, h, i, j) radial flakes; (c) scraper; (d) bifacial point; (e) microblade; (f) bidirectional core; (g) radial core; (k) burin; (l) retouched flake.*

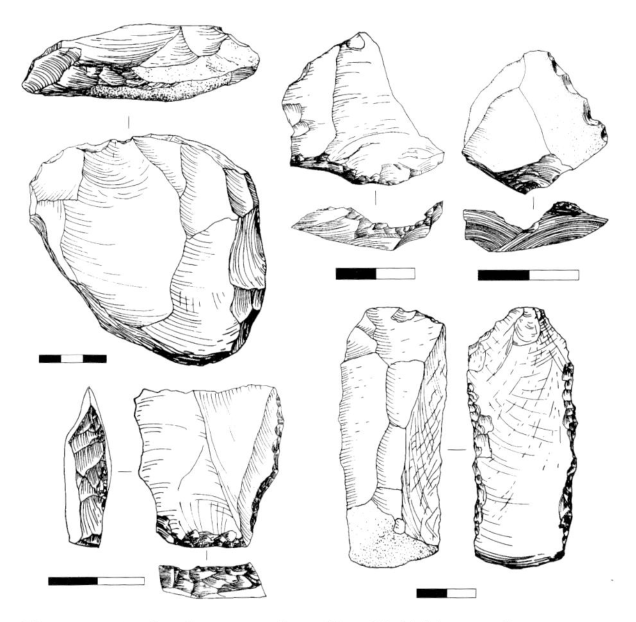

Figure 7.8. *Jwalapuram Locality 22 lithic artefacts. Clockwise from top left: Levallois core; two radial flakes with facetted platforms; blade with possible retouch;* éclat debordant. *Scales are in centimetres.*

Table 7.3. *Comparison of frequencies of different platform types on complete flakes and proximal fragments, below-ash Jwalapuram Localities 22 and 3.*

Locality	% cortical only	% cortical and conchoidal	% crushed	% dihedral	% focalized	% multiple conchoidal	% single conchoidal
3	14	0	9	9	3	12	54
22	9	2	8	9	0	33	40
Total no.	**20**	**2**	**15**	**16**	**2**	**43**	**84**

Table 7.4. *Frequencies of platform preparation on complete flakes and proximal fragments from below-ash Jwalapuram Localities 22 and 3.*

Locality	% none	% both	% facetting	% overhang removal	Total
3	51.81	1.20	2.41	44.58	**139**
22	65.47	5.04	15.83	13.67	**83**
Total	**134**	**8**	**24**	**56**	**222**

Table 7.5. *Percentages of raw materials used at Jwalapuram Localities 22 and 3, below-ash.*

Locality	Chalcedony	Chert	Dolerite	Limestone	Quartz	Quartzite
22	0.3	1.7	8.2	89.5	0.0	0.3
3	6.1	10.7	1.0	59.8	0.9	21.5

bidirectional and non-proximal orientations) found on the dorsal surfaces of flakes shows no significant difference between Localities 22 and 3 (χ^2 = 0.476, p = 0.495). This means the two samples likely derive from the same or very similar populations in terms of methods of core preparation and flake removal.

From this comparison we can conclude that flakes discarded at Jwalapuram Localities 22 and 3 tend to be quite heavy, relatively stout but not thick, with parallel to gently tapering lateral margins, relatively large and low-angled platforms, and dorsal scar patterns that tend to show slightly more unidirectional alignments from the proximal end than radial and non-proximal orientations.

Some significant differences between the assemblages nevertheless exist. Flakes from Locality 3, for example, possess higher proportions of single conchoidal, cortical and crushed platforms, whereas Locality 22 flakes have higher proportions of single conchoidal and cortical, and cortical and multiple conchoidal platforms, with small differences in the proportions of dihedral platforms (χ^2 = 32.7, p = <0.0005) (Table 7.3).

Another difference is present in the proportions of different forms of platform preparation (Table 7.4). Jwalapuram Locality 22 flakes show a much higher proportion of platform facetting and unprepared platforms, but much lower proportions of overhang removal. These differences are significant (χ^2 = 32.06, p = <0.0005) and clearly reflect different emphasis in core preparation between locations, but given the strong similarity in flake form, they may simply represent specific responses to working individual cores, or different emphasis on different reduction strategies that yield flakes of much the same form. Furthermore, these differences in platform type and preparation may partially reflect the vagaries of sampling, since our own experiments show that a wide range of platform types and preparation can be generated by various types of radial and bidirectional core reduction. Given small sample size, different platform types will be present in varying proportions.

Raw materials

The most marked difference between the Locality 22 and Locality 3 assemblages lies in the proportions of raw materials used (see Table 7.5). While stone artefacts at both localities are predominantly made from limestone, Locality 3 shows higher proportions of chert, chalcedony and quartzite, whereas Locality 22 has significantly more dolerite. This suggests that a greater range of raw material type was more frequently reduced at Locality 3 than Locality 22, although the actual raw materials employed were much the same at both locations.

All of the raw materials found at both localities can be obtained on the foothills surrounding the valley, with quartzite easily obtained from the escarpment a short distance to the north, limestone, chert, quartz and chalcedony from hillslopes to the north and south, and dolerite from a dyke that snakes its way along the southern hillslopes of the Jurreru Valley. Raw material acquisition is therefore likely to be predominantly local, with all materials available within a few kilometres (or less) of the sites. The greater raw material richness at Jwalapuram Locality 3, with greater emphasis on chert and chalcedony, could indicate that this assemblage accumulated within the context of a base camp from which foraging trips around the surrounding landscape were mounted. Locality 22, on the other hand, seems to represent more concentrated use of limestone, perhaps representing the accumulation of debris from short-term occupation and single procurement episodes. The limestone utilized at Locality 22 is identical in appearance and fracture patterns to that seen outcropping in the Jurreru Valley at nearby Jwalapuram Locality 20 (Petraglia *et al.* 2009b).

Site formation
High rates of artefact breakage and edge damage suggest destructive site-formation processes were active at Jwalapuram Localities 22 and 3. A total of 83 per cent of the artefacts from Locality 22 are broken in some way, and 67 per cent from Locality 3. At Locality 22, breaks are mostly caused by manufacturing errors in the form of longitudinal splits. At Locality 3, the majority of breakage is transverse snapping, which may have resulted from either manufacturing errors or trampling/retouch/use. Again this difference may reflect the emphasis at Locality 22 on primary flaking and at Locality 3 of a wider range of activities.

Edge damage is difficult to separate from retouch on many flakes with modified lateral margins at both Jwalapuram Locality 3 and Locality 22. The possibility cannot yet be ruled out that retouch is rarer than the percentages in Table 7.1 suggest, and that heavy trampling and rolling of artefacts can account for much of the damage. Retouch often appears isolated, haphazard and alternating between dorsal and ventral, without any appearance of having created a continuous edge. Heavy abrasion and edge rounding are also present on some pieces, and aeolian weathering appears to have affected many artefacts. Retouch on other pieces appears quite deliberate, forming marked concavities, or continuous retouched margins. Further examination of tool edges under high magnification is required to determine the extent of site-weathering and site-formation processes, as well as use-wear, on the creation of the large 'retouched' assemblage.

Discussion

The two localities described here provide an initial perspective on the activities and movements of the hominin occupants of the Jurreru valley immediately prior to the YTT ash-fall. They also comprise the first well-dated and detailed record yet discovered in India of hominin activities at the OIS-5/-4 transition, and contribute valuable data to ongoing debates over the presence or otherwise of modern humans in South Asia at this time. Both sites employed a range of core-reduction systems that produced extremely similar flakes in terms of size, shape and dorsal morphology. However, significant differences in the abundance of certain core types with resulting differences in the proportions of platform types suggest greater emphasis on radial flaking at Jwalapuram Locality 22, and more flaking of rotated cores at Locality 3, resulting in slightly more Levallois flakes at the former and slightly more blades at the latter. The proportions of different raw materials are also markedly different between the two sites, with Locality 3 exhibiting greater use of a range of material types, and Locality 22 showing much more focus on the use of limestone. The range of scraper forms is very alike at both localities, but Locality 22 is unique owing to the appearance of tanged and unifacial and bifacial points. These are very rare in the Indian Middle Palaeolithic (James 2007), but could indicate the use of hafted projectiles, something rarely documented among pre-modern humans (Shea 1988; Shea *et al.* 2001).

The Jurreru Valley Toba ash-fall isochron has allowed us to identify a relatively intact and extensive palaeosurface occupied by Middle Palaeolithic hominins. While the below-YTT assemblages from Jwalapuram Localities 3 and 22 may not be precisely coeval given their stratigraphic positioning relative to the ash, they are tightly constrained by chronometric dating and the Toba tephra, and the strong similarities in flake form and the reliance on local raw materials at both sites point to clear continuities in both the desired flake products and the nature of procurement and movement within the valley. We can therefore emphasize that predominantly local procurement of lithic resources, a flexible approach to core-reduction techniques (perhaps with a chronological component) and a regularity of artefact morphological production are characteristic of the technological industry present at the two sites. These trends should serve as hypotheses to be tested against the records of other late OIS-5/early OIS-4 assemblages recovered from within and peripheral to South Asia (e.g. Zuraina 2003).

The 'Middle–Upper Palaeolithic transition' in South Asia: an inappropriate template

What are the implications of the Indian Middle Palaeolithic record for our understanding of the nature of modern human dispersal out of Africa, and the Middle to Upper Palaeolithic transition as seen in Europe? For the moment we can leave aside the possibility that now-extinct *Homo sapiens* lineages extended their initial movement out of Africa some 90–120,000 years ago beyond the known occupation of the Levant (Shea 2003) into India, although this cannot be ruled out on present evidence. There is, however, broad acceptance that initial modern human dispersal into the Indian subcontinent occurred before 50 kyr BP, and perhaps prior to 70 kyr BP, based on dates for Australasian colonization and genetic coalescence (Roberts *et al.* 1994; Oppenheimer 2009). Unequivocally, at no time during or close to this period do the South Asian lithic data provide evidence of a technological 'revolution' to parallel the changes seen at the European Middle/Upper Palaeolithic boundary (Mellars 2007). Instead, the data suggest that the earliest humans entering South Asia used technologies based on the African Middle Stone Age, similar to those previously created by other hominins on the subcontinent from the late Middle Pleistocene onwards (Misra 1995). The inappropriateness of transposing expectations derived from the European transition to other regions has been noted before (e.g. McBrearty & Brooks 2000; James & Petraglia 2005), and we reaffirm it here.

The fact that humans arrived in South Asia with very similar lithic propensities to the existing non-*sapiens* groups is unsurprising, as commonalities in the lithic output of different hominins have been recorded elsewhere. The Levantine Middle and Later Middle Palaeolithic assemblages were considered for some time to be similar enough to be evidence of continuity, even though the former was created by *Homo sapiens* and the latter by Neanderthals (Shea 2008). Close similarities have also recently been documented between the technological strategies of *Homo floresiensis* and *Homo sapiens* at Liang Bua in Indonesia (Moore *et al.* 2009). The lithic industries of these sites could be parsimoniously assigned to a single tradition, were it not for accompanying skeletal remains, and in this regard the continuing lack of South Asian Late Pleistocene hominin remains before *c.* 35–30 kyr BP (Deraniyagala 1992) is a persistent barrier.

Only one major Late Pleistocene technological shift — from Middle Palaeolithic to systematic microlithic technologies — has been identified within South Asia. Recent postulation has suggested that this may in fact be a signature of initial modern human dispersal, linking the South Asian microliths typologically to those found in the South African Howiesons Poort and various East African sites (Mellars 2006). This speculation, however, unravels in the face of genetic and dating evidence. The earliest microliths in South Asia appear at around 35 kyr BP, and correspond to environmental deterioration and a demographic spike seen in reconstructed genetic histories of human occupation of the Indian subcontinent (Clarkson *et al.* 2009; Petraglia *et al.* 2009a). Importantly, this spike is clearly not a marker of new populations entering the region, but an autochthonous differentiation and expansion, as revealed by a thorough review and modelling of available South Asian genetic sequences (e.g. Kivisild *et al.* 2003; Sahoo *et al.* 2006; Sengupta *et al.* 2006). In any case, the appearance of microlithic technologies in this region is much too late to allow for subsequent colonization of Australasia by 45–60 kyr BP (O'Connell & Allen 2004), recognizing the human propensity for rapid occupation of large landmasses. Indeed, entry into South Asia 35–30 kyr BP may have been prohibited by desertification of the region's northwest at the time (Petraglia *et al.* 2009a, fig. 1 and supplementary material).

Conclusion

The lithic data presented here add much-needed empirical focus to discussions of the technology and ultimately taxonomic identity of the hominins in southern India *c.* 74,000 years ago. Comparable data sets are nonexistent for any other region of South Asia, and the emerging Jurreru Valley results therefore establish a behavioural baseline for this time period. Continued exploration of the Jurreru palaeolandscape holds significant promise for constructing the first chronologically constrained and technologically comprehensive record for the South Asian Late Pleistocene. In the absence of fossil evidence, the lithic data set, in combination with genetic coalescence ages and chronometric dating, offers the most productive means currently available to accurately identify the timing, manner and behavioural correlates of the modern human dispersal into India.

Acknowledgements

Our thanks to Arawazhi, Kate Connell, Alison Crowther, Clair Harris, Jinu Koshy, Bert Roberts and Ceri Shipton, as well as Hari, Ramesh, and the villagers of Jwalapuram. For permission and assistance we acknowledge the Archaeological Survey of India and the American Institute for Indian Studies. This research is supported by the British Academy, the Leverhulme Trust, the Natural Environment Research Council (NERC) (Environmental Factors in the Chronology of Human Evolution and Dispersal program), the Leakey Foundation, the NERC Arts and Humanities Research Council Oxford Radiocarbon Accelerator Dating Service, the McDonald Institute for Archaeological Research and the Australian Research Council.

References

Acharyya, S.K. & P.K. Basu, 1993. Toba ash on the Indian subcontinent and its implications for correlation of late Pleistocene alluvium. *Quaternary Research* 40(1), 10–19.

Allchin, B. & R. Allchin, 1997. *Origins of a Civilization: the Prehistory and Early Archaeology of South Asia.* New York (NY): Viking.

Allchin, B., A. Goudie & K. Hedge, 1978. *The Prehistory and Palaeogeography of the Great Indian Desert.* London: Academic Press.

Ambrose, S.H., 1998. Late Pleistocene human population bottlenecks, volcanic winter, and differentiation of modern humans. *Journal of Human Evolution* 34(6), 623–51.

Ambrose, S.H., 2003. Did the super-eruption of Toba cause a human population bottleneck? Reply to Gathorne-Hardy and Harcourt-Smith. *Journal of Human Evolution* 45(3), 231–7.

Atkinson, Q.D., R.D. Gray & A.J. Drummond, 2008. MtDNA variation predicts population size in humans and reveals a major southern Asian chapter in human prehistory. *Molecular Biology and Evolution* 25(2), 468–74.

Blumenschine, R., S. Brandt & J. Clark, 1983. Excavations and analysis of Middle Palaeolithic artifacts from Patpara, Madhya Pradesh, in *Palaeoenvironments and Prehistory in the Middle Son Valley*, eds. G. Sharma & J. Clark. Allahabad: Abinash Prakashan, 39–114.

Clarkson, C., M. Petraglia, R. Korisettar *et al.*, 2009. The oldest and longest enduring microlithic sequence in India: 35 000 years of modern human occupation and change at the Jwalapuram Locality 9 rockshelter. *Antiquity* 83, 326–48.

Deraniyagala, S.U., 1992. *The Prehistory of Sri Lanka: an Ecological Perspective.* (Memoir 8.) Colombo: Department of the Archaeological Survey, Government of Sri Lanka.

Endicott, P., M. Metspalu & T. Kivisild, 2007. Genetic evidence on modern human dispersals in South Asia: Y chromosome and mitochondrial DNA perspectives. The world through the eyes of two haploid genomes, in *The Evolution and History of Human Populations in South Asia,* eds. M.D. Petraglia & B. Allchin. Dordrecht: Springer Netherlands, 229–44.

Field, J. & M.M. Lahr, 2005. Assessment of the southern dispersal: GIS-based analyses of potential routes at Oxygen Isotopic Stage 4. *Journal of World Prehistory* 19(1),1–45.

Field, J., M. Petraglia & M.M. Lahr, 2007. The southern dispersal hypothesis and the South Asian archaeological record: examination of dispersal routes through GIS analysis. *Journal of Anthropological Archaeology* 26(1), 88–108.

Gathorne-Hardy, F.J. & W.E. Harcourt-Smith, 2003. The super-eruption of Toba, did it cause a human bottleneck? *Journal of Human Evolution* 45(3), 227–30.

Grayson, D.K. & S.C. Cole, 1998. Stone tool assemblage richness during the Middle and Early Upper Palaeolithic in France. *Journal of Archaeological Science* 25(9), 927–38.

Hiscock, P., 2001. Sizing up prehistory: sample size and composition of Australian artefact assemblages. *Australian Aboriginal Studies* 1, 48–62.

James, H.V.A., 2007. The emergence of modern human behavior in South Asia: a review of the current evidence and discussion of its possible implications, in *The Evolution and History of Human Populations in South Asia,* eds. M.D. Petraglia & B. Allchin. Dordrecht: Springer Netherlands, 201–27.

James, H.V.A. & M.D. Petraglia, 2005. Modern human origins and the evolution of behavior in the Later Pleistocene record of South Asia. *Current Anthropology* 46, S3–S27.

Jones, S.C., 2007. The Toba supervolcanic eruption: tephra-fall deposits in India and paleoanthropological implications, in *The Evolution and History of Human Populations in South Asia,* eds. M.D. Petraglia & B. Allchin. Dordrecht: Springer Netherlands, 173–200.

Kivisild, T., K. Kaldma, M. Metspalu, J. Parik, S. Papiha & R. Villems, 1999. The place of the Indian mtDNA variants in the global network of maternal lineages and the peopling of the Old World, in *Genomic Diversity: Applications in Human Population Genetics*, eds. S. Papiha, R. Deka & R. Chakraborty. London: Kluwer Academic/Plenum, 135–52.

Kivisild, T., S. Rootsi, M. Metspalu *et al.*, 2003. The genetic heritage of the earliest settlers persists both in Indian tribal and caste populations. *American Journal of Human Genetics* 72, 313–32.

Kuhn, S., 1990. A geometric index of reduction for unifacial stone tools. *Journal of Archaeological Science* 17, 583–93.

Lee, M.-Y., C.-H. Chen, K.-Y. Wei, Y. Iizuka & S. Carey, 2004. First Toba supereruption revival. *Geology* 32(1), 61–4.

McBrearty, S. & A.S. Brooks, 2000. The revolution that wasn't: a new interpretation of the origin of modern human behavior. *Journal of Human Evolution* 39, 453–563.

Mellars, P., 2006. Going east: new genetic and archaeological perspectives on the modern human colonization of Eurasia. *Science* 313, 796–800.

Mellars, P., 2007. Rethinking the human revolution: Eurasian and African perspectives, in *Rethinking the Human Revolution: New Behavioural and Biological Perspectives on the Origin and Dispersal of Modern Humans*, eds. P. Mellars, K. Boyle, O. Bar-Yosef & C. Stringer. (McDonald Institute Monographs.) Cambridge: McDonald Institute for Archaeological Research, 1–11.

Mishra, S., 1995. Chronology of the Indian Stone Age: the impact of recent absolute and relative dating attempts. *Man and Environment* 20(2), 11–16.

Misra, V.N., 1985. The Acheulean succession at Bhimbetka, Central India, in *Recent Advances in Indo-Pacific Prehistory*, eds. V.N. Misra & P. Bellwood. New Delhi: Oxford I-B-H, 35–47.

Misra, V.N., 1989. Stone Age India: an ecological perspective. *Man and Environment* 14, 17–64.

Misra, V.N., 1995. Geoarchaeology of the Thar Desert, North West India, in *Quaternary Environment and Geoarchaeology of India*, eds. S. Wadia, R. Korisettar & V.S. Kale. Bangalore: Geological Society of India Press, 210–30.

Moore, M.W., T. Sutikna, Jatmiko, M.J. Morwood & A. Brumm, 2009. Continuities in stone flaking technology at Liang Bua, Flores, Indonesia. *Journal of Human Evolution* 57(5), 503–26.

Ninkovich, D., N.J. Shackleton, A.A. Abdel-Monem, J.D. Obradovich & G. Izett, 1978. K-Ar age of the late Pleistocene eruption of Toba, north Sumatra. *Nature* 276, 574–7.

O'Connell, J.F. & J. Allen, 2004. Dating the colonization of Sahul (Pleistocene Australia-New Guinea): a review of recent research. *Journal of Archaeological Science* 31, 835–53.

Oppenheimer, C., 2002. Limited global change due to the

largest known Quaternary eruption, Toba ≈74 kyr BP? *Quaternary Science Reviews* 21(14–15), 1593–609.
Oppenheimer, S., 2009. The great arc of dispersal of modern humans: Africa to Australia. *Quaternary International* 202(1–2), 2–13.
Pappu, S., 2001. Middle Palaeolithic stone tool technology in the Kortallayar Basin, South India. *Antiquity* 75, 107–17.
Petraglia, M., 2007. Mind the gap: factoring the Arabian peninsula and the Indian subcontinent into Out of Africa models, in *Rethinking the Human Revolution: New Behavioural and Biological Perspectives on the Origin and Dispersal of Modern Humans*, eds. P. Mellars, K. Boyle, O. Bar-Yosef & C. Stringer. (McDonald Institute Monographs.) Cambridge: McDonald Institute for Archaeological Research, 383–94.
Petraglia, M., J. Schuldenrein & R. Korisettar, 2003. Landscapes, activity, and the Acheulean to Middle Paleolithic transition in the Kaladgi Basin, India. *Eurasian Prehistory* 1(2), 3–24.
Petraglia, M., R. Korisettar, N. Boivin *et al.*, 2007. Middle Paleolithic assemblages from the Indian subcontinent before and after the Toba super-eruption. *Science* 317, 114–16.
Petraglia, M., C. Clarkson, N. Boivin *et al.*, 2009a. Population increase and environmental deterioration correspond with microlithic innovations in South Asia ca. 35,000 years ago. *Proceedings of the National Academy of Sciences of the USA* 106(30), 12,261–6.
Petraglia, M., R. Korisettar, M.K. Bai *et al.*, 2009b. Cave and rockshelter records, the Toba super-eruption and forager-farmer interactions in the Kurnool District, India. *Eurasian Prehistory* 6(1–2), 119–66.
Rampino, M.R. & S. Self, 1993. Climate-volcanism feedback and the Toba eruption of ~74,000 years ago. *Quaternary Research* 40(3), 269–80.
Rao, C.V. & C.V.N.K. Rao, 1992. Palaeontological studies of the cave fauna of Kurnool District, Andhra Pradesh. *Records of the Geological Survey of India* 135(5), 240–41.
Roberts, R.G., R. Jones, N.A. Spooner, M.J. Head, A.S. Murray & M.A. Smith, 1994. The human colonization of Australia: optical dates of 53,000 and 60,000 bracket human arrival at Deaf Adder Gorge, Northern Territory. *Quaternary Science Reviews* 13(5–7), 575–83.
Sahoo, S., A. Singh, G. Himabindu *et al.*, 2006. A prehistory of Indian Y chromosomes: evaluating demic diffusion scenarios. *Proceedings of the National Academy of Sciences of the USA* 103(4), 843–8.
Sali, S.A., 1989. *The Upper Palaeolithic and Mesolithic Cultures of Maharashtra*. Pune: Deccan College Post-Graduate and Research Institute.
Sengupta, S., L.A. Zhivotovsky, R. King *et al.*, 2006. Polarity and temporality of high-resolution Y-chromosome distributions in India identify both indigenous and exogenous expansions and reveal minor genetic influence of Central Asian pastoralists. *American Journal of Human Genetics* 78, 202–21.
Shane, P., J. Westgate, M. Williams & R. Korisettar, 1995. New geochemical evidence for the Youngest Toba Tuff in India. *Quaternary Research* 44(2), 200–204.
Shea, J.J., 1988. Spear points from the Middle Paleolithic of the Levant. *Journal of Field Archeology* 15, 441–50.
Shea, J.J., 2003. Neandertals, competition, and the origin of modern human behavior in the Levant. *Evolutionary Anthropology* 12, 173–87.
Shea, J.J., 2008. Transitions or turnovers? Climatically-forced extinctions of *Homo sapiens* and Neanderthals in the east Mediterranean Levant. *Quaternary Science Reviews* 27, 2253–70.
Shea, J.J., Z. Davis & K. Brown, 2001. Experimental tests of Middle Paleolithic spear points using a calibrated crossbow. *Journal of Archaeological Science* 28, 807–16.
Stock, J.T., M.M. Lahr & S. Kulatilake, 2007. Cranial diversity in South Asia relative to modern human dispersals and global patterns of human variation, in *The Evolution and History of Human Populations in South Asia*, eds. M.D. Petraglia & B. Allchin. Dordrecht: Springer Netherlands, 245–68.
Sun, C., Q.-P. Kong, M. Palanichamy *et al.*, 2006. The dazzling array of basal branches in the mtDNA macrohaplogroup M from India as inferred from complete genomes. *Molecular Biology and Evolution* 23, 83–690.
Westgate, J.A., P.A.R. Shane, N.J.G. Pearce *et al.*, 1998. All Toba tephra occurrences across peninsular India belong to the 75,000 yr B.P. eruption. *Quaternary Research* 50(1), 107–12.
Zuraina, M., 2003. *Archaeology in Malaysia*. Penang: Universiti Sains Malaysia.

Chapter 8

The Middle to Upper Palaeolithic Transition in Western Asia

Ofer Bar-Yosef & Anna Belfer-Cohen

The debate concerning the colonization of Eurasia by modern humans and the ensuing demise of the Neanderthals revolves around the major cultural change defined as the 'transition' from the Middle to the Upper Palaeolithic. The main disagreements among scholars concerning the processes, results and implications of the Middle–Upper Palaeolithic transition stem from fuzzy archaeological terminologies, disregard of the meaning of learnt behaviour in traditional societies, ambiguities in the interpretation of radiocarbon chronology within the range of 50–40 ka BP, and uncritical acceptance of visual observations made in the field. Nonetheless, though some of these malaises haunt the relevant archaeological record of western Asia, any discussion pertaining to the Middle–Upper Palaeolithic transition has to comprise the available evidence derived from this vast region as it is of crucial importance to the understanding of that transformation. In the following pages we will briefly summarize the evidence from several areas within western Asia. Times of geo-political unrest, and limitations imposed by laws of antiquities in several countries hindered obtaining a clear picture of the geographic distribution of prehistoric sites in this topographically, climatically and vegetationally variable mega-region.

Today the climate of western Asia is generally, but not exclusively, dominated by two seasons: cool, rainy winters and hot, dry summers. Winter temperatures are milder in the coastal ranges and more severe inland or at higher elevations. Precipitation is affected by distance from the sea and by altitude, with the central Anatolian and Iranian plateaux, the Syro-Arabian desert and Mesopotamia being the driest zones. The current complex climatic system of western Asia makes it difficult to reconstruct the patterns of past climates. Most of the precipitation of the region is brought by storm tracks following various paths across Europe and the Mediterranean Sea, except for the southern edges at the shores of the Indian Ocean where the monsoon is the main vehicle for summer rains. Deep-sea cores from the Mediterranean and isotope studies of spelaeothems in Israel have demonstrated that the Upper Pleistocene rainfall distributions were similar to those of today although higher or lower at certain times than the current averages (Bar-Matthews *et al.* 2003). Rather than temperature changes, the decadal and centennial fluctuations in the amount of precipitation were responsible for the expansion and contraction of vegetation belts and therefore dictated the distribution of the available plant and animal sources for Middle and Upper Palaeolithic humans.

The definition of the Middle–Upper Palaeolithic boundary is fraught with methodological problems. First among these is the recurrent absence of a common definition of how this cultural boundary is to be recognized (e.g. McBrearty & Brooks 2000; Bar-Yosef 2002). In the present paper we follow the original contention that it is the typological change which determines the cultural change, namely the dominance of end-scrapers, burins and blade forms (including points or backed pieces) over the characteristic late Middle Palaeolithic types such as side scrapers and flake points. The second issue relates to the term 'transition' which actually implies some sort of cultural continuity. As the first Upper Palaeolithic industry emerged from a Mousterian background it comprised a relatively large number of elements and technological attributes of the former industries. If the carriers of this type of Initial Upper Palaeolithic (IUP) had moved relatively quickly across land masses, the presence of such an IUP industry in many localities along their route, hundreds or thousands kilometres away from the 'core area', would be falsely interpreted as 'endemic evolution'. The tempo of the time-transgressive movement of people on a continental scale can be resolved when the ambiguities of radiocarbon

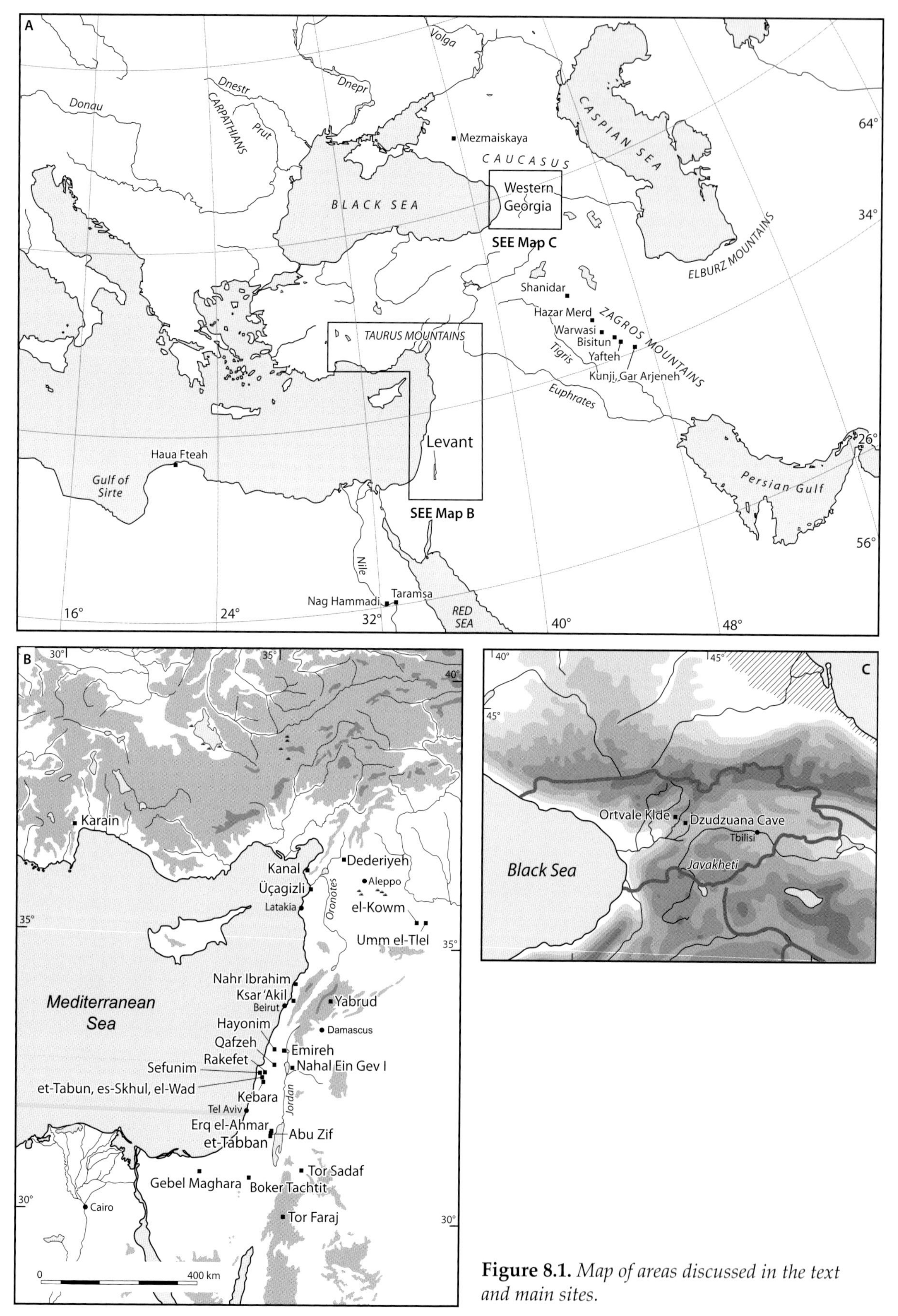

Figure 8.1. *Map of areas discussed in the text and main sites.*

dating for the span of 50–40 ka BP are finally clarified. The first step has already been taken (Mellars 2006), although by employing a different calculation only 3000–5000 years were needed for modern humans to move from the Levant to western Europe (Bar-Yosef 2007). The cultural changeover is obviously more safely recognized as such when the typological and technological components of the local Late Middle Palaeolithic *vis à vis* those of the IUP are entirely different, as in the case of the local Mousterian and the IUP Bohunician in Moravia (Svoboda & Skrdla 1995; Svoboda & Bar-Yosef 2003 and papers therein).

The information concerning Late Middle and Early Upper Palaeolithic cultures derives from the Mediterranean Levant, a few cave sites in the Zagros Mountains and a few sites in the Caucasus, both north and south (Fig. 8.1) (Bar-Yosef & Pilbeam 2000 and papers therein; Bar-Yosef 2007; Goring-Morris & Belfer-Cohen 2003 and papers therein; Belfer-Cohen & Goring-Morris 2007). From the archaeological records of these different sites we offer a summary of what is known and what is missing concerning the Middle to Upper Palaeolithic transition, as well as a description of Early Upper Palaeolithic industries to about 25/20 ka BP. For clarity we have rounded the available dates, uncalibrated, and the interested reader should consult the cited original publications. The archaeological data of the different regions are presented geographically.

The Levant

The Initial Upper Palaeolithic

Historically, the Middle and Upper Palaeolithic entities were first described on the stratigraphic basis of et-Tabun, es-Skhul and el-Wad caves in Mt Carmel (Garrod & Bate 1937) and Abu Zif, et-Tabban and Erq el-Ahmar in the Judean desert, with a short addition from Qafzeh Cave (Neuville 1951). The cultural scheme, from 'Levalloiso-Mousterian' through 'Aurignacian', was originally defined using terminology derived from west European contexts based on broad similarities with the latter. The pioneer researchers, given their technique of excavations, considered the lithic assemblages of the Early Upper Palaeolithic as composed of both Mousterian elements (e.g. Levallois products, side scrapers), and new tool types heralding the full-fledged Upper Palaeolithic entities (e.g. blades, end-scrapers, burins). Later studies by Copeland (1975) of some of the relevant material brought her to suggest that the typological change to Upper Palaeolithic forms still involved the use of blanks produced by Levallois technique.

Garrod (1955) reclassified Turville-Petre's finds from Emireh Cave and suggested labelling the first post-Mousterian entity from both this cave and el-Wad layer E as the first Early Upper Palaeolithic culture, the 'Emiran' (the same culture Neuville considered as 'Upper Palaeolithic Phase I' in his six-part scheme of the local Upper Palaeolithic). The re-assessment of the Upper Palaeolithic layers at Ksar 'Akil led to the coining of the term 'Transitional Industry'. It applied to assemblages deriving from layers overlying the Mousterian occupations and displaying both Upper Palaeolithic typology and partial Middle Palaeolithic technology (Azoury 1986; Copeland 1975; Ohnuma 1988). The notion of 'transitional' characteristics was reinforced with the excavations at Boker Tachtit in the Negev highlands (Marks 1983; 1993), where morphologically Mousterian artefacts (i.e. Levallois points) were shown to be produced by Upper Palaeolithic bi-directional blade technology. Thus the use of the term 'Transitional Industry', implying *in situ* cultural transition, was in accord with the notion that the Upper Palaeolithic of the Levant emerged from a local Middle Palaeolithic industry.

The profile of the pan-Levantine earliest IUP industries is currently rather vague, mainly due to paucity of sites, chronological ambiguities and the presence of particular local tool types. For example, the chamfered blades and flakes — the *chanfreins* — found in north Levantine sites such as Ksar 'Akil, Abri Antelias (next to each other) (Newcomer 1970) and Üçagizli on the coast of southeast Turkey (Kuhn 2002; 2004a) (Fig. 8.1). The active edge of the chamfered pieces was shaped by a side blow, on the retouched end of the piece, forming a rounded edge that is assumed to have served as a scraper (Newcomer 1970). It should be noted that the chamfered tool type is common at Haua Fteah, Cyrenaica, in the Dabban layer, the IUP culture at that site (McBurney 1967). It was also found in Nag Hamadi in Egypt (McBurney 1967), an undated open-air site. The absence of chamfered pieces from stratified sites in the southern Levant is intriguing and only one surface site was found (Goring-Morris & Rosen 1989). Emireh points, although not abundant, prevail in the southern Levant (Copeland 2001), and in particular at Boker Tachtit Levels 1 and 2 (Volkman & Kaufman 1983).

The earlier series of excavations at Ksar 'Akil provided prominent assemblages of the Levantine IUP. The lowermost layers (XXV–XXIV) were rather poor, but they were clearly characterized by opposed platform cores with parallel sides, one of the attributes of a major change in volumetric concept (Ohnuma 1988; Ohnuma & Bergman 1990; Meignen & Bar-Yosef 2002) ascribed to the Upper Palaeolithic. In the layers above (XXIII–XXI/XX), there were triangularly shaped cores which indeed produced convergent blades and Levallois points. However, Ohnuma & Bergman (1990, 202) note that 'the Levallois points in levels XXIII–XXI/XX are far more numerous than Levallois cores probably because many of them were detached from prismatic cores during continuous production of blades'. They draw this conclusion in light of the Boker Tachtit evidence (Marks 1983; 1993; Volkman 1983), where the refitted cores clearly demonstrate the change in the knapper's conception of the volume of the nodule. The Levallois points in these assemblages are shaped by bi-directional removals as documented by the scars (the Y-pattern), and thus are different from the locally common Late Mousterian Levallois points (e.g. Meignen & Bar-Yosef 1991; Kerry & Henry 2000). Similar observations concerning the techno-typological attributes of an early Upper Palaeolithic industry in Tor Sadaf (Wadi Hassa, Jordan) were made by Fox (2003). Although Emireh points were missing, the basic core technology, with blades and elongated triangular blanks with facetted platforms (referred to as 'Levallois points'), resemble the early Ksar 'Akil assemblages and those from Umm el-Tlel (and see below).

These observations raise the question as to why the artisans continued with the production of Levallois points when other tool forms were already being made. The answer lies perhaps in the continued use of thrusting spears at the time the first bows or spear throwers (e.g. atlatl) were invented (Bar-Yosef 1994; Shea 2006; Knecht 1997 and papers therein). Undoubtedly, the production of an old tool form through a new core-reduction strategy testifies to continuity. However, whether this was continuity from the local Late Mousterian (Copeland 1975), an idea borrowed from a neighbouring entity where it was first observed, such as the Nile Valley Mousterian (e.g. Taramsa: Van Peer 1998; 2004; Vermeersch 2001), or it indicates the arrival of new people, remains to be further tested.

Additional IUP assemblages were uncovered in Üçagizli and Kanal caves. These industries are blade-based, with facetted striking platforms, and the tool types comprise the so-called Umm el-Tlel points (see below), a few chamfered pieces, end-scrapers, burins and retouched blades (Kuhn 2004a,b; Kuhn *et al.* 2004; Kuhn pers. comm.). Noteworthy are the shell beads made of marine molluscs similar to those found in Ksar 'Akil (Kuhn *et al.* 2001). Unit V contains retouched Mousterian points, a tool-type missing in earlier layers. Similar points were recorded in the Final Middle Palaeolithic layer at Dederiyeh Cave (Akazawa & Muhesen 2002).

The dating of the Middle–Initial Upper Palaeolithic transition in the Levant is currently based on the TL readings for the latest Mousterian and the radiocarbon dates for the IUP assemblages (for a comprehensive list of the latter see appendix in Goring-Morris & Belfer-Cohen 2003). The TL dates in Kebara Cave (Valladas *et al.* 1987) indicate that the Late Mousterian therein (Unit VI) is *c.* 48 kya, while there are yet no secure dates for Unit V, the very last Mousterian layer in the cave. The phase with Emireh points is missing in Kebara or was destroyed by the erosion that marked the beginning of the Upper Palaeolithic period, following the subsidence and tilt of the Mousterian deposits. Hence, the IUP and the following layers are radiocarbon dated to 43/42 ka BP (Bar-Yosef *et al.* 1996). It seems that the gap in the sequence of Kebara lasted from *c.* 47/45 ka BP to *c.* 43/42 ka BP. This proposal lends credence to the first radiocarbon dates of Boker Tachtit Level 1, 47 ka and 46 ka BP (Marks 1983).

The Middle Palaeolithic–IUP boundary in the Ksar 'Akil sequence (Mellars & Tixier 1989), based on 11 AMS radiocarbon readings of the upper part of the Upper Palaeolithic sequence, three conventional dates for the Aurignacian assemblages, as well as a calculated rate of sedimentation, is assigned to *c.* 50 kya. U-series disequilibrium dates on two bone samples produced somewhat similar readings (van der Plicht *et al.* 1989).

Further north, the AMS charcoal dates from Üçagizli of *c.* 41–39 ka BP place the earliest assemblage in this cave within the range of the IUP industries (Kuhn 2002; 2004a,b; Kuhn *et al.* 2004).

An additional interesting site, located in the steppic belt of the Levant, in the el-Kowm basin (northeast Syria) is Umm el-Tlel, which illustrates another 'facies' of the IUP industries. Layers II Base and III2A, on top of a long Mousterian sequence, are characterized by what is described as a *chaîne opératoire* that follows the Levallois concept (Boëda & Muhesen 1993; Bourguignon 1996; 1998). Many of the cores are volumetrically flat, producing numerous blades which mostly resemble narrow and elongated Levallois points, with uni-directional scar patterns. These points and pointed blades, which grade into regular blades, are named Umm el-Tlel points. Perhaps it would be more appropriate, given the grading of formal points (triangular flakes) into real blades, to name this *chaîne opératoire* the Umm el-Tlel technique or method. Other products, in particular the retouched pieces, are definitely of Upper Palaeolithic character, with numerous burins and end-scrapers, as well as some Middle Palaeolithic elements such as the Nahr Ibrahim technique, notches and denticulates. An AMS date of 34 ka BP and a TL date of 36 ka BP for III2A indicate the age of the industry, but perhaps additional dating would bring them closer to other early dates in the Levant.

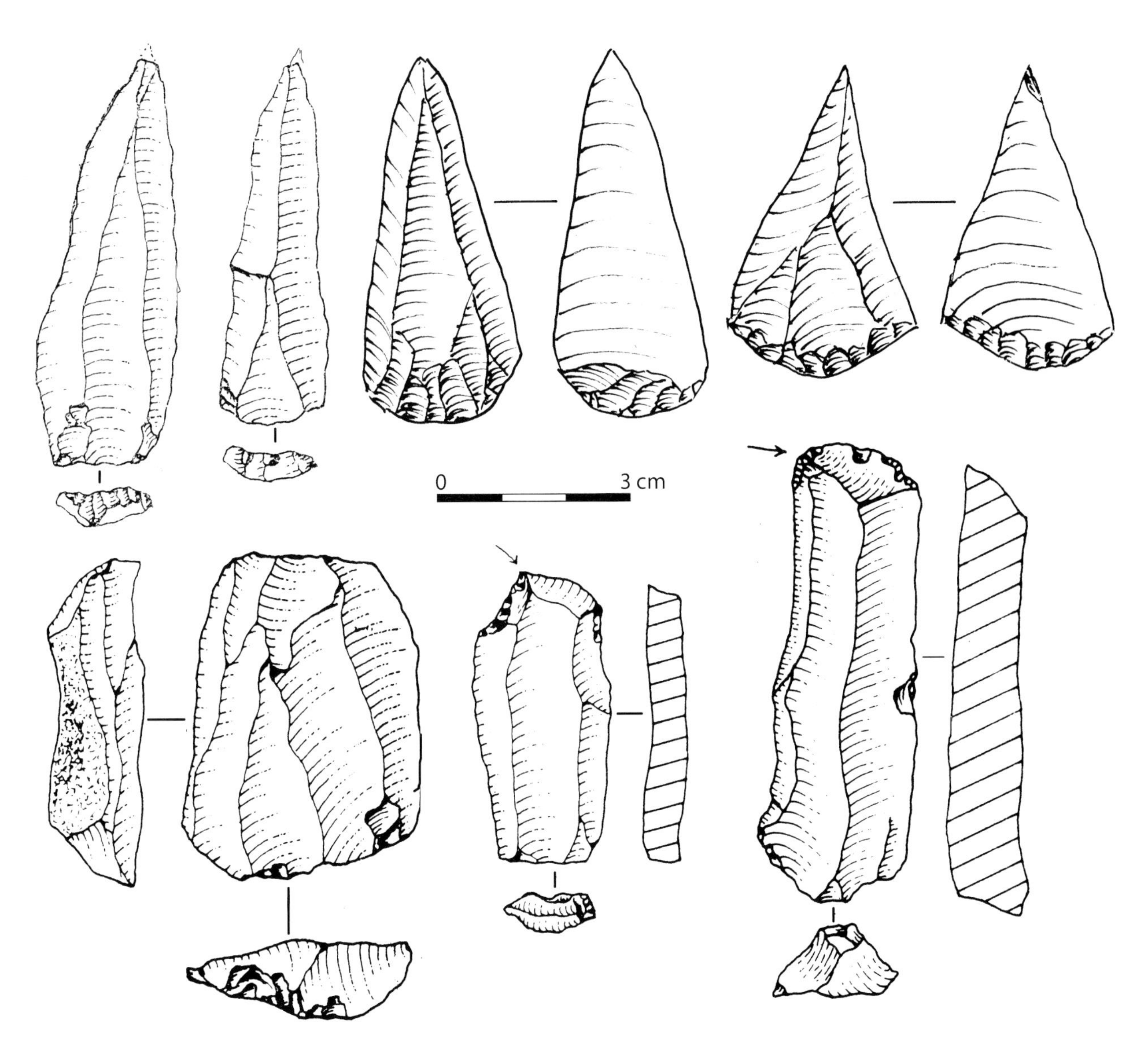

Figure 8.2. *Early Upper Palaeolithic artefacts from the Levant — Emiran and other 'Transitional Industries' (redrawn from Bar-Yosef 2000).*

The earliest Levantine IUP with its two 'facies' (the southern with Emireh points and the northern with chamfered pieces) (Fig. 8.2) is followed by a blade industry, often called the Early Ahmarian. It has been reported from Kebara Units IV and III, Boker A (*c.* 38 ka BP), Boker Tachtit Level 4 (35 ka BP) and is techno-typologically similar to Ksar 'Akil layers XXI–XVII, among others.

The Ahmarian (Fig. 8.3)

The Ahmarian tradition was defined some two and half decades ago (Gilead 1981; 1991; Marks 1981) when the EUP blade/bladelet-dominated assemblages were recognized as different from the Levantine Aurignacian. Already Neuville (1951) realized that in Erq el-Ahmar rockshelter in layers B through F the production of blades exceeded that of flakes. In a later examination Copeland (1976) noted the rarity or absence of carinated and nosed scrapers, commonly seen as typical of the Aurignacian, in local assemblages ascribed on the whole to the Levantine Aurignacian facies. The Ahmarian was later subdivided into Early (38/37 –28/27 ka BP) and Late Ahmarian (*c.* 28/27–20 ka BP). The latter is beyond the scope of this paper and will not be described (see Goring-Morris & Belfer Cohen 2003 and papers therein).

A proliferation of core reduction aimed at obtaining laminar blanks, often with facetted striking platforms was already found in IUP contexts such as Ksar 'Akil layers XXV–XXI (Azoury 1986; Ohnuma 1988) as

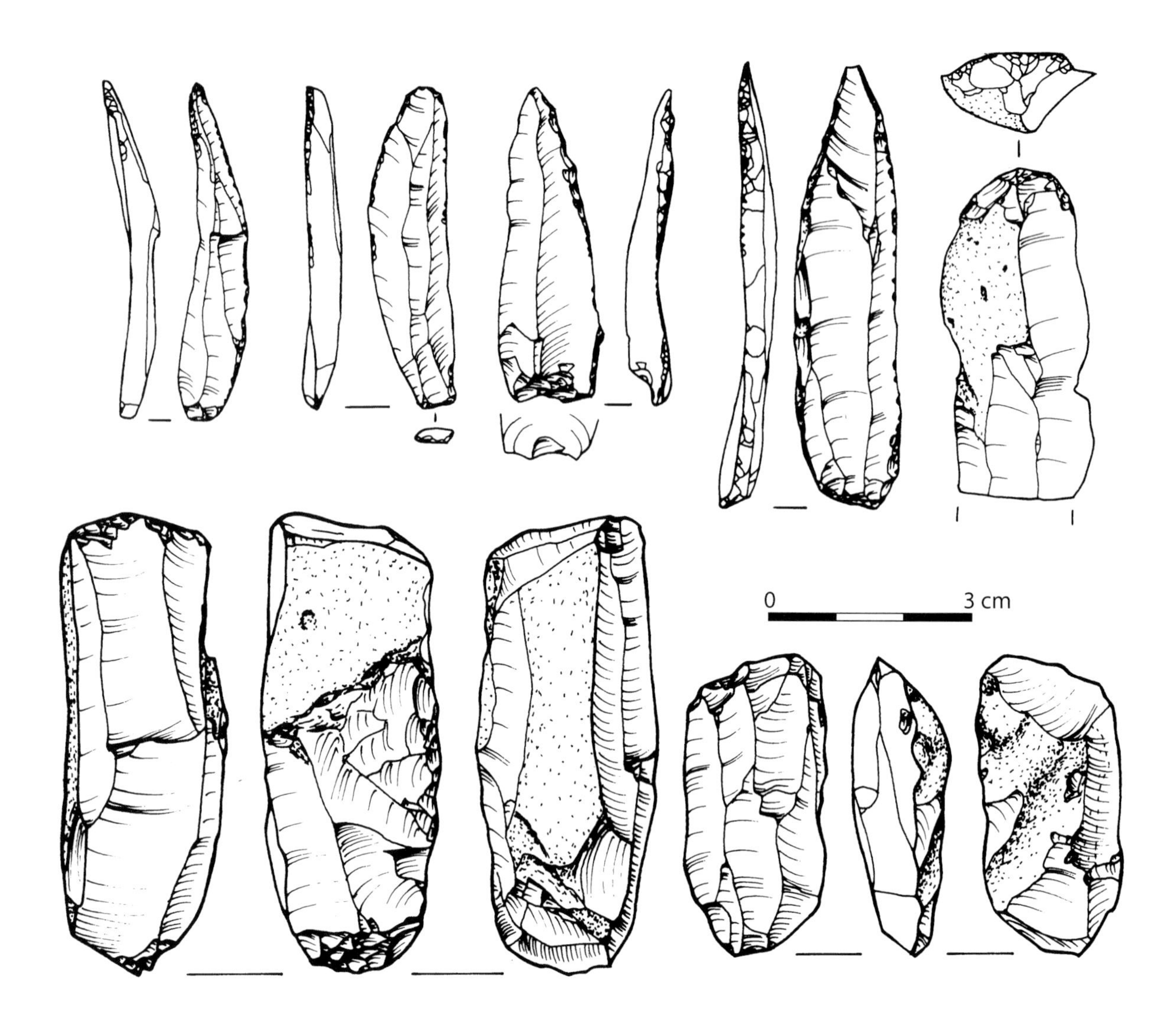

Figure 8.3. *Ahmarian artefacts from the Levant.*

well as in Kebara (at *c.* 42–38/36 ka BP: Bar-Yosef *et al.* 1996). The full production of blades got high in Boker A (*c.* 38 ka BP), Qafzeh Cave (*c.* 31–28 ka BP) as well as in numerous sites in the Negev (e.g. Boker Tachtit level 4, *c.* 35 ka BP), in southern Jordan, and the Sinai peninsula, around 36–34 ka BP (Marks 1983; 1993; Belfer-Cohen & Goring-Morris 2003; Monigal 2003; Fox 2003; Baruch & Bar-Yosef 1986; Bar-Yosef & Belfer 1977; Bar-Yosef & Belfer-Cohen 2005; Becker 1999; Coinman & Henry 1995; Gilead & Bar-Yosef 1993; Phillips 1988; 1994). All these assemblages often retrieved through systematic excavations, exhibit fully-fledged blade/bladelet core-reduction strategies. Blank reduction was carried out by manipulating cores with one or two platforms.

The major tool-groups consist of retouched and backed blades and bladelets and include el-Wad points. End scrapers are quite common, but burins are rare. There is definitely a high degree of typological variability among sites that were assigned to the Early Ahmarian. One example is the localities from Gebel Maghara in northern Sinai, dated to 34–30 ka BP, named also as 'Lagaman' (Bar-Yosef & Belfer 1977).

The ephemeral nature of many of the EUP sites suggests that these were temporary camps of mobile foragers. It seems that their social, economic and cultural frameworks were entirely different from those of the Late Mousterian. The larger exploitable territories of these EUP hunter-gatherers would have required improved communication systems in order to maintain contact with other groups and to ensure that their mating networks remained viable.

The Aurignacian (Figs. 8.4 & 8.5)

The Levantine Aurignacian was originally defined on the basis of the same typological criteria as in Europe.

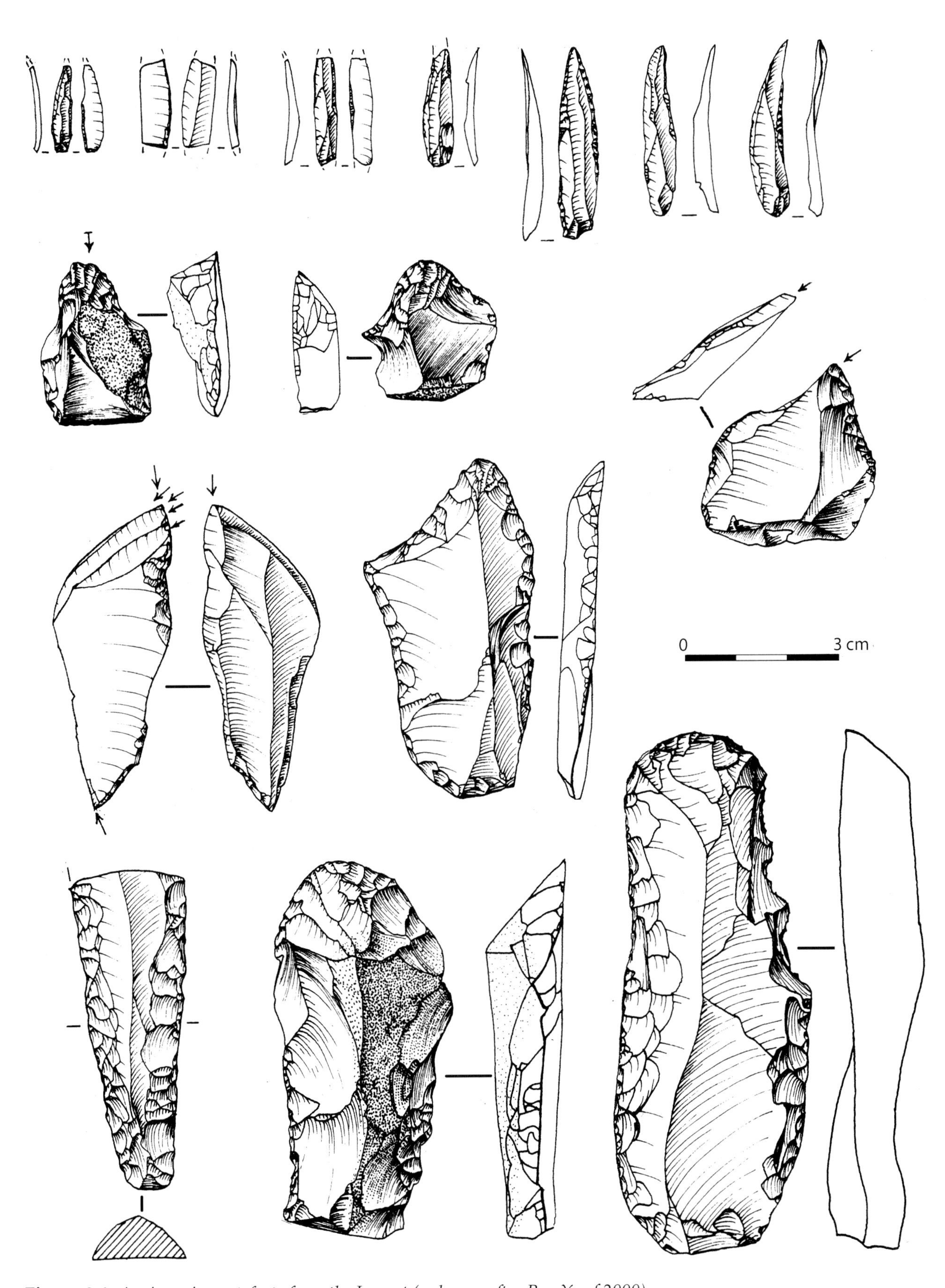

Figure 8.4. *Aurignacian artefacts from the Levant (redrawn after Bar-Yosef 2000).*

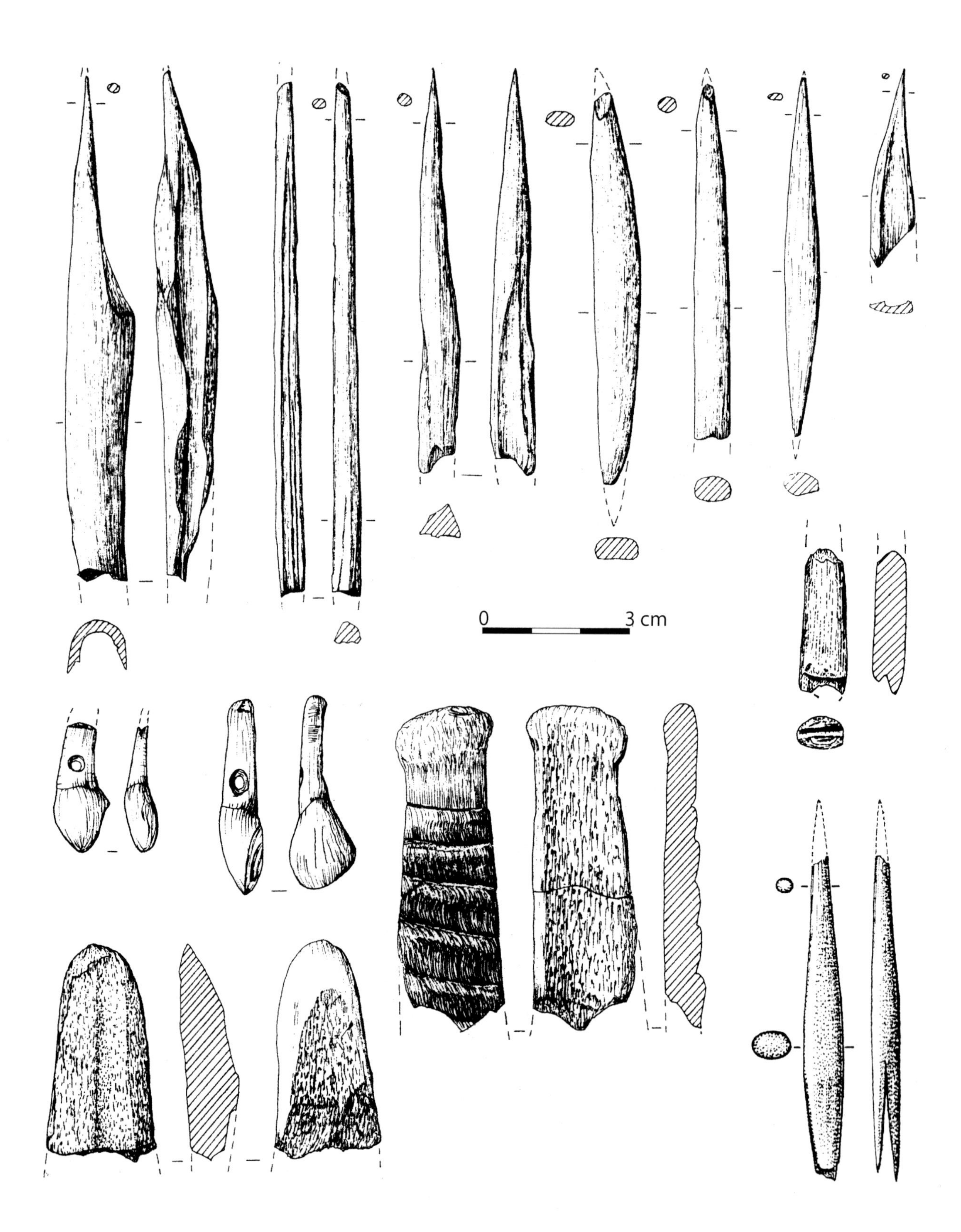

Figure 8.5. *Aurignacian bone, horn and teeth artefacts from the Levant.*

Prominent among the lithic types are the nosed and carinated scrapers, which often outnumber the burins, Dufour bladelets, and the presence of the el-Wad points (considered first by Garrod as a variant of the west European Font Yves or Krems points though she remarked that perhaps they are the local contribution to the basically European variety). There is a wide range of morphological variability among the el-Wad points. In some cases the retouch is fine to semi-abrupt, and seemingly 'corrects' rather than modifies the shape of a given blade or bladelet blank. In rare cases the retouch is abrupt (Bergman 1981). This kind of shaping was necessary since many points served as projectiles (Bergman & Newcomer 1983). The presence of the Dufour bladelets was noticed only when careful sieving was practised during fieldwork. This entity was conceived as rich in bone and antler tools and two split-base points, the hallmark of the Early Aurignacian in western Europe were reported from the Aurignacian levels at Kebara and Hayonim caves (Bar-Yosef & Belfer-Cohen 1996; Belfer-Cohen & Bar-Yosef 1999).

Since the 1970s, the definition of the 'Levantine Aurignacian' has followed the Ksar 'Akil sequence with 'Levantine Aurignacian A' (layers 13–11), 'Levantine Aurignacian B' (layers 10–8), and 'Levantine Aurignacian C' (layers 7–6) (Azoury 1986; Bergman 1987; Copeland 1975; Bergman & Goring-Morris 1987). Later, techno-typological studies demonstrated that the elevated percentages of simple flake scrapers and burins and lack of Aurignacian characteristics clearly indicate that the assemblages assigned to phase C should be removed from the Levantine Aurignacian sequence. In addition these previously called 'Levantine Aurignacian C' assemblages are known not only from the Negev and Sinai but also from the Jordan Valley (i.e. Nahal Ein Gev I: Belfer-Cohen *et al.* 2004). Indeed, it was recognized that the 'classic' Aurignacian assemblages were recovered from cave and rockshelter contexts including Ksar 'Akil, Hayonim, el-Wad, Kebara, Sefunim, Yabrud II, and Rakefet caves.

In sum, there is undoubtedly a dichotomy between flake-dominated and blade-dominated assemblages within the Levantine Upper Palaeolithic that led Gilead (1981) and Marks (1981) to recognize two lithic traditions. This view was held until recently by most scholars in the Near East. However, as the overall picture broadens geographically, it demands the return to the more basic definitions of the known assemblages, not only in terms of the practised core-reduction strategies, but also by taking into account the typological variability as well as the presence of bone and antler artefacts (Bar-Yosef & Belfer-Cohen 1996; Belfer-Cohen & Goring-Morris 2003; Goring-Morris & Belfer-Cohen 2006).

The bone and antler assemblages referred to above were actually reported only from cave sites. Aurignacian typical items such as the split-base bone point were recovered only in two assemblages (see above) while animal teeth pendants (deer, equid and carnivore teeth, polished after the removal of the enamel), identical to those from European Aurignacian contexts were found only at Hayonim Cave (Belfer-Cohen & Bar-Yosef 1981). Of the rare finds which resemble those of the European Aurignacian are two engraved limestone slabs from Hayonim Cave. One bears a series of fine, intentional, incisions and the other an incised horse (Belfer-Cohen & Bar-Yosef 1981; Marshack 1997).

The few dates of the Levantine Aurignacian range between *c.* 36/34–28/27 ka BP, but probably additional dates will shorten this time span. In comparison with the persistence of laminar Ahmarian industries in the region (42/38–18 ka BP), the Levantine Aurignacian *senso stricto* is a short-lived archaeological phenomenon. It justifies earlier proposals (e.g. Kozłowski 1992) to interpret the presence of the Aurignacian in the Levant as evidence for population movement from southeast Europe into the Near East. It also supports the proposal to see the classical Aurignacian as a culture that developed in western Europe and not in the eastern Mediterranean or elsewhere in central Asia (Bar-Yosef 2000).

The Taurus Mountains

The Mousterian sequence in both the Taurus and the Zagros regions is still poorly dated and least known (Kozłowski 1998; Yalçinkaya 1995; Yalçinkaya *et al.* 1993; Rink *et al.* 1994). The main source of information is the excavation of Karain E (the main hall of the cave) where the Mousterian was ESR dated to the Last Interglacial (Rink *et al.* 1994; Kuhn 2002), comprising plenty of side and convergent scrapers, sometimes heavily reduced, as well as retouched points. The amount of debitage present at the site cannot account for all the blanks that were used for modifying the retouched pieces, and therefore one expects this site to be only a seasonal camp into which stone tools were brought, often in already complete form. There is no evidence for the presence of IUP or even EUP industries in this site, and the blade-/bladelet-dominated assemblages in Karain B (a smaller hall in this cave) date to the late Upper Palaeolithic (Atici 2007).

There are other Middle Palaeolithic sites in Anatolia (excluding southeast Turkey, which is part of the northern Levant), containing Levallois-dominated assemblages (e.g. Yalçinkaya 1995). Upper Palaeolithic sites are not yet reported.

The Zagros Mountains

Several cave sites, some with only Late Mousterian industry, were excavated in the Zagros Mountains (Dibble & Holdaway 1993; Solecki & Solecki 1993; Smith 1988; Olszewski & Dibble 2006; Baumler & Speth 1993) including Shanidar, Hazar Merd, Bisitun, Warwasi, Houamian and Kunji. The cave of Shanidar with Neanderthal burials in a 6 m deep occupational sequence produced a Mousterian industry similar to the one published in details from Kunji Cave. In Warwasi, the thickness of the Middle Palaeolithic layers reached about 2.5 m. Bisitun and Kunji caves had shallower deposits. Some of these cave sites are rather far away from each other when compared to the situation in the Levant. For example, the distance between Shanidar and Kunji is about 800 km, without known excavated sites in between. In the Levant, the *c.* 1200 km between el-Kowm (Syria) and Tor Faraj (Jordan), comprise over 20 cave and open-air sites which have been reported in detail.

Among the Late Mousterian sites, Shanidar and Warwasi are located in low altitude (about 350 m above sea level), while Kunji Cave is situated at about 1300 m above sea level. The first two probably served as base camps, while Kunji was an ephemeral camp into which artefacts were brought in from the lowland sites, as indicated by the rarity of cores and the great abundance of retouched pieces on site (Baumler & Speth 1993).

The industry of layer D in Shanidar (with Neanderthal human remains) and that from Kunji Cave compare reasonably well with that of the Karain Mousterian sequence, 1600 km away. These heavily retouched, non-Levalloisian, Mousterian assemblages are rich in side scrapers and retouched points. Even if the radiocarbon readings of 46 ka BP and 50 ka BP from Shanidar upper D are only minimal dates, they do indicate a Late Mousterian age.

The scarcity of field research in this region causes disagreements concerning the recognition of IUP or EUP industries. The lithic assemblages of layer C in Shanidar served as the basis for the definition of the Baradostian, a laminar Upper Palaeolithic industry dated to 37–28 ka BP (Henry & Servello 1974). Similar assemblages were recovered from Warwasi, Gar Arjeneh and Yafteh caves (Olszewski 1999; Olszewski & Dibble 1994). The radiocarbon dates, even if considered as minimal readings, do not support currently temporal correlation between the Baradostian and the IUP of the Levant. Based on the available counts and drawings, it seems that the Baradostian lithic assemblages from Shanidar would correlate at best with the Ahmarian. The attribution of the Warwasi laminar industry to the Aurignacian (Olszewski & Dibble 1994) is rather problematic as the assemblage lacks the main Aurignacian characteristics, e.g. nosed and carinated scrapers on flakes, bone and antler objects, etc. The presence of bladelets defined as Dufour bladelets is not sufficient to call the industry Aurignacian. On the other hand it is possible that the Aurignacian is represented in Arjeneh or Yafteh caves (Otte & Biglari 2004). In the latter cave recent excavations exposed an industry with a few carinated pieces, but mostly regular end-scrapers, burins and el-Wad (or Krems) points. The bone industry, including the *sagie* type and the few beads are not different from those found in the Caucasus Upper Palaeolithic and the radiocarbon dates (both the earlier and later series) fall within the range of 35–28 ka BP (Otte *et al.* 2007). However a clearer Aurignacian assemblage was reported from Afghanistan (Davis 2004). The discussion concerning the origins of the Aurignacian culture is beyond the scope of this paper, but as written above there is currently a consensus that the classical, culturally rich Aurignacian emerged as a west European culture and expanded eastward (e.g. Kozłowski 1992; Mellars 2006). How far east the bearers of this industry dispersed is yet to be determined.

In sum, the rarity of early Upper Palaeolithic stratified and dated assemblages from the Taurus–Zagros region and the rather limited number of known late Mousterian sites from this vast area makes any far-reaching conclusions merely speculations based on long-distance interpolations. It is quite possible that the beginning of the Upper Palaeolithic in the Zagros was relatively late when compared to adjacent regions such as the Levant, a situation that resembles the cultural history of the Caucasus.

The Caucasus region

The better-known area of the Middle to Upper Palaeolithic occupations at the Caucasus is in the western region, on both sides of this high mountain range. Numerous deep rivers flowing north and south create a mosaic of ecological niches populated by a diverse flora and fauna, and provide access to the higher altitudes. Human movements in a south to north direction and *vice versa* were facilitated by the coastal plain along the eastern seaboard of the Black Sea during the Upper Pleistocene. While the cold period of OIS-4 had seen the development of mountain glaciers, OIS-3 witnessed variable warming and cooling conditions. Good-quality lithic raw materials such as flint and radiolarite are widely distributed and were accessible during the entire year. Obsidian was available in the Javakheti region about 80–100 km to the southeast

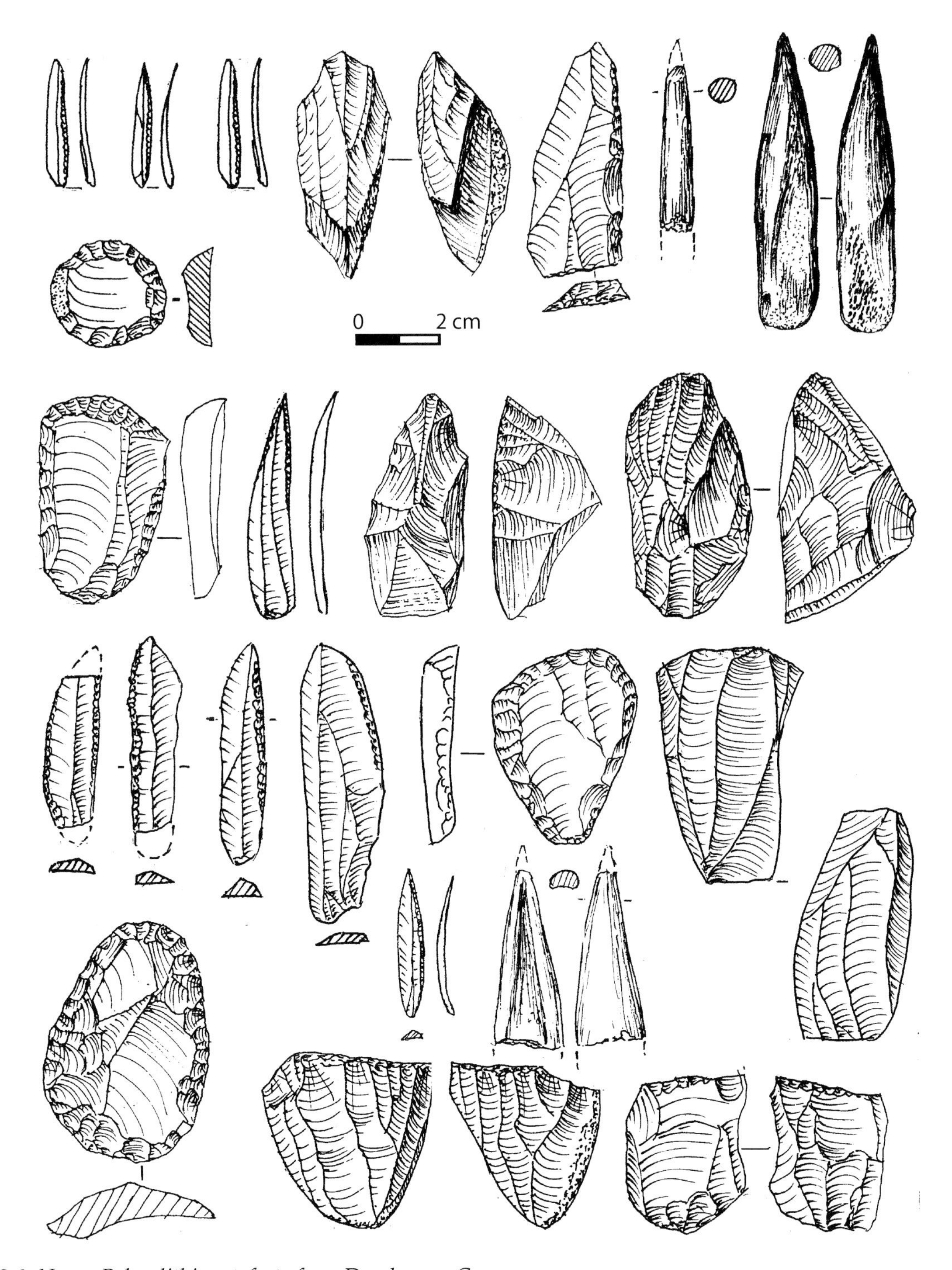

Figure 8.6. *Upper Palaeolithic artefacts from Dzudzuana Cave.*

(Blackman *et al.* 1998) and was obtained by both Middle and Upper Palaeolithic humans.

Recent fieldwork conducted by a joint team of Georgian, American and Israeli researchers from various institutions (Tushabramishvili *et al.* 2002; Meshveliani *et al.* 2004; Bar-Oz *et al.* 2002; Adler *et al.* 2006a,b) focused on new explorations at Ortvale Klde rockshelter and Dzudzuana Cave, both originally excavated by D. Tushabramishvili. The sites are located some 5 km apart, about 550 m above sea level. Modern excavation techniques were employed including water sieving in fine mesh.

The stratigraphic evidence at Ortvale Klde demonstrates a clear break between the Middle Palaeolithic layers (7–5) and the Upper Palaeolithic (4–2) ones. The Late Mousterian industry of this site is not different from other Late Mousterian assemblages from the southern slopes of the Caucasus. It is generally shaped by a Levallois technique, heavily retouched, with side scrapers and points (Lubin 1989; Golovanova & Doronichev 2003; Adler *et al.* 2006a,b; Golovanova *et al.* 2006). This industry resembles the ones known from the Zagros and the Taurus Mountains (Baumler & Speth 1993; Yalçinkaya *et al.* 1993).

Conversely, the Mousterian of the northern slopes of the Caucasus is a variant of the Eastern Micoquian, an entity recorded across eastern Europe and essentially not much different from other Mousterian industries characterized by dominance of foliates and handaxes (Golovanova & Doronichev 2003). Apparently, the Caucasus Mountains during the Late Mousterian formed an important social barrier between two, most probably Neanderthal, populations.

The Upper Palaeolithic sequence in Ortvale Klde begins in layer 4 which contains uni-directional blade cores, end-scrapers on blades, rounded flake scrapers, burins on truncation, numerous retouched (some 2–3 mm wide) and backed bladelets. Noteworthy are some bone/antler points, polished bone/antler abraders, and a polished bone implement with parallel linear incisions of unknown function (Adler *et al.* 2006b).

In addition the two cultural entities used obsidian in different ways. Within the Mousterian occupations obsidian artefacts appear as small debitage items and heavily reduced tools while in the IUP (Layer 4a–d) obsidian is represented by full reduction sequences, including cores and debitage. The dates of sub-layers 4d and 4c range from 34 to 30 ka BP (Adler *et al.* 2006a).

The excavations at Dzudzuana Cave, an elongated karstic cavity with an indoor small creek running towards the entrance revealed the following:

The lower layers (Unit D) contained an industry similar to that of layer 4 in Ortvle Klde characterized by the production of short blades and small bladelets from unidirectional cores. The most distinctive tool-types are microliths, mostly minute, finely retouched bladelets often less than 4 mm wide. Among the larger pieces there are typical end-scrapers on flakes and blades, including an oval type that occurs also in Ortvale Klde. The radiocarbon readings from Unit D, arranged in stratigraphic order and spread over a thickness of 0.6 m, provide ages of 32 ka through 27 ka BP (Meshveliani *et al.* 2004; Fig. 8.6).

Unit C dates to 23–20 ka BP and is dominated by the production of small blades and bladelets detached predominantly from carinated narrow cores. Among several thousands of retouched pieces there are simple end-scrapers on flakes and blades, burins and rare borers.

The industry from Unit C throws light on one aspect of the confusing Upper Palaeolithic terminology which we referred to in the introduction. The issue is the carinated core type. This shape is formed during the preparation of a nodule for the detachment of blanks when the knapper first shaped it into a quasi-biface and then removed one of the thinner (or narrow) sides in order to form a striking platform. From the narrow end of this platform, which is 'nose-shaped', another ridge blade is removed. In order to keep a standard length (and thus avoid maximizing the curvature of the bladelet), the edge opposite the platform (or the 'keel' of the core), was continuously shaped into a 'notched-form', either by retouch or bifacial removal of small flakes. The chosen bladelets were modified into tools by fine retouch. This type of core was described by one of us (Bar-Yosef 1991) under the term 'narrow cores' in reporting the Kebaran in the Levant (dated to *c.* 18–14.5 ka BP) and was discussed in detail by Goring-Morris *et al.* (1998). Later they were more appropriately called 'carinated cores' (Belfer-Cohen & Grosman 2007).

Carinated cores were recognized almost a century ago by Bourlon & Bouyssonie (1912) in the French Aurignacian and had been defined as a '*rabot*', intuitively interpreted as a push-plane. The Aurignacian of western Europe was and still is considered by many scholars as the first cultural manifestation of the modern humans, the Cro-Magnons. Thus it has become a norm to use the term 'Aurignacian' for every first appearance of blade/bladelet industry that was produced of carinated cores. Attributing the term 'Aurignacian' to industries that had nothing to do with this particular west European early Upper Palaeolithic culture is nicely demonstrated in Unit C at Dzudzuana Cave and should serve as a cautionary note for future deliberation.

Industries similar to the Upper Palaeolithic assemblages from Dzudzuana and Ortvale Klde were recovered from the northern side of the Caucasus, such as in Mezmaiskaya Cave (Golovanova *et al.* 2006), dated by radiocarbon and ESR to the same period (35–34 ka BP) as those in Ortvale Klde (Skinner *et al.* 2005).

Given the major differences between the Late Mousterian entities from both sides of the Caucasus Mountains and the similarity in the EUP industries, it is quite clear that modern humans succeed to expand into new territories. The issue whether the Neanderthals were already extinct or were actually replaced by the advancing groups of migrating foragers is unresolved. The dispersal of these colonizers continued into the Crimea area where the Upper Palaeolithic appears even later. The Crimea refugium could have been swept by migrants from both sides of the Black Sea. The crossing of the Caucasus boundary and the dates of the earliest Upper Palaeolithic occurrences on both sides of the mountain range indicate that modern humans entered these regions later than the Levant (Cohen & Stepanchuck 1999).

Discussion

The techno-typological variability among IUP and EUP assemblages from western Asia portrays, in our opinion, individuality of the various groups of modern humans. It is also apparent in the production of particular bone tools or the use of items of personal adornment. When the small migrating groups moved into territories occupied by local Neanderthals, a certain degree of interaction, whether friendly or aggressive, occurred, though for sure at times they could simply ignore each other. The frequency and form of interactions between the newcomers and the indigenous populations are the subject of continued research, but additional methodologies are needed in order to identify the archaeological markers signalling the interactions and the nature of the encounters. Evidence for physical conflicts is missing owing to the rarity of skeletal remains of both populations. The advantages of the incoming migrants could have been the result of better long-distance communication systems (both verbal and non-verbal), improved weaponry (i.e. reduced physical risk to hunters and increased kill rates), or improved reproductive conditions (i.e. decreased infant mortality or shortened inter-birth intervals). Another intriguing option is a better social organization, with role differentiation of the various group members by gender (Kuhn & Stiner 2006) and age which could aid in strategic decisions of the group as a whole.

The dispersal patterns and tempo varied across western Asia as in other parts of the Old World. The route into Europe through Anatolia was the faster one, following both the Danube corridor (Conard & Bolus 2003) and a southern venue through the Mediterranean lands. The move northward through the confluence of the Taurus–Zagros and the Caucasus, took place much later. We may speculate that the dispersal into northern and central Asia preceded the move into the Caucasus. Little is known archaeologically about the Asian southern route into Southeast Asia (Forster 2004), except for the colonization of Australia.

When we consider the interactions between local Middle Palaeolithic groups across Eurasia, including the Neanderthals, and the expanding foreign populations, it is possible that the locals retreated of their own accord or died out as a result of pandemics. Yet the final outcome recorded in the history of humanity was the same, whether after a few hundred years or some thousands years later: the Middle Palaeolithic, industries and humans alike was replaced by Upper Palaeolithic cultures, people and life-ways.

References

Adler, D.S., G. Bar-Oz, A. Belfer-Cohen & O. Bar-Yosef, 2006a. Ahead of the game: Middle and Upper Palaeolithic hunting behaviors in the southern Caucasus. *Current Anthropology* 47(1), 89–118.

Adler, D.S., A. Belfer-Cohen & O. Bar-Yosef, 2006b. Between a rock and a hard place: Neanderthal–modern human interactions in the southern Caucasus, in *When Neanderthals and Modern Humans Met*, ed. N.J. Conard. (Tübingen Publications in Prehistory.) Tübingen: Kerns Verlag, 165–88.

Akazawa, T. & S. Muhesen (eds.), 2002. *Neanderthal Burials: Excavations of the Dederiyeh Cave, Afrin, Syria*. Kyoto: International Research Center for Japanese Studies.

Atici, A.L., 2007. Before the Revolution: a Comprehensive Zooarchaeological Approach to Terminal Pleistocene Forager Adaptations in the Western Taurus Mountains, Turkey. Unpublished PhD thesis, Harvard University.

Azoury, I., 1986. *Ksar Akil, Lebanon: a Technological and Typological Analysis of the Transitional and Early Upper Palaeolithic Levels of Ksar Akil and Abu Halka*. (British Archaeological Reports, International Series 289.) Oxford: BAR.

Bar-Matthews, M., A. Ayalon, M. Gilmour, A. Matthews & C. Hawkesworth, 2003. Sea–land oxygen isotopic relationships from planktonic foraminifera and speleothems in the eastern Mediterranean region and their implication for paleorainfall during interglacial intervals. *Geochimica et Cosmochimica Acta* 67(17), 3181–99.

Bar-Oz, G., D.S. Adler, A. Vekua *et al.*, 2002. Faunal exploitation patterns along the southern slopes of the Caucasus during the Late Middle and Early Upper Palaeolithic, in *Colonisation, Migration and Marginal Areas: a Zooarchaeological Approach*, eds. M. Mondini, S. Munoz & S. Wickler. Oxford: Oxbow Books, 46–54.

Bar-Yosef, O., 1991. The search for lithic variability among Levantine Epi-Paleolithic industries, in *25 ans d'études technologiques en préhistoire, XIe Rencontre Internationales d'Archéologie et Histoire d'Antibes*, ed. L. Meignen. Juan-les-Pins: Éditions APDCA, 319–36.

Bar-Yosef, O., 1994. The contributions of southwest Asia to the study of the origin of modern humans, in *Origins of Anatomically Modern Humans*, eds. M.H. Nitecki & D.V. Nitecki. New York (NY): Plenum Press, 23–66.

Bar-Yosef, O., 2000. The Middle and Early Upper Palaeolithic in southwest Asia and neighbouring regions, in *The Geography of Neandertals and Modern Humans in Europe and the Greater Mediterranean*, eds. O. Bar-Yosef & D. Pilbeam. Cambridge (MA): Peabody Museum, Har-

vard University, 107–56.
Bar-Yosef, O., 2002. The Upper Paleolithic Revolution. *Annual Review of Anthropology* 31, 363–93.
Bar-Yosef, O., 2007. The dispersal of modern humans in Eurasia: a cultural interpretation, in *Rethinking the Human Evolution: New Behavioural and Biological Perspectives on the Origin and Dispersal of Modern Humans*, eds. P. Mellars, K. Boyle, O. Bar-Yosef & C.B. Stringer. (McDonald Institute Monographs.) Cambridge: McDonald Institute for Archaeological Research, 207–18.
Bar-Yosef, O. & A. Belfer, 1977. The Lagaman industry, in *Prehistoric Investigations in Gebel Maghara, Northern Sinai*, eds. O. Bar-Yosef & J.L. Phillips. (Qedem 7, Monographs of the Institute of Archaeology.) Jerusalem: The Hebrew University of Jerusalem, 42–84.
Bar-Yosef, O. & A. Belfer-Cohen, 1996. Another look at the Levantine Aurignacian, in *Proceedings of the XIIIe International Union of Prehistoric and Protohistoric Sciences, Colloquium XI*, eds. A. Montet-White & A. Palma di Cesnola. Forlì: A.B.A.C.O. Edizioni, 139–50.
Bar-Yosef, O. & A. Belfer-Cohen, 2005. The Qafzeh Upper Paleolithic assemblages: 70 years later. *Eurasian Prehistory* 2(1), 145–80.
Bar-Yosef, O. & D. Pilbeam (eds.), 2000. *The Geography of Neandertals and Modern Humans in Europe and the Greater Mediterranean*. Cambridge (MA): Peabody Museum, Harvard University.
Bar-Yosef, O., M. Arnold, N. Mercier *et al.*, 1996. The dating of the Upper Paleolithic layers in Kebara Cave, Mt Carmel. *Journal of Archaeological Science* 23(2), 297–306.
Baruch, U. & O. Bar-Yosef, 1986. Upper Paleolithic assemblages from Wadi Sudr, western Sinai. *Paléorient* 12(2), 69–84.
Baumler, M.F. & J.D. Speth, 1993. A Middle Paleolithic assemblage from Kunji Cave, Iran, in *The Paleolithic Prehistory of the Zagros-Taurus*, eds. D.I. Olszewski & H.L. Dibble. (University Museum Symposium series 5.) Philadelphia (PA): University of Pennsylvania, 1–74.
Becker, M.S., 1999. Reconstructing Prehistoric Hunter-Gatherer Mobility Patterns and the Implications for the Shift to Sedentism: a Perspective from the Near East. Unpublished PhD Thesis, University of Colorado.
Belfer-Cohen, A. & O. Bar-Yosef, 1981. The Aurignacian at Hayonim Cave. *Paléorient* 7(2), 19–42.
Belfer-Cohen, A. & O. Bar-Yosef, 1999. The Levantine Aurignacian: 60 years of research, in *Dorothy Garrod and the Progress of the Palaeolithic*, eds. W. Davies & R. Charles. Oxford: Oxbow Books, 118–34.
Belfer-Cohen, A. & A.N. Goring-Morris, 2003. Current issues in Levantine Upper Palaeolithic research, in *More Than Meets the Eye: Studies on Upper Palaeolithic Diversity in the Near East*, eds. A.N. Goring-Morris & A. Belfer-Cohen. Oxford: Oxbow Books, 1–12.
Belfer-Cohen, A. & A.N. Goring-Morris, 2007. From the beginning: Levantine Upper Palaeolithic cultural continuity, in *Rethinking the Human Evolution: New Behavioural and Biological Perspectives on the Origin and Dispersal of Modern Humans*, eds. P. Mellars, K. Boyle, O. Bar-Yosef & C.B. Stringer. (McDonald Institute Monographs.) Cambridge: McDonald Institute for Archaeological Research, 199–206.
Belfer-Cohen, A. & L. Grosman, 2007. Tools or cores? Carinated artifacts in Levantine Late Upper Paleolithic assemblages and why does it matter?, in *Tools versus Cores: Alternative Approaches to Stone Tool Analysis*, ed. S.P. McPherron. Cambridge: Cambridge Scholars Publishing, 143–63.
Belfer-Cohen, A., A. Davidzon, A.N. Goring-Morris, D. Lieberman & M. Spears, 2004. Nahal Ein Gev I: A late Upper Palaeolithic site by the Sea of Galilee, Israel. *Paléorient* 30(1), 25–46.
Bergman, C.A., 1981. Point types in the Upper Palaeolithic sequence at Ksar Akil, Lebanon, in *Préhistoire du Levant*, eds. J. Cauvin & P. Sanlaville. Paris: CNRS, 319–30.
Bergman, C.A., 1987. *Ksar Akil, Lebanon: a Technological and Typological Analysis of the Later Palaeolithic Levels of Ksar Akil*, vol. II: *Levels XIII–VI*. (British Archaeological Reports, International Series 329.) Oxford: BAR.
Bergman, C.A. & A.N. Goring-Morris, 1987. Conference: the Levantine Aurignacian with special reference to Ksar Akil, Lebanon, March 27–28, 1987. *Paléorient* 13, 140–45.
Bergman, C.A. & M.H. Newcomer, 1983. Flint arrowhead breakage: examples from Ksar Akil, Lebanon. *Journal of Field Archaeology* 10, 238–43.
Blackman, J., R. Badaljan, Z. Kikodze & P. Kohl, 1998. Chemical characterization of Caucasian obsidian — geological sources, in *L'obsidienne au Proche et Moyen Orient*, eds. M.-C. Cauvin, A. Gourgaud, B. Gratuze *et al.* (British Archaeological Reports, International Series 738.) Oxford: BAR, 205–31.
Boëda, E. & S. Muhesen, 1993. Umm el-Tlel (El Kowm, Syrie): étude préliminaire des industries lithiques du Paléolithique moyen et supérieur 1991–1992. *Cahiers de l'Euphrate* 7, 47–92.
Bourguignon, L., 1996. Un Moustérien tardif sur le site d'Umm el-Tlel (Bassin d'El Khowm, Syrie)? Exemples des Nivaux II Base' et III2A', in *The Last Neandertals, the First Anatomically Modern Humans*, eds. E. Carbonell & M. Vaquero. Tarragona: Universitat Rovira i Virgili, 317–36.
Bourguignon, L., 1998. Les industries du Paléolithique intermédiaire d'Umm el Tlel: nouveaux éléments pour le passage entre Paléolithique moyen et supérieur dans le Bassin d'El Khowm, in *Préhistoire d'Anatolie: Genèse de deux mondes*, ed. M. Otte. Liège: ERAUL 85, 709–30.
Bourlon, J. & A. Bouyssonie, 1912. Grattoirs carénés, rabots et grattoirs nucléiformes: essai de classification des grattoirs. *Revue Anthropologique* 22(12), 475–86.
Cohen, V.Y. & V.N. Stepanchuck, 1999. Late Middle and Early Upper Paleolithic evidence from the East European plain and Caucasus: a new look at variability, interactions, and transitions. *Journal of World Prehistory* 13(3), 265–319.
Coinman, N. & D.O. Henry, 1995. The Upper Paleolithic sites, in *Prehistoric Cultural Ecology and Evolution: Insights from Southern Jordan*, ed. D.O. Henry. New York (NY): Plenum Press, 133–214.

Conard, N.J. & M. Bolus, 2003. Radiocarbon dating the appearance of modern humans and timing of cultural innovations in Europe: new results and new challenges. *Journal of Human Evolution* 44, 331–71.

Copeland, L., 1975. The Middle and Upper Paleolithic of Lebanon and Syria in the light of recent research, in *Problems in Prehistory: North Africa and the Levant*, eds. F. Wendorf & A.E. Marks. Dallas (TX): SMU Press, 317–50.

Copeland, L., 1976. Terminological correlations in the Early Upper Palaeolithic of Lebanon and Syria, in *Deuxième Colloque sur la Terminologie de la Préhistoire au Proche Orient*, ed. F. Wendorf. Nice: UISPP, IXe Congrès, 35–48.

Copeland, L., 2001. Forty-six Emireh points from the Lebanon in the context of the Middle to Upper Paleolithic transition in the Levant. *Paléorient* 26(1), 73–92.

Davis, R.S., 2004. Kara Kamar in northern Afghanistan: Aurignacian, Aurignacoid, or just plain Upper Paleolithic? in *Archaeology and Paleoecology of Eurasia: Papers in Honor of Vadim Ranov*, eds. A.P. Derevianko & T.I. Nokhrina. Novosibirsk: Institute of Archaeology and Ethnography, SBRAS Press, 210–17.

Dibble, H.L. & S.J. Holdaway, 1993. The Middle Paleolithic industries of Warwasi, in *The Paleolithic Prehistory of the Zagros-Taurus*, eds. D.I. Olszewski & H.L. Dibble. (University Museum Symposium series 5.) Philadelphia (PA): University of Pennsylvania, 75–100.

Forster, P., 2004. Ice ages and the mitochondrial DNA chronology of human dispersals: a review. *Philosophical Transactions of the Royal Society, London* Series B 359(1442), 255–64.

Fox, J.R., 2003. The Tor Sadaf lithic assemblages: a technological study of the Early Upper Palaeolithic in the Wadi al-Hasa, in *More than Meets the Eye: Studies on Upper Palaeolithic Diversity in the Near East*, eds. A.N. Goring-Morris & A. Belfer-Cohen. Oxford: Oxbow Books, 80–94.

Garrod, D.A.E., 1955. The Mugharet el Emireh in Lower Galilee: type station of the Emiran Industry. *Journal of the Royal Anthropological Institute* 85, 141–62.

Garrod, D.A.E. & D.M. Bate, 1937. *The Stone Age of Mount Carmel*. Oxford: Clarendon Press.

Gilead, I., 1981. Upper Palaeolithic tool assemblages from the Negev and Sinai, in *Préhistoire du Levant*, eds. J. Cauvin & P. Sanlaville. Paris: CNRS, 331–42.

Gilead, I., 1991. The Upper Paleolithic in the Levant. *Journal of World Prehi*story 5(2), 105–54.

Gilead, I. & O. Bar-Yosef, 1993. Early Upper Paleolithic sites in the Qadesch Barnea area, northeast Sinai. *Journal of Field Archaeology* 20, 265–80.

Golovanova, L.V. & V.B. Doronichev, 2003. The Middle Paleolithic of the Caucasus. *Journal of World Prehistory* 17(1), 71–140.

Golovanova, L.V., N.E. Cleghorn, V.B. Doronichev, J.F. Hoffecker, G.S. Burr & L.D. Sulergizkiy, 2006. The early Upper Paleolithic in the northern Caucasus (new data from Mezmaiskaya Cave, 1997 excavation). *Eurasian Prehistory* 4(1–2), 43–78.

Goring-Morris, A.N. & A. Belfer-Cohen (eds.), 2003. *More than Meets the Eye: Studies on Upper Palaeolithic Diversity in the Near East*. Oxford: Oxbow Books.

Goring-Morris, A.N. & A. Belfer-Cohen, 2006. A hard look at the 'Levantine Aurignacian': How real is the taxon? in *Towards a Definition of the Aurignacian*, eds. O. Bar-Yosef & J. Zilhão. (Trabalhos de Arqueologia 45.) Lisbon: American School of Prehistoric Research/Instituto Português de Arqueologia, 297–314.

Goring-Morris, A.N. & S.A. Rosen, 1989. An Early Upper Palaeolithic assemblage with chamfered pieces from the central Negev, Israel. *Mitekufat Haeven* (22), 31*–40*.

Goring-Morris, A.N., O. Marder, A. Davidzon & F. Ibrahim, 1998. Putting Humpty Dumpty together again: preliminary observations on refitting studies in the eastern Mediterranean, in *From Raw Material Procurement to Tool Production: the Organisation of Lithic Technology in Late Glacial and Early Postglacial Europe*, ed. S. Milliken. (British Archaeological Reports International Series 700.) Oxford: BAR, 149–82.

Henry, D. & F. Servello, 1974. Compendium of C-14 determinations derived from Near Eastern prehistoric sites. *Paléorient* 2(1), 19–44.

Kerry, K.W. & D.O. Henry, 2000. Conceptual domains, competence and *chaîne opératoire* in the Levantine Mousterian, in *The Archaeology of Jordan and Beyond: Essays in Honor of James A. Sauer*, eds. L.E. Stager, J.A. Greene & M.D. Coogan. Winona Lake (IN): Eisenbrauns, 238–54.

Knecht, H. (ed.), 1997. *Projectile Technology*. New York (NY): Plenum Press.

Kozłowski, J.K., 1992. The Balkans in the Middle and Upper Palaeolithic: the gateway to Europe or a cul-de-suc? *Proceedings of the Prehistoric Society* 58, 1–20.

Kozłowski, J.K., 1998. The Middle and the Early Upper Paleolithic around the Black Sea, in *Neandertals and Modern Humans in Western Asia*, eds. T. Akazawa, K. Aoki & O. Bar-Yosef. New York (NY): Plenum Press, 461–82.

Kuhn, S.L., 2002. Paleolithic archeology in Turkey. *Evolutionary Anthropology* 11, 198–210.

Kuhn, S.L., 2004a. From Initial Upper Paleolithic to Ahmarian at Üçagizli Cave, Turkey. *Anthropologie* (Brno) XLII(3), 249–62.

Kuhn, S.L., 2004b. Upper Palaeolithic raw material economies at Üçagizli cave, Turkey. *Journal of Anthropological Archaeology* 23, 431–48.

Kuhn, S.L. & M.C. Stiner, 2006. What's a mother to do? The division of labor among Neandertals and modern humans in Eurasia. *Current Anthropology* 47(6), 953–80.

Kuhn, S.L., M.C. Stiner, D.S. Reese & E. Gulec, 2001. Ornaments of the earliest Upper Paleolithic: new insights from the Levant. *Proceedings of the National Academy of Sciences of the USA* 98(13), 7641–6.

Kuhn, S.L., M.C. Stiner & E. Gulec, 2004. New perspectives on the Initial Upper Paleolithic: the view from Üçagizli Cave, Turkey, in *The Early Upper Paleolithic Beyond Western Europe*, eds. P.J. Brantingham, S.L. Kuhn & K.W. Kerry. Berkeley (CA): University of California Press, 113–28.

Lubin, V.P., 1989. The Palaeolithic of the Caucasus, in *The Palaeolithic of the Caucasus and Northern Asia*, ed. P.I. Boriskovski. Moscow: Institute of Archaeology, 9–142.

Marks, A., 1981. The Upper Paleolithic of the Negev, in *Préhistoire du Levant*, eds. J. Cauvin & P. Sanlaville. Paris: CNRS, 299–304.

Marks, A.E. (ed.), 1983. *Prehistory and Paleoenvironments in the Central Negev, Israel*, vol. III: *The Avdat/Aqev Area*, part 3. Dallas (TX): SMU Press.

Marks, A.E., 1993. The Early Upper Paleolithic: the view from the Levant, in *Before Lascaux: the Complex Record of the Early Upper Paleolithic*, eds. H. Knecht, A. Pike-Tay & R. White. London: CRC Press, 5–21.

Marshack, A., 1997. Paleolithic image making and symboling in Europe and the Middle East: a comparative review, in *Beyond Art: Pleistocene Image and Symbol*, eds. M.W. Conkey, O. Soffer, D. Stratmann & N.G. Jablonski. (Memoirs 23.) San Francisco (CA): California Academy of Sciences, 53–91.

McBrearty, S. & A.S. Brooks, 2000. The revolution that wasn't: a new interpretation of the origin of modern human behavior. *Journal of Human Evolution* 39(5), 453–563.

McBurney, C.B.M.,1967. *The Haua Fteah (Cyrenaica) and the Stone Age of the South-east Mediterranean*. Cambridge: Cambridge University Press.

Meignen, L. & O. Bar-Yosef, 1991. Les outillages lithiques mousteriens de Kebara (Fouilles 1982–1985), in *Le Squelette Mousterien de Kebara 2*, eds. O. Bar-Yosef & B. Vandermeersch. Paris: CNRS, 49–76.

Meignen, L. & O. Bar-Yosef, 2002. The lithic industries of the Middle and Upper Paleolithic of the Levant: continuity or break? *Archaeology, Ethnology & Anthropology of Eurasia* 3(11), 12–21.

Mellars, P., 2006. A new radiocarbon revolution and the dispersal of modern humans in Eurasia. *Nature* 439, 931–5.

Mellars, P. & J. Tixier, 1989. Radiocarbon-accelerator dating of Ksar Akil (Lebanon) and the chronology of the Upper Palaeolithic sequence in the Middle East. *Antiquity* 63, 761–8.

Meshveliani, T., O. Bar-Yosef & A. Belfer-Cohen, 2004. The Upper Paleolithic of western Georgia, in *The Early Upper Paleolithic Beyond Western Europe*, eds. P.J. Brantingham, S.L. Kuhn & K.W. Kerry. Berkeley (CA): University of California Press, 129–43.

Monigal, K., 2003. Technology, economy, and mobility at the beginning of the Levantine Upper Palaeolithic, in *More Than Meets the Eye: Studies on Upper Palaeolithic Diversity in the Near East*, eds. A.N. Goring-Morris & A. Belfer-Cohen. Oxford: Oxbow Books, 118–33.

Neuville, R., 1951. *Le Paléolithique et le Mésolithique de Désert de Judée*. (Archives de L'Institut de Paléontologie Humaine Mémoire 24.) Paris: Masson et Cie.

Newcomer, M.H., 1970. The chamfered pieces from Ksar Akil. *Bulletin of the Institute of Archaeology* 8–9, 177–91.

Ohnuma, K., 1988. *Ksar Akil, Lebanon: a Technological Study of the Earlier Upper Palaeolithic Levels at Ksar Akil*, vol. III: *Levels XXV–XIV*. (British Archaeological Reports, International Series 426.) Oxford: BAR.

Ohnuma, K. & C.A. Bergman, 1990. A technological analysis of the Upper Palaeolithic Levels (XXV–VI) of Ksar Akil, Lebanon, in *The Emergence of Modern Humans: an Archaeological Perspective*, eds. P. Mellars & C. Stringer. Edinburgh: Edinburgh University Press, 91–138.

Olszewski, D.I., 1999. The Early Upper Palaeolithic in the Zagros Mountains, in *Dorothy Garrod and the Progress of the Palaeolithic*, eds. W. Davies & R. Charles. Oxford: Oxbow Books, 167–80.

Olszewski, D.I. & H.L. Dibble, 1994. The Zagros Aurignacian. *Current Anthropology* 35(1), 68–75.

Olszewski, D.I. & H.L. Dibble, 2006. To be or not to be Aurignacian: the Zagros Upper Paleolithic, in *Towards a Definition of the Aurignacian*, eds O. Bar-Yosef & J. Zilhão. (Trabalhos de Arqueologia 45.) Lisbon: American School of Prehistoric Research/Instituto Português de Arqueologia, 355–73.

Otte, M. & F. Biglari, 2004. Témoins Aurignaciens dans le Zagros, Iran. *Anthropologie* (Brno) XLII(3), 243–7.

Otte, M., F. Biglari, D. Flas *et al.*, 2007. The Aurignacian in the Zagros region: new research at Yafteh cave, Lorestan, Iran. *Antiquity* 81, 82–96.

Van Peer, P., 1998. The Nile Corridor and the Out-of-Africa model: an examination of the archaeological record. *Current Anthropology* 39 (Supplement), S115–S140.

Van Peer, P., 2004. Did Middle Stone age Moderns of sub-Saharan African descent trigger an Upper Palaeolithic Revolution in the Nile Valley. *Anthropologie* (Brno) XLII(3), 215–26.

Phillips, J.L., 1988. The Upper Paleolithic of the Wadi Feiran, southern Sinai. *Paléorient* 14(2), 183–99.

Phillips, J.L., 1994. The Upper Paleolithic chronology of the Levant and the Nile Valley, in *Late Quaternary Chronology and Paleoclimates of the Eastern Mediterranean*, eds. O. Bar-Yosef & R.S. Kra. Tucson (AZ): Radiocarbon, 169–76.

van der Plicht, J., A. van der Wijk & G.J. Bartstra, 1989. Uranium and thorium in fossil bones: activity ratios and dating. *Applied Geochemistry* 4, 339–42.

Rink, W.J., H.P. Schwarcz, R. Grun *et al.*, 1994. ESR dating of the Last Interglacial Mousterian at Karain Cave, southern Turkey. *Journal of Archaeological Science* 21, 839–49.

Shea, J.J., 2006. The origins of lithic projectile point technology: evidence from Africa, the Levant and Europe. *Journal of Archaeological Science* 33(6), 823–47.

Skinner, A.R., B.A.B. Blackwell, S. Martin *et al.*, 2005. ESR dating at Mezmaiskaya Cave, Russia. *Applied Radiation and Isotopes* 62, 219–24.

Smith, P.E.L., 1988. *Palaeolithic Archaeology in Iran*. Philadelphia (PA): The American Institute of Iranian Studies, The University Museum, University of Pennsylvania.

Solecki, R.S. & R.L. Solecki, 1993. The pointed tools from the Mousterian occupations of Shanidar Cave, northern Iraq, in *The Paleolithic Prehistory of the Zagros-Taurus*, eds. D.I. Olszewski & H.L. Dibble. (University Museum Symposium series 5.) Philadelphia (PA): University of Pennsylvania, 119–46.

Svoboda, J. & O. Bar-Yosef, 2003. *Stranska Skala: Origins of the Upper Paleolithic in the Brno Area, Moravia, Czech Republic*. Cambridge (MA): Peabody Museum, Harvard University.

Svoboda, J. & P. Skrdla, 1995. Bohunician technology, in *The Definition and Interpretation of Levallois Technology*, eds.

H. Dibble & O. Bar-Yosef. Madison (WI): Prehistory Press, 432–8.

Tushabramishvili, N., D.S. Adler, O. Bar-Yosef & A. Belfer-Cohen, 2002. Current Middle and Upper Palaeolithic research in the southern Caucasus. *Antiquity* 76, 927–8.

Valladas, H., J.-L. Joron, G. Valladas *et al.*, 1987. Thermoluminescence dates for the Neanderthal burial site at Kebara in Israel. *Nature* 330, 159–60.

Vermeersch, P.M., 2001. 'Out of Africa' from an Egyptian point of view. *Quaternary International* 75, 103–12.

Volkman, P., 1983. Boker Tachtit: core reconstructions, in *Prehistory and Paleoenvironments in the Central Negev, Israel*, vol. III: *The Avdat/Aqev Area*, part 3, ed. A.E. Marks. Dallas (TX): SMU Press, 127–88.

Volkman, P.W. & D. Kaufman, 1983. A reassessment of the Emireh point as a possible type fossil for the technological shift from the Middle to the Upper Paleolithic in the Levant, in *The Mousterian Legacy, Human Biocultural Change in the Upper Pleistocene*, ed. E. Trinkaus. (British Archaeological Reports, International Series 167.) Oxford: BAR, 35–51.

Yalçinkaya, I., 1995. Thoughts on Levallois technique in Anatolia, in *The Definition and Interpretation of Levallois Technology*, eds. H. Dibble & O. Bar-Yosef. Madison (WI): Prehistory Press, 399–412.

Yalçinkaya, I., M. Otte, O. Bar-Yosef, J. Kozłowski, J.M. Léotard & H. Taskiran, 1993. The excavations at Karain Cave, south-western Turkey: an interim report, in *The Paleolithic Prehistory of the Zagros-Taurus*, eds. D.I. Olszewski & H.L. Dibble. (University Museum Symposium series 5.) Philadelphia (PA): University of Pennsylvania, 100–106.

Chapter 9

The Middle to Upper Palaeolithic Transition in Southern Siberia and Mongolia

Anatoly P. Derevianko

The transition from the Middle to Upper Palaeolithic is one of the most crucial topics of modern archaeology, and an issue with which Paul Mellars has been deeply concerned through-out his career. During the past twenty to thirty years, many symposia, dozens of monographs and hundreds of papers have explored various aspects of the problem of modern human origins and the transition from the Middle to the Upper Palaeolithic. Apparently such attention to the resolution of this multidisciplinary problem has had a positive role. New data have been circulated within the scientific community, and available data have been analysed from new viewpoints. New geochronological techniques have been employed, and specialized laboratories for the biomolecular analysis of the mtDNA of early humans have been established.

Abundant material obtained in the course of fieldwork in southern Siberia and central Asia provides new insights into the problem. Here we will discuss Palaeolithic evidence from the Altai Mountains. Early lithic industries from the Altai are among the best studied in eastern Eurasia. Researchers from the Institute of Archaeology and Ethnography SB RAS have pursued a targetted strategy in Palaeolithic studies during the past 25 years. Along with fieldwork carried out in Siberia, Mongolia, Kazakhstan, Kyrgyzstan, Uzbekistan and elsewhere, fieldwork in the neighbouring valleys of the Anui and Ursul rivers in the Altai Mountains has always been a priority (Fig. 9.1). Since 1983, six caves and more than ten open-air sites have been investigated within an area no more than 150 km from Denisova Cave. The majority of these sites had sequences of 20 or more lithological layers, most of which yielded cultural remains. Research has been carried out following an interdisciplinary approach combining the efforts of experts in archaeology, geology, geomorphology, palaeobotany, palaeontology, physical anthropology, geochronology, palaeoecology and other disciplines working for the Institute of Archaeology and Ethnography as well as other research institutions in Novosibirsk, Moscow, St Petersburg and elsewhere in Russia. Foreign specialists have also participated in this interdisciplinary research.

The Altai caves and open-air sites have revealed well-stratified sedimentary sequences providing sufficient material for the reconstruction of palaeoclimatic and palaeoenvironmental conditions through evidence of floral associations and animal populations during the final Middle to Late Pleistocene. The majority of established horizons have yielded remains of the cultural activity of early humans, allowing reconstructions of human culture and the recognition of the development of lithic industries during the last 800,000 years. A number of books and hundreds of scientific papers have been published describing the results of various analyses. Here I would mention just one such publication dedicated to the results of the complex studies at a site located in the vicinity of Denisova Cave: *Prirodnaya sreda i chelovek v paleolite Gornogo Altaia*, which I edited with numerous colleagues and was published in 2003. To my regret, many publications are not known to the wider public for a variety of reasons.

In 2001–2004, the international scientific journal *Archaeology, Ethnology and Anthropology of Eurasia* published a discussion of the Middle to Upper Palaeolithic transition. All the papers in the journal relating to this discussion have been included in a separate volume, published in both Russian and English. The collection of papers entitled 'The Middle to Upper Palaeolithic Transition in Eurasia. Hypotheses and Facts' (Derevianko 2005c,d) was launched at the opening of the Symposium held at the Denisova Cave field research centre in August 2005. The present paper highlights results of many years of research carried out by archaeological teams headed by the present author in the Altai mountains, Mongolia, Kazakhstan, Uzbekistan and other countries.

Figure 9.1. *Map of the Altai showing the locations of Palaeolithic sites.*

Lower and Middle Palaeolithic industries in the Altai Mountains

Available evidence suggests that there were two global migrations of archaic hominins during the Early Palaeolithic (Derevianko 2001; 2004; 2005a; 2006a). The first migration wave of *Homo ergaster–Homo erectus* left Africa 2–1.8 Mya and inhabited a considerable part of Eurasia over the course of several hundred thousand years. It was the first great human migration during which humans expanded beyond the boundaries of Africa and populated other regions of the Earth. It is most likely that the first migration did not represent a uniform phenomenon. The earliest industries in Eurasia can be subdivided into two major types: the Oldowan and the 'small tool industry' (Derevianko 2006a). Such evidence supports the hypothesis of at least two groups of early hominins from Africa: populations manufacturing tools of Oldowan type (pebble tools) and those with small or micro-tools (Derevianko 2006a, 2).

Karama is the earliest (pebble tool) Palaeolithic site discovered in northern Asia. It is located at the confluence of the Karama and Anui rivers, 15 km downstream from the well-known archaeological site of Denisova Cave. Excavations at the site began in 2001, revealing artefacts which turned out to be earlier than those from any other site in northern Asia. In excavation trench 2, at an elevation of 51 m above the river surface, the profile exposed up to 8 m of Pleistocene deposits. Four horizons bearing archaic lithic implements are recognized. Two culture-bearing horizons have been established within the middle and lower portions of the red-coloured deposits; another two horizons with artefacts have been recognized within the alluvial deposits. Primary reduction strategy is mostly illustrated by pebble cores — striking platforms retain natural crust. From such cores, shortened non-facetted spalls of varying size were removed. Within the tool kit, the most common tool types are those of longitudinal and transverse side-scrapers, nosed tools, choppers and chopping tools, notches and denticulates.

Red-coloured sediments forming the two upper culture-bearing horizons are well correlated with the red deposits in the Cherny Anui profile located 20 m upstream from the Karama site. The age of these sediments has been estimated through RTL analysis to be about 542 kya. Available palaeontological material suggests their attribution to the upper border of the early Middle Pleistocene (Derevianko *et al.* 1992a,c). Flood-plain deposits of the ancient brook have also yielded artefacts and have been attributed to an earlier period — the early Middle Pleistocene. On the basis of all available data, the age of the lowest culture-bearing horizon has been estimated as no later than 800 kya (Derevianko *et al.* 2005). The Karama site is the oldest such site in northern Asia. Its northern location compared to other known sites allows us to hypothesize that populations of the first migration wave (*Homo erectus*?) disseminated over considerable territories in Eurasia, with the possible exclusion of areas located to the north of 52°N.

The second wave of migration most likely originated in the Near East within the chronological range of 450–350 kya and is associated with the distribution of the Late Acheulean culture. The major diagnostic features are Levallois primary lithic reduction and the presence of bifaces. In many territories the new human populations met representatives of the early migration wave, resulting in subsequent mixture of industries: the older local and the Late Acheulean. The mixing of the two industrial bases demonstrates different features in various regions of central Asia: some areas are mostly characterized by Acheulean features (in primary and secondary working techniques), while others show predominantly pebble-tool techniques. Acheulean features are best illustrated by the archaeological collections of Turkmenistan and Kazakhstan. More than ten sites with bifaces and Levallois reduction features have been reported from Kazakhstan (Medoev 1964; 1970; Voloshin 1987; 1988; 1990). New sites with Acheulean features were discovered in the Elba river basin in the Mugodjary Mountains, Kazakhstan in 1999–2000 (Derevianko *et al.* 1999; 2001a,b). The relevant collections comprise bifaces, side-scrapers of various modifications, notch-denticulate tools, Levallois- and disc-cores.

The earliest lithic artefacts in Turkmenistan have been discovered in the Yangadja-Karatengir complex of sites with bifaces (Liubin 1984; Liubin & Vishniatsky 1990; Vishniatsky 1996). In my view, the earliest of these complexes are attributable to the Late Acheulean and have much in common with those from Mugodjary.

In Tajikistan, lithic industries from the late Lower and Middle Palaeolithic mostly belong to the pebble-tool tradition. Similar attribution may be assigned to broadly contemporary sites in Kyrgyzstan and Uzbekistan.

More than one thousand Stone Age sites have been discovered in Mongolia during the last 20 years. Available lithic collections clearly show a new trend in stone-tool reduction demonstrating emergence of the Levallois reduction strategy, bifaces and other features of the Late Acheulean industry (Derevianko *et al.* 1990; 2000a,c).

Archaeological sites of the Altai in southern Siberia provide indisputable evidence of occupation of this area by human populations practising the Late Acheulean tradition of stone-tool manufacturing. Representatives of the earlier populations with pebble tools might have abandoned this area due to deterioration in environmental and climatic conditions. Available archaeological materials attest to a 'gap' occurring in the history of human habitation of the Altai. Altai artefact collections attributable to the Late Acheulean–early Middle Palaeolithic do not suggest any possible superimposition of the Late Acheulean industry on the pebble-tool tradition. The Altai Late Acheulean/Middle Palaeolithic industries show closest similarities of their primary and secondary working strategies with the industries of the Mugaranian and Yabrudian of the Near East (Derevianko 2001). Recent investigations have established considerably older chronometrical ranges for the Mugaranian tradition: layers Ed–Ea dated to 350–270 kya are worth noting (Jelinek 1992; Bar-Yosef 1995; Schwarcz & Rink 1998).

Altai caves (Denisova, Kaminnaya, Okladnikov, Maloyalomanskaya, Biyke etc.) and open-air sites (Ust-Karakol-1 & -2, Anui 1–3, Kara-Bom, Tiumechin-1–4 etc.) have revealed well-stratified sedimentary sequences illustrating the development of Middle Palaeolithic and transitional industries (Derevianko 1990; Derevianko & Markin 1992; 1998; 2000; Derevianko & Shunkov 1992; 2004; Derevianko & Zenin 1990; Derevianko *et al.* 1987; 1991; 1998a; 2000d; 2003).

The earliest evidence for the emergence of human populations of the second migration wave in the Altai Mountains has been obtained from the lowermost culture-bearing horizon 22 at Denisova Cave. This layer has been established within the lowermost soft sediments in the cave and is dated to 282±56 ka BP (RTL-548). Two geomagnetic excursions have been recognized within this layer: Biwa 1 (176–220 kya) within sub-layer 22.1 and Biwa 2 (266–300 kya) within sub-layer 22.2 (Derevianko *et al.* 1998b).

Settlement by early human migrants from the west in this area about 300–250 kya prompted further development of human culture in this territory. Local

Middle Palaeolithic sites have yielded a variety of techno-typological complexes. The noted differences reflect palaeoecological changes as well as behavioural and adaptive strategies rather than migration processes. For instance, collections from 70–60 kya contain greater proportions of denticulate, notched and similar forms than earlier ones. According to palynological data, this period witnessed an increase of the proportion of coniferous trees in the vegetation. The higher percentage of notch-denticulate tools suggests certain changes in adaptation strategies in response to ecological changes, in particular the use of wooden implements. Occurrences of bifacial tools in culture-bearing horizons dated to 90–100 kya at some sites (Ust-Karakol-1 and Anui-3) should be interpreted in a similar way.

The Middle to Upper Palaeolithic transition in the Altai Mountains

The early Middle Palaeolithic human habitation zone can be subdivided into three major parts (African, Eurasian and Sino-Malaysian), each demonstrating lithic industries dramatically different from the others (Derevianko 2005a,b). The differences are well expressed within Middle to Upper Palaeolithic transitional industries and others attributed to the Initial Upper Palaeolithic. According to the present state of knowledge, we can recognize three major models for the transition from the Middle to the Upper Palaeolithic on the basis of available lithic materials: the African (Aterian tradition), the Eurasian (characterized by the standardization of blade production and the manufacturing of blade tools) and the Sino-Malaysian (flake tools, and no traces of Levallois primary reduction) trends. Bordering zones produce industries illustrating acculturation processes, e.g. Cyrenaica and the lower reaches of the Nile in northeastern Africa.

In terms of both chronology and typology, the Altai facies of the Middle to Upper Palaeolithic transition has much in common with the Eurasian developmental trend and the Near East variant in particular. This similarity can be explained by the fact that the Eurasian trend originated in a single industry attributable to the ancient human populations associated with the second major migration wave.

During the final Middle Palaeolithic, the Karakol and the Kara-Bom major trends, or variants, of development of the Upper Palaeolithic industry appeared in the Altai Mountains (Derevianko 2001; Derevianko & Shunkov 2004). The Karakol trend is illustrated by the artefacts from the Denisova Cave and the Ust-Karakol-1 open-air sites. At Denisova Cave we see archaeological material illustrating development of the Middle Palaeolithic industry between 90 and 50 kya from horizons 18–12 in the Main Chamber of the cave and from horizons 10 and 9 in the entrance zone. The density of artefacts varies across horizons. The tool kit comprises Middle Palaeolithic artefacts homogeneous in technical and typological features. Insignificant variations representing various techniques of primary and secondary reduction strategies have been noted across horizons, a fact which has been interpreted in terms of alternating adaptive strategies occurring due to environmental changes rather than as a result of changes and shifts of cultural-historical communities. Primary reduction is characterized by irregular, radial and Levallois technologies, but the proportions of parallel cores and Upper Palaeolithic tools increase upwards in the profile.

The final stage of development of the Denisova industry is illustrated by the assemblage from stratum 11. An infinite date of more than 37,235 cal. BP (SOAN-2504) was generated on bone from this stratum. Another date of 48,630+2380/–1840 cal. BP has been generated for stratum 11 in the Department for Genetics of the Max Planck Institute of Evolutionary Anthropology.

The major characteristic feature of the industry in stratum 11 is the equal proportions of Middle and Upper Palaeolithic tools in the assemblage. Within the collection of typologically distinct retouched tools, the share of Mousterian points and side-scrapers is 22.5 per cent. This category is dominated by longitudinal side-scrapers with one cutting edge. Along with clear Mousterian tools, a few typologically distinct Levallois points are also present (IL_{ty} = 2.5). The share of denticulate tools is not great (12.3 per cent), while the value of the denticulate, notched and beak-shaped tool index is twice as high. The share of Upper Palaeolithic tools is the highest within the collection (29.7 per cent). This category includes such typologically distinct types as end-scrapers, burins, borers and backed blades. These tool types can be considered diagnostic for this industry. Stratum 11 also yielded a few foliate bifaces. In general, the recognized tool types and the noted proportions of various tool categories allow us to interpret the collection of stratum 11 as a new cultural-chronological stage transitional to the initial Upper Palaeolithic.

Another important argument in favour of the Upper Palaeolithic attribution of lithic artefacts from stratum 11 is the occurrence of bone tools and personal body decorations made of bone, shell, mammoth tusk and animal teeth (Fig. 9.2). The set of bone implements recovered from stratum 11 includes more than 60

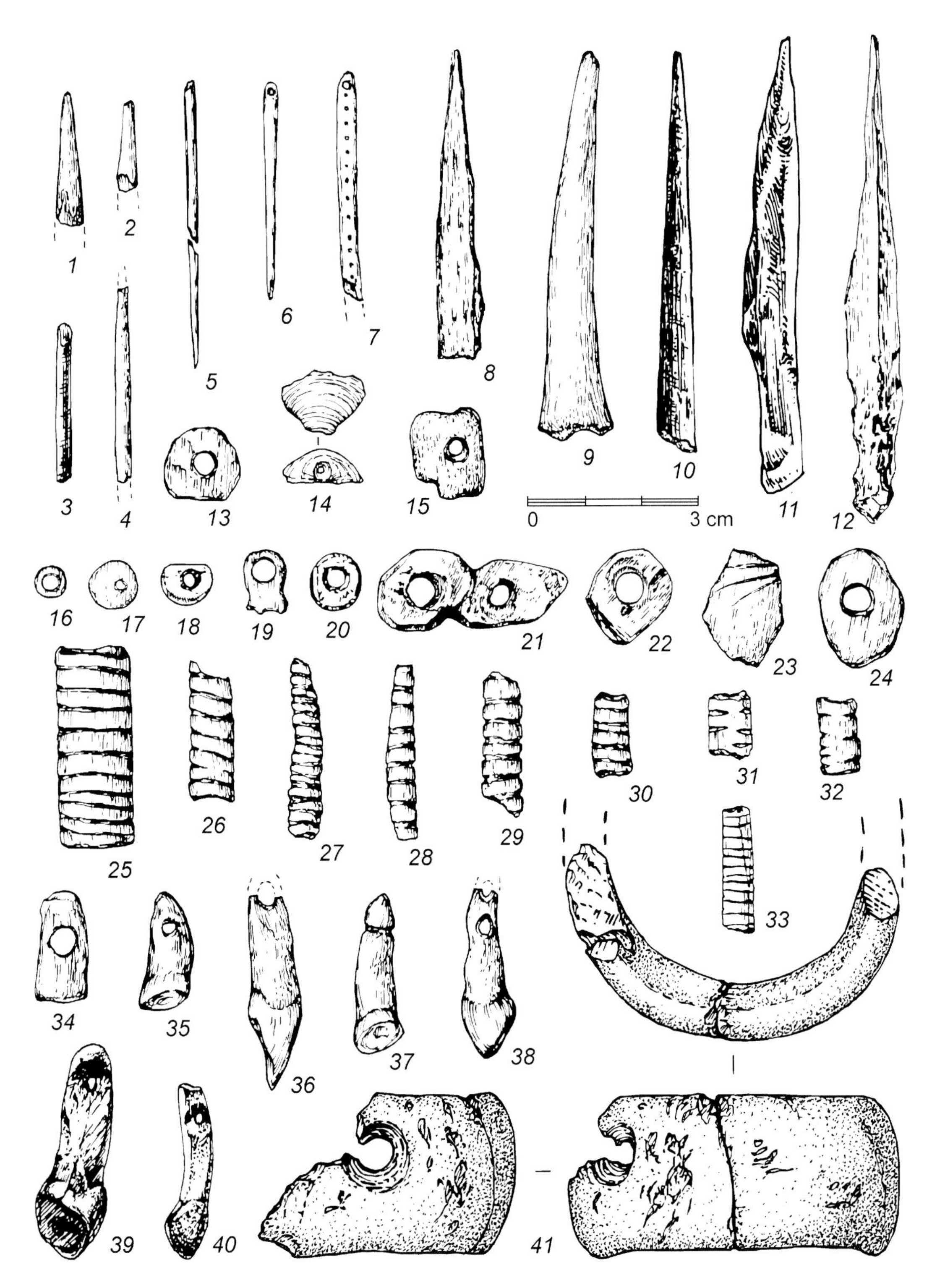

Figure 9.2. *Early Upper Palaeolithic adornments and implements made of bone (1–12, 17, 19, 21–33), stone (13, 15, 18, 41), animal teeth (34–40), mammoth ivory (20), eggshell (16) and mollusc shell (14) from the Main Chamber of Denisova Cave. 1, 2, 8–12) bone awls/borers; 3–7) bone eyed needles; 13, 15, 18) stone beads; 14) ornament of mollusc shell; 16) bead made of ostrich eggshell; 17) bead made of fractured long bones; 19, 21, 22, 24) bead blanks (?); 20) bead made of mammoth ivory; 23) fragment of a rib of a large ungulate with incised decorative motifs; 25–33) cylinder beads with ornamentation; 34–6, 38–40) animal-tooth pendants with a biconical drilled hole; 37) animal-tooth pendant with linear incisions; 41) stone bangle.*

items (Derevianko & Shunkov 2004). The collection comprises:

- miniature eyed needles including a flat tool with a broken tip bearing lines of incised dots on both surfaces;
- awls/borers made of fractured tubular bones of large mammals;
- pendants of fox, bison and deer teeth with bi-conically drilled-out holes or alternatively with linear incisions encircling the root;
- sets of cylinder beads made of tubular bones including specimens decorated with symmetrical linear incisions encircling the bone;
- a fragment of a ring made of a mammoth tusk;
- a mammoth tusk bead bearing natural 'decoration' on the thoroughly polished surface and biconically drilled hole;
- a mammoth tusk fragment with two drilled openings and a transversal groove at the narrow portion in the middle, probably representing a bead blank (?) and sub-ovoid fragments of a mammoth tusk and a tubular bone with well-polished surfaces and wide openings in the centre;
- a ring with thin walls representing a transversally cut fragment of a tubular bone belonging to a large bird;
- a fragment of a rib of a large ungulate with three fan-shaped incised decorative motifs; and
- bars made of the diaphyses of mammal limb bones including medial fragments with polished surfaces and one distal fragment with a flattened tip, as well as fragments of large mammal bones with drilled holes.
- a flat bead-ring made of ostrich eggshell, a material unique for the Altai, is also noteworthy.

In addition, a set of adornment pieces made of gemstones and mollusc shells has been recovered. This set includes fragmented pendants made of kaolinitic agalmatolite and talc/steatite showing a bi-conical drilled opening on one of the narrow sides, beads made of various stones (talc, serpentine and shale) as well as ornaments with drilled holes made of the shells of fresh-water mollusc *Corbicula tibetensis*.

A very interesting artefact was recovered from the lowermost portion of culture-bearing horizon 11 in 2005. It has been identified as a fragmented bracelet consisting of two pieces of dark green serpentine. The bracelet is 27 mm wide and 9 mm thick, the diameter of the whole piece being about 7 mm. The bracelet shows signs of grinding and polishing and an opening made through bifacial drilling.

In summary, the Denisova archaeological assemblage demonstrates development of the lithic industry from the initial Middle Palaeolithic to the early Upper Palaeolithic. Despite some variations across culture-bearing horizons, the industry demonstrates technological and typological homogeneity.

The artefact collection from the lowermost culture-bearing horizons 19–13 at Ust-Karakol (133±33 ka BP (RTL-661) for stratum 19; 100±20 ka BP (RTL-636) for stratum 18A; 90±18 ka BP (RTL-638) for stratum 18B) illustrates two major techniques of primary reduction: Levallois and the parallel detachment of blades and lamellar flakes. These lowermost culture-bearing horizons yielded typical Middle Palaeolithic tools such as side-scrapers of various modifications, notch-denticulate tools, points and others along with Upper Palaeolithic tools such as burins, end-scrapers, borers and truncation tools, which constitute 10 per cent of the total collection.

An industry attributable to the early Upper Palaeolithic has been recovered from culture-bearing horizons 11–9. This industry is characterized by a parallel reduction strategy aimed at detaching elongated blanks from cores with single and double platforms, including prismatic varieties. The production of micro-laminar products is one of the most important features of this techno-complex. The collection comprises wedge-shaped and cone-shaped cores, carinated end-scrapers fashioned using micro-laminar flaking and a set of distinct micro-blades. The tool kit is dominated by longitudinal side-scrapers, end-scrapers and carinated scrapers, knives with natural backs and backed edges, angle burins and borers, spear-shaped tools and notched implements fashioned from retouched *encoches*. There are also points on blades, angular points, beak-shaped and denticulate forms, large retouched blades and micro-blades with backed edges. Isolated specimens of tool types such as typical and atypical Levallois points, a fragment of an ovoid biface and a medial fragment of a foliate biface, a truncation flake and blade, and fragments of a serpentine body decoration piece bearing a bi-conical drilled hole have been noted in the collection.

A series of radiocarbon dates has been generated on charcoal and humic acids from fireplaces in strata 10, 9C and 5 suggesting a Karginian age for the sediments: 35,200±2850 cal. BP (SOAN-3259) for the uppermost portion of stratum 10; 33,400±1285 cal. BP (SOAN-3257), 29,860±355 cal. BP (SOAN-3358) and 29,720±360 cal. BP (SOAN-3359) for stratum 9C; 27,020±435 cal. BP (SOAN-3356) and 26,305±280 cal. BP (SOAN-3261) for stratum 5. An RTL date of 50±12 ka BP (RTL-660) has been obtained on a sample of burnt soil from stratum 9C.

In sum, the evidence from culture-bearing horizons at multi-layered and the well-stratified sites of Ust-Karakol-1 and Denisova Cave allows us to recognize the development of a Middle Palaeolithic industry into an Upper Palaeolithic one on the basis of the Levallois facies. The process of transition seems to have started around 60–50 kya and ended with the formation of an Upper Palaeolithic industry around 50–40 kya or possibly earlier.

The Kara-Bom site is located in the Ursul valley, 150 km from Denisova Cave. Two Middle Palaeolithic and six Upper Palaeolithic habitation horizons have been recognized at Kara-Bom (Derevianko *et al.* 1998e; Derevianko 2001). The Middle Palaeolithic horizons yielded parallel cores and Levallois point and flake cores. All the cores are flat-face. A significant proportion of the tools were made on blades. Middle Palaeolithic horizon 2, underlying the lithological layer with an ESR date of 62.2 ka BP, has yielded a set of Levallois Mousterian tools constituting 32 per cent of the total tool-kit, while the proportion of Upper Palaeolithic tools is 16 per cent. The proportion of Upper Palaeolithic tools from the overlying Middle Palaeolithic habitation horizon rises to 21 per cent. It attests to a gradual increase in the proportion of Upper Palaeolithic tools compared with that of Middle Palaeolithic ones.

The combination of a mature Levallois primary reduction technique with the presence of narrow-face cores, from which narrow and long blades were removed and modified into burins, end-scrapers and other implements, can be regarded as an important indicator of an Upper Palaeolithic industry. Pressure-flaking technique may also have been used (Derevianko & Volkov 2004).

Habitation horizons 5 and 6, attributable to the early Upper Palaeolithic, have yielded mainly parallel cores dominated by narrow-face forms, from which long and narrow blades were removed. Tools are mostly made on blades. For instance, the share of tools on flakes is 19.5 per cent of the total number of tools from horizon 6, while the share of tools on blades in 70.6 per cent and that of tools on pointed flakes is 6.9 per cent. Tools were mostly fashioned on thick blades. The tools on blades comprise end-scrapers, multi-facetted burins, points, blade-knives made on blades up to 25 cm long and combination tools. The share of various retouched notch-denticulate tools is also considerable. The major tool types are Upper Palaeolithic forms. Comparative analyses have shown continuity and similarities within major stone tool types from the Mousterian and Upper Palaeolithic horizons. Apparently, the Upper Palaeolithic industry was formed on the basis of the preceding Middle Palaeolithic. Available radiocarbon dates are 43,200±1500 cal. BP (Gx-17597) for horizon 6 and 43,300±1600 cal. BP (Gx-17596) for horizon 5. These two Kara-Bom horizons have yielded a well-developed Upper Palaeolithic industry, an attribution which is supported by evidence for symbolic behaviour (Derevianko & Rybin 2003).

The Kara-Bom industry differs considerably from the Upper Palaeolithic industry of Ust-Karakol-1 in terms of major features of both primary and secondary reduction. As soon as the early Upper Palaeolithic, industries vary across the sites in the Gorny Altai although all of them are based on the earlier Acheulean-Levallois tradition.

Thus, the Ust-Karakol and Kara-Bom trends of the Upper Palaeolithic industry in the Altai Mountains began to form around 60–50 kya, maturing by 50–40 kya. This inference is supported by occurrences of bone implements within some collections as well as evidence of new subsistence strategies and symbolic behaviour.

Upper Palaeolithic industry formation processes are well illustrated by the archaeological materials recovered from other Altai sites, e.g. Anui-3 (Derevianko & Shunkov 2002), Anui-1 (Derevianko & Zenin 1990), Strashanya Cave site (Derevianko & Zenin 1997) and Ushlep-6 (Kungurov *et al.* 2003) among others. Materials from Kara-Tenesh are attributable to the Kara-Bom variant of the Upper Palaeolithic (Derevianko *et al.* 1998d). Lithological stratum 3 at Kara-Tenesh has been radiocarbon dated to between 42,165±4170 cal. BP (SOAN-2485) and 26,875±625 cal. BP (SOAN-2134) and has yielded an industry mostly based on large blades. Assemblages from the Ust-Kan and Maloyalomanskaya caves illustrate early stages in the development of the Upper Palaeolithic.

In sum, studies of stratified cave and open-air sites in the Altai provide evidence of the formation of two Upper Palaeolithic traditions: the Kara-Bom blade-based tradition and the Karakol tradition based on blades and micro-blades, both of them on a Middle Palaeolithic substrate.

The Middle to Upper Palaeolithic transition in Mongolia and the Lake Baikal region

Archaeological materials from regions in eastern Siberia and Mongolia contiguous to the Altai show both similar and different features in the development of local final Middle Palaeolithic industries into Upper Palaeolithic forms.

In Mongolia, a few well-stratified sites have been studied. These provide evidence for the gradual development of Early, Middle and Transitional Upper Palaeolithic industries. Archaeological material from the lower horizons 13–10 at the Tsagaan Agui Cave demonstrates the use of orthogonal and radial primary reduction techniques (Derevianko *et al.* 1998c;

2000c). Stratum 12 has yielded a date of 520±130 ka BP (RTL-805), and stratum 11 has been dated to 450±117 ka BP (RTL-806). Levallois cores have been recovered in stratum 9. It is likely that the industry from the lower horizons is linked with humans of the first global migration wave, while the subsequent industry associated with horizons 9 and above is related to the second wave. Culture-bearing horizons 4 and 3 have yielded archaeological materials illustrating the process of formation of the Upper Palaeolithic industry. Several dates have been generated for horizon 4: 66±9 ka BP (RU), 57±7 ka BP (LU), 49±6 ka BP (EU) while the dates for horizon 3 are 33,840±640 cal. BP (AA-23158), 33,777±585 cal. BP (AA-26587), 33,497±600 cal. BP (AA-26588), 32,960±670 cal. BP (AA-23159) and 30,942±478 cal. BP (AA-26589). Regrettably, the assemblage recovered from these culture-bearing horizons is too small for us to recognize particular features in the development of the lithic industry within this chronological period. The collection from horizon 3 represents a mature Upper Palaeolithic culture of the Kara-Bom variant (Derevianko 2001).

Important data for the Middle to Upper Palaeolithic transition in Mongolia have been obtained during excavations at the multi-layered and stratified sites of Orkhon-1 and -7, located in the Orkhon river basin 3 km from Khara-Khorin (Derevianko *et al.* 1992b; Derevianko & Petrin 1993). Several culture-bearing horizons at these sites have been dated to 70–38 ka BP. In the Orkhon floodplain alluvium, Middle Palaeolithic horizons with hearths have been recognized. Relevant artefact collections are dominated by Levallois cores, from which large blades and lamellar flakes were detached, as well as by implements fashioned on large blades. Overlying horizons (40–38 kya) yielded mainly proto-prismatic and prismatic cores as well as implements on blades. The Orkhon sites have yielded a considerable number of notch-denticulate tools as at Kara-Bom. The assemblages from Orkhon and Kara-Bom share many techno-typological features. It seems that the Upper Palaeolithic industry in the Altai and Mongolia formed from a single base at approximately the same time.

The technocomplexes from Orok-Noor I and II in the Gobi Altai also demonstrate many features similar to the Middle Palaeolithic industry of Orkhon (Derevianko & Petrin 1990), both complexes being manufactured on raw material of similar quality; pebbles were used predominantly as blanks in tool production. Cores, especially Levallois cores, demonstrate closely similar techno-morphological features, while tool kits are also similar. The techno-typological characteristics of the transition from Middle to Upper Palaeolithic industries in the Mongolian and Gobi Altai are similar to those of the Kara-Bom trend. Other early Upper Palaeolithic sites in Mongolia (late complexes of the Flint Valley, Chikhen-1, Tuin-Gol etc.) also yield cores for detaching large blades, various implements on blades and lamellar flakes.

One of the most interesting sites of the western Baikal region is that of Makarovo-4, located in the upper reaches of the Lena River. The industry is characterized by parallel and sub-parallel reduction. The tool kit is rich and includes side-scrapers of various types, end-scrapers, transverse burins, knives and borers. Especially interesting is a set of leaf-shaped points made on long spalls with bifacially flaked bases (Aksenov & Shunkov 1978). Similar points have been recognized in industries of the Kara-Bom variant. Two infinite dates are available for culture-bearing horizons at Makarovo-4: earlier than 38 ka BP (AA-8879) and earlier than 39 ka BP (AA-8880) (Goebel & Aksenov 1995). Given the features of sedimentation at the site and aeolian corrosion of the artefacts, some authors have argued in favour of earlier dates for the Makarovo sites (Medvedev 1983; Aksenov *et al.* 1987).

In the eastern Baikal region, sites such as Tolbaga (stratum 4), Varavrina Gora (stratum 3), Kamenka (complex A), Khotyk (layer 3) and Podzvonkaya have yielded archaeological material chronologically and technically close to the artefacts of the Kara-Bom industry (Konstantinov 1994; Lbova 2000; Tashak 2003). Deposition of culture-bearing horizons at these sites occurred during the Karginian period, judging from their stratigraphic position within the profiles and available radiocarbon dates. The earliest dates have been generated for Podzvonkaya (43,900±960 cal. BP (SOAN-44445): Tashak 2002) and Kamenka (complex A) (40,500±3800 cal. BP (AA-26743): Lbova 2002).

The technology of the eastern Baikal industries was based on parallel flaking of flat face and proto-prismatic cores, and some narrow face flaking. Various types of blades were used as tool blanks. The majority of tools were fashioned on long, large- and medium-sized blanks. The tool kit comprises sets of points and burins, end-scrapers and borers, chisel-like tools and retouched blades. The characteristic feature of these complexes is the presence of non-utilitarian implements, such as pendants, small and cylindrical beads made of bird long bones, ostrich eggshells and soft gemstones (Lbova *et al.* 2002; Tashak 2002) as well as a unique figurine with a bear head made from the tooth of the axis vertebra of a woolly rhinoceros (Konstantinov *et al.* 1983).

Palaeolithic sites located in Mongolia and in the Lake Baikal region dating from 40–30 ka BP share many characteristics (tool kits, exploitation and adaptive strategies, symbolic behaviour etc.), with sites attributable to the Kara-Bom Upper Palaeolithic trend in the Altai (Derevianko 2001; Derevianko & Shunkov 2004). Two explanations may be offered. Firstly, in the late Middle Pleistocene, human populations of the second migration wave spread out over the vast territories of the Altai, Mongolia and Baikal region. In the final stage of the Middle Palaeolithic, industries of the early Upper Palaeolithic developed independently in each of these areas; features of primary reduction technique and stone tool type shared common roots. Secondly, around 60–50 kya, populations from the Altai migrated into the territories of Mongolia and the Baikal region and gave birth to the formation of Upper Palaeolithic technocomplexes of the Kara-Bom and Karakol affinity.

Based on our current knowledge, the second hypothesis seems to be more reliable as no stratified Middle Palaeolithic site east of the Altai provides evidence of the initial stage of formation of blade- and micro-blade-based industries.

Conclusions

In southern Siberia and Mongolia, the Upper to Middle Palaeolithic transition began before 50 kya. Around 50–40 kya, two types of Upper Palaeolithic industry, the Kara-Bom and Karakol, were formed. The Altai facies of the Middle to Upper Palaeolithic transition has much in common with the Eurasian developmental trend characterized by the standardization of blade production and the manufacturing of blade tools. Within the vast Eurasian zone, the Altai and the Levant show closest similarity in techniques of stone knapping: the Karakol type and the Ksar Akil technocomplex and the Kara-Bom and Boker Tachtit share major features (Derevianko 2001). A possible explanation of this similarity is that Altai Upper Palaeolithic industries developed from a common Middle Palaeolithic substrate originating in the Near East.

Quite a different developmental trajectory of human culture can be traced in the Sino-Malaysian zone. Eastern and southeastern Asia have not produced evidence of a Middle Palaeolithic industry of the same types as those seen in other parts of Eurasia. Levallois technology was not present and local Middle and early Upper Pleistocene assemblages show considerable proportions of tools on flakes and special blanks. In China and Korea, parallel reduction strategy and tools on knife-like blades emerged only around 30 kya. This industry seems to have been imported into China from southern Siberia and Mongolia either with migrating populations or through the borrowing of new technologies (Derevianko 2005b). New technologies did not replace older ones. Technology developed through acculturation. Collections dating from 25–18 kya have yielded flake tools together with considerable proportions of blade tools and later even micro-blade-tools.

Acknowledgements

The author expresses his gratitude to Elena Pankeyeva for translating this paper from the original Russian.

References

Aksenov, M.P. & M.V. Shunkov, 1978 Novoe v paleolite Verhnei Leny, in *Drevnaia istoria narodov yuga Vostochnoi Sibiri.* Irkutsk: Izd. Irkutsk Gos. Univ, iss. 4, 22–4.

Aksenov, M.P., M.A. Berdnikov, G.I. Medvedev, S.N. Perzhakov & A.B. Fedorenko, 1987. Morfologia i arheologicheskii vozrast kamennogo inventaria 'makarovskogo paleoliticheskogo plasta', in *Problemy antropologii i arheologii kamennogo veka Evrazii.* Irkutsk: Izd. Irkutsk Gos. Univ., 26–30.

Bar-Yosef, O., 1995. The Lower and Middle Paleolithic in the Mediterranean Levant: chronology and cultural entities, in *Man and Environment in the Paleolithic,* ed. H. Ullrich. (ERAUL 62.) Liège: Université de Liège, 247–63.

Derevianko, A.P., 1990. *Paleolithic of Northern Asia and the Problem of Ancient Migrations.* Novosibirsk: Academy of Sciences of the USSR. Siberian Division. Inst. of History, Philology and Philosophy.

Derevianko, A.P., 2001. The Middle to Upper Paleolithic transition in the Altai. *Archaeology, Ethnology and Anthropology of Eurasia* 3(7), 70–103.

Derevianko, A.P., 2004. Problemy antropogeneza i zaselenia chelovekom vostochnoi chasti Evrazii, in *Sovremennye problemy nauki: Materialy nauchnoi sessii 25–26 noiab. 2003 g.: Proishozhdenie i evoliutsia zhizni na Zemle.* Novosibirsk: Izd. SO RAN, 52–72.

Derevianko, A.P., 2005a. The earliest human migrations in Eurasia and the origin of the Upper Paleolithic. *Archaeology, Ethnology and Anthropology of Eurasia* 2(22), 22–36.

Derevianko, A.P., 2005b. Formation of blade industries in eastern Asia. *Archaeology, Ethnology and Anthropology of Eurasia* 4(24), 2–29.

Derevianko, A.P., 2005c. The Middle to Upper Paleolithic transition: a view from northern Asia, in *The Middle to Upper Paleolithic Transition in Eurasia: Hypotheses and Facts,* ed. A.P. Derevianko. Novosibirsk: Izd. IAE SO RAN.

Derevianko, A.P. (ed.), 2005d. *The Middle to Upper Paleolithic Transition: Hypotheses and Facts.* Novosibirsk: Izd. IAE SO RAN.

Derevanko, A.P., 2006a. The Lower Paleolithic small tool industry in Eurasia: migration or convergent evolution? *Archaeology, Ethnology and Anthropology of Eurasia* 1(25), 2–32.

Derevianko, A.P., 2006b. *Paleolit Kitaia: Itogi i nekotorye problemy v izuchenii.* Novosibirsk: Izd. IAE SO RAN.

Derevianko, A.P., 2006c. *Perehod ot srednego k verhnemu paleolitu v Vostochnoi Azii (Kitai, Koreiskii poluostrov).* Novosibirsk: Izd. IAE SO RAN.

Derevianko, A.P. & S.V. Markin, 1992. Predvaritelnye opredelenia musterskih industrii Altaia, in *Valihanovskie chtenia: Tez. respubl. konf. 25–28 marta 1992 g.* AN Kazahstana. Institute istorii i entologii im. CH.CH. Valihanova. Kokshetau, 18–20.

Derevianko, A.P. & S.V. Markin, 1998. Paleolit severo-zapada Altae-Saian. *Rossiiskaya Arheologia* 4, 17–34.

Derevianko, A.P. & S.V. Markin, 2000. Prostranstvenno-vremennye izmenenia paleoliticheskih industrii severo-zapada Altae-Saianskoi gornoi strany, in *Paleogeografia kamennogo veka.* Krasnoyarsk: Izd. Krasnoyarsk Gos. Ped. Univ., 32–4.

Derevianko, A.P. & V.T. Petrin, 1990. *Svoeobraznaia kamennaia industria s severnogo poberezhia Doliny Ozer, in Arheologicheskie, etnograficheskie i antropologicheskie issledovania v Mongolii.* Novosibirsk: Nauka, 3–39.

Derevianko, A.P. & V.T. Petrin, 1993. The Levallois of Mongolia, in *The Definition and Interpretation of Levallois Technology*, eds. H. Dibble & O. Bar-Yosef. Madison (WI): Prehistory Press, 1–30.

Derevianko, A.P. & E.P. Rybin, 2003. The earliest representations of symbolic behavior by Paleolithic humans in the Altai Mountains. *Archaeology, Ethnology and Anthropology of Eurasia* 3(15), 27–50.

Derevianko, A.P. & M.V. Shunkov, 1992. Archaeological investigations in the Anui River basin. *Altaica* 1, 8–12.

Derevianko, A.P. & M.V. Shunkov, 2002. Middle Paleolithic industries with foliate bifaces in Gorny Altai. *Archaeology, Ethnology and Anthropology of Eurasia* 1(9), 16–42.

Derevianko, A.P. & M.V. Shunkov, 2004. Formation of the Upper Paleolithic traditions in the Altai. *Archaeology, Ethnology and Anthropology of Eurasia* 3(19), 12–40.

Derevianko, A.P. & P.V. Volkov, 2004 Evolution of lithic reduction technology in the course of the Middle to Upper Paleolithic transition in the Altai Mountains. *Archaeology, Ethnology and Anthropology of Eurasia* 2(18), 21–35.

Derevianko, A.P. & A.N. Zenin, 1990. Paleoliticheskoe mestonahozhdenie Anui-1, in *Kompleksnoe issledovanie paleoliticheskih obektov basseina r. Anui.* Novosibirsk: Izd. IAE SO RAN, 31–42.

Derevianko, A.P. & A.N. Zenin, 1997. The Mousterian to Upper Paleolithic transition though the example of the Altai cave and open air site, in *Suyanggae and Her Neighbours: the 2nd International Symposium.* Chungbuk: Chungbuk National University, 241–54.

Derevianko, A.P., V.I. Molodin & S.V. Markin, 1987. *Arheologicheskie issledovania na Altae v 1986 g. (Predvarit. itogi sov.-japon. ekspeditsii).* Novosibirsk. (Preprint AN SSSR. Sib.otd-nie. In-t istorii, filologii i filosofii).

Derevianko, A.P., D. Dorj, R.S. Vasilevskii *et al.*, 1990. *Kamennyi vek Mongolii: Paleolit i neolit Mongolskogo Altaia.* Novosibirsk: Nauka.

Derevianko, A.P., V.I. Molodin, Y.V. Grichan *et al.*, 1991. Arheologia i paleoekologia paleolita Gornogo Altaia, 1990, in *Hronostratigrafia paleolita Severnoi, Tsentralnoi, Vostochnoi Azii i Ameriki (paleoekolog. aspekt): Putevoditel Mezhdunar.* symp. k XIII congress INQUA, China, AN SSSR. Sib. Otd. Institute istorii, filologii i filosofii; Komis. po izuch. chetvertich. perioda. Novosibirsk.

Derevianko, A.P., S.A. Laukhin, E.M. Malaeva, O.A. Kulikov & M.V. Shunkov, 1992a. Nizhnii pleistotsen na severo-zapade Gornogo Altaia, Doklady AN SSSR. *Seria geologia* 323, 509–13.

Derevianko, A.P., S.V. Nikolayev & V.T. Petrin, 1992b. *Geologia, stratigrafia, paleogeografia paleolita Yuzhnogo Hangaia.* Preprint. Novosibirsk: IAE SO RAN.

Derevianko, A.P., S.M. Popova, E.M. Malaeva, S.A. Laukhin & M.V. Shunkov, 1992c. Paleoklimat severo-zapada Gornogo Altaija v eopleistotsene, in *Doklady AN SSSR. Seria geologia* 324, 842–6.

Derevianko, A.P., A.K. Agadjanian, G.F. Baryshnikov, *et al.* 1998a. *Arheologia, geologia i paleogeografia pleistotsena i golotsena Gornogo Altaia.* Novosibirsk: Izd. IAE SO RAN.

Derevianko, A.P., Z.N. Gnibidenko & M.V. Shunkov, 1998b. Srednepleistocenovye ekskursy geomagnitnogo polia v otlozheniah Denisovoi peschery (Gornyi Altai). *Doklady AN* 360, 511–13.

Derevianko, A.P., J. Olsen, D. Tseveendorj *et al.*, 1998c. *Archaeological Studies carried out by the Joint Russian-Mongolian-American Expedition in Mongolia in 1996.* Novosibirsk: Izd. IAE SO RAN.

Derevianko, A.P., V.T. Petrin, S.V. Nikolaev *et al.*, 1998d. Stoianka Kara-Tenesh-pamiatnik nachalnoi pory pozdnego paleolita, in *Problemy paleoekologii, geologii i arheologii paleolita Altaia.* Novosibirsk: Izd. IAE SO RAN, 205–38.

Derevianko, A.P., V.T. Petrin & E.P. Rybin, 1998e. *Paleoliticheskie kompleksy stratifitsyrovannoi chasti stoianki Kara-Bom.* Novosibirsk: Izd. IAE SO RAN.

Derevianko, A.P., V.T. Petrin, J.K. Taimagambetov, A.P. Zenin & S.A. Gladyshev, 1999. Paleoliticheskie kompleksy poverhnostnogo zalegania Mugodjarskih gor, in *Problemy arheologii, etnografii, antropologii Sibiri i sopredelnyh territorii: Mat-ly VII Godovoi itogovoi sessii IAEt SO RAN. Dekabr 1999 g.* Novosibirsk: Izd. IAE SO RAN, vol. V, 50–55.

Derevianko, A.P., V.T. Petrin, D. Tseveendorj *et al.*, 2000a. *Kamennyi vek Mongolii: Paleolit i neolit severnogo poberezhia Doliny Ozer.* Novosibirsk: Izd. IAE SO RAN.

Derevianko, A.P., J. Olsen, D. Tseveendorj *et al.*, 2000b. *Archaeological Studies carried out by the Joint Russian-Mongolian-American Expedition in Mongolia in 1997–1998.* Novosibirsk: Izd. IAE SO RAN.

Derevianko, A.P., J., Olsen, D. Tseveendorj *et al.*, 2000c. The stratified cave site of Tsagaan-Agui in the Gobi Altai (Mongolia), *Archaeology, Ethnology and Anthropology of Eurasia* 1, 23–36.

Derevianko, A.P., M.V. Shunkov & V.A. Ulianov, 2000d. Izuchenie paleoliticheskoi stoianki v doline r. Anui, in *Problemy arheologii, etnografii i antropologii Sibiri i sopredelnyh territorii: Materialy Godovoi sessii IAEt SO RAN. Dekabr 2000 g.* Novosibirsk: Izd. IAE SO RAN, vol. 6, 99–104.

Derevianko, A.P., V.T. Petrin, S.A. Gladyshev, A.N. Zenin & J.K. Taimagambetov, 2001a. Acheulian complexes from the Mugodjari Mountains (northwestern Asia). *Archaeology, Ethnology and Anthropology of Eurasia* 2(6), 20–36.

Derevianko, A.P., V.T. Petrin, S.A. Gladyshev, A.N. Zenin & J.K. Taimagambetov, 2001b. *Ashelskie kompleksy Mugodjarskih gor (Severo-Zapadnaia Aziia).* Novosibirsk: Izd. IAE SO RAN.

Derevianko, A.P., M.V. Shunkov, A.K. Agadjanian *et al.*, 2003. *Prirodnaya sreda i chelovek v paleolite Gornogo Altaia.* Novosibirsk: Izd. IAE SO RAN

Derevianko, A.P., M.V. Shunkov, N.S. Bolihovskaia *et al.*, 2005. *Stoianka rannego paleolita Karama na Altae.* Novosibirsk: Izd. IAE SO RAN.

Goebel, T. & M. Aksenov, 1995. Accelerator radiocarbon dating of the initial Upper Paleolithic in southeast Siberia. *Antiquity* 69, 349–57.

Jelinek, A.J., 1992. Problems in the chronology of the Middle Paleolithic and the first appearance of early modern *Homo sapiens* in southwest Asia, in *The Evolution and Dispersal of Modern Humans in Asia*, eds. T. Akazawa, K. Aohi & T. Kimura. Tokyo: Hokusen-sha, 253–75.

Konstantinov, M.V., 1994. *Kamennyi vek vostochnogo regiona Baikalskoi Azii: K Vsemir. arheol. inter-kongr.* Ulan-Ude: Chita, Izd. BION SO RAN; CHGPI.

Konstantinov, M.V., V.B. Sumarokov, A.K. Filippov & N.M. Ermolova, 1983. Drevneishaia skulptura Sibiri. *KSIA* 173, 78–81.

Kungurov, A.L., M.M. Markin & V.P. Semibratov, 2003. Vosmoi kulturnyi sloi mnogosloinoi paleoliticheskoi stoianki Ushlep-6, in *Problemy arheologii, etnografii, antropologii Sibiri i sopredelnyh territorii.* Novosibirsk: Izd. IAE SO RAN, vol. 9, part 1, 159–62.

Lbova, L.V., 2000. *Paleolit severnoi zony Zapadnogo Zabaikalia.* Ulan-Ude: Izd. BNC SO RAN.

Lbova, L.V., 2002. The transition from the Middle to Upper Paleolithic in western Trans-Baikal. *Archaeology, Ethnology and Anthropology of Eurasia* 1(9), 59–75.

Lbova, L.V., P.V. Volkov & B.A. Bazarov, 2002. Semanticheskii aspekt nahodok 3-go urovnia pamiatnika Hotyk (Zapadnoe Zabaikale), in *Istoria i kultura vostoka Azii: Materialy Mezhdunar. nauch. konf.* Novosibirsk: Izd. IAE SO RAN, vol. 2, 101–3.

Liubin, V.P., 1984. Paleolit Turkmenii. *Sovetskaya Arheologia* 1, 26–45.

Liubin, V.P. & L.B. Vishniatsky, 1990. Otkrytie paleolita v Vostochnoi Turkmenii. *Sovetskaya Arheologia* 4, 5–15.

Medoev, A.G., 1964. Kamennyi vek Sary-Arka v svete noveishih issledovanii. *Izv. AN KazSSR. Ser. obscestv. Nauk* 6, 90–98.

Medoev, A.G., 1970. Arealy paleoliticheskih kultur Sary-Arka, in *Po sledam drevnih kultur Kazahstana.* Alma-Ata: Nauka KazSSR, 200–216.

Medvedev, G.I., 1983. Paleolit Yuzhnogo Priangaria. Unpublished doctoral dissertation, Novosibirsk.

Schwarcz, H.P. & W.J. Rink, 1998. Progress in ESR and U-series chronology of the Levantine Paleolithic, in *Neanderthals and Modern Humans in Western Asia*, eds. T. Akazawa, K. Aoki & O. Bar-Yosef. New York (NY): Plenum Press, 57–68.

Tashak, V.I., 2002. Obrabotka skorlupy iaits strausov v verhnem paleolite Zabaikalia, in *Istoria i kultura Vostoka Azii: Materialy Mezhdunar. nauch. konf. k 70-letiiu V.E. Laricheva (Novosibirsk, 9–11 dekabria 2002 g.).* Novosibirsk: Izd. IAE SO RAN, vol. 2, 159–64.

Tashak, V.I., 2003. Hearths at the Podzvonkaya Paleolithic site: evidence suggestive of the spirituality of early populations of the Trans-Baikal region. *Archaeology, Ethnology and Anthropology of Eurasia* 3(15), 70–78.

Vishniatsky, L.B., 1990. Otkrytie paleolita v Vostochnom Turkmenistane. *Sovetskaya Arheologia* N4, 5–15.

Vishniatsky, L.B., 1996. *Paleolit Srednei Azii i Kazahstana.* St Petersburg: Izd. "Evropeiskii Dom".

Voloshin, V.S., 1987. Voprosy hronologii i periodizatsii paleolita Tsentralnogo Kazahstana, in *Voprosy periodiztsii arheologicheskih pamiatnikov Tsentralnogo i Severnogo Kazahstana.* Karaganda: Izd. KarGU, 3–13.

Voloshin, V.S., 1988. Ashelskie bifasy iz mestonahozhdenia Vishnevka-3 (Tsentralnyi Kazahstan). *Sovetskaya Arheologia* 4, 19–203.

Voloshin, V.S., 1990. Stratigrafia i periodizatsia paleolita Tsentralnogo Kazahstana, in *Hronostratigrafia paleolita Severnoi, Centralnoi i Vostochnoi Azii i Ameriki:* Dokl. Mezhdunar. simp. Novosibirsk: SO AN SSSR. IIFiF, 99–106.

Chapter 10

The Transition to Upper Palaeolithic Industries in the Korean Peninsula

Kidong Bae

One of the most critical issues in Korean Palaeolithic archaeology at present is how Upper Palaeolithic industries first appeared in the Korean Peninsula. Dozens of Palaeolithic sites have been identified as Upper Palaeolithic since the first discovery at the Sokchangni site in the early 1960s. However, the process involved in the development of Upper Palaeolithic industries in Korea is not yet clear. It is generally understood that blade industries using specially selected raw material are found at some sites in the peninsula dated back to 32,000 BP. At the time, some specialized tool types appeared in the peninsula, a fact which is presumably considered to be an indication of the appearance of new technology and possible new adaptive strategies. While local development of the Upper Palaeolithic is being considered by some archaeologists, diffusion from Siberia has been a more common explanation. At present, there may not be enough evidence to make the developmental process clear. As few sites have been properly dated in the Korean Peninsula and in northeastern Asia, it is difficult to establish the chronology of the initial stage of development of the Upper Palaeolithic in this region. The lack of extensive comparative analysis of stone industries in this area could be one reason for the difficulty in understanding regional movement of Upper Palaeolithic cultures. Current palaeoanthropological data is not sufficient to enable us to understand hominin dispersal to this region in the Upper Pleistocene, especially for understanding the appearance of anatomically modern humans.

History and major sites

The Upper Palaeolithic industry from the Sokchangni site was the first Palaeolithic find in the Korean Peninsula (Sohn 1967). Blades and microblades were found in the uppermost layers above fluvial deposits of the Kum river. The Upper Palaeolithic industry from the Sokchangni site was dated to around 30,000 BP, for sometime believed to be the lower boundary of the Upper Palaeolithic age. Few Upper Palaeolithic sites were reported until an industry which includes tanged points was found at the Suyanggae site in the early 1980s, which is located on the upper reach of the Nam Han river (Lee 1983). Tanged points and microcores and blades were recovered from a lower layer at the site. It was the first find of tanged point in Korea and is considered a typical Upper Palaeolithic tool type because they were found together with microblades. Radiocarbon dates from the site are younger than 18,000 BP. Until now, the earliest came from the Yongho-dong site near Daejeon city on a river terrace of the Kum river, which was radiocarbon-dated 38,000 BP (Han 2002).

Blades, along with many blade cores, were recovered from the Gorye-ri site in southeastern part of the Korean Peninsula in early 1990s. The site is situated on a gentle slope in a small valley in a mountain range. Long thick blades of volcanic tuff with large prismatic cores were recovered. The industry is considered to represent an early stage of development in blade technology in the Upper Palaeolithic because no microblades were observed. The geographical location of the Gorye-ri site provides some information on how Upper Palaeolithic people used the land, i.e. resource exploitation. In spite of a claim of AT tephra (tephra from Aira-Tanzawa volcano in Japan) at the site, no absolute date is available for the industry.

The Sinbuk site in Jeonnam Province yielded ground-stone axes in association with microblades possibly dated back to 20,000 BP, the oldest ground-stone artefacts in the Korean Peninsula (Lee 2004). Partially ground stone artefacts were found at the Jangheung site in Gyeongsang Province, although somewhat later than those from the Sinbuk site. Ground axes from the two sites in the southern part of the Korean Peninsula are considered possible

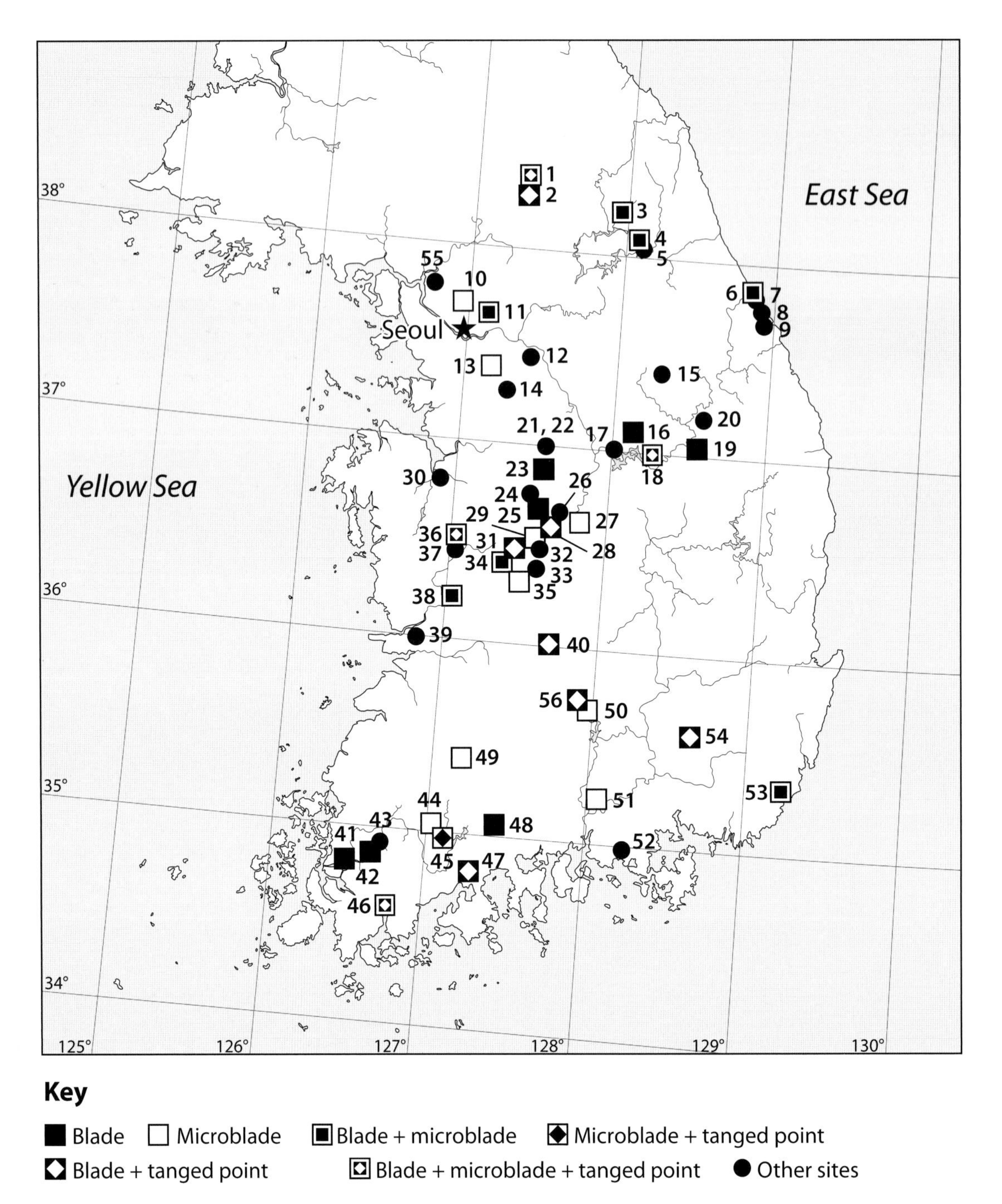

Figure 10.1. *Upper Palaeolithic sites in the Korean Peninsula: 1) Janghung-ri; 2) Hwadae-ri; 3) Sangmuryong-ri; 4) Hahwagye-ri; 5) Yeonbong-2; 6) Gigok; 7) Nobong; 8) Balhan-dong; 9) Gumi-dong; 10) Millak-dong; 11) Hopyeong-dong; 12) Byeongsan-ri; 13) Pyeongchang-ri; 14) Sam-ri; 15) Sangsi Rockshelter; 16) Sagi-ri, Changnae; 17) Chommal; 18) Suyanggae; 19) Geum Cave; 20) Gunang Cave; 21) Songdu-ri; 22) Janggwan-ri; 23) Soro-ri; 24) Yullyang-dong; 25) Nosan-ri; 26) Durubong Cave; 27) Saemgol; 28) Yongho-dong; 29) Sindae-dong; 30) Yonggok-dong; 31) Yongsan-dong; 32) Dunsan-dong; 33) Bongmyoung-dong; 34) Noeun-dong; 35) Daejeong-dong; 36) Sokchangni; 37) Maam-ri; 38) Sinmak; 39) Naeheung-dong; 40) Jin Rockshelter; 41) Danghasan; 42) Chongok-ri; 43) Dangga; 44) Daejeon, Hwasun; 45) Juksan-ri; 46) Sinbuk; 47) Wolpyeong, Suncheon; 48) Jungnae-ri; 49) Okgwa; 50) Imbul-ri; 51) Jangheung-ri, Wolpyeong; 52) Igeum-dong; 53) Haeundae; 54) Gorye-ri; 55) Kumpa-ri; 56) Jeongjang-ri.*

evidence of adaptation to woody environments. However, it may not be easy to reconstruct the environment of the glacial maximum in the southern part of the Korean Peninsula, because some argue that a dry continental climate similar to modern-day Siberia prevailed during the period (Kim *et al.* 2002).

At present, Upper Palaeolithic sites have been found across most of the country, often on riverine terraces and colluvial deposits on hill slopes. Some new types of tools appeared in the early stage of the Upper Palaeolithic; however the transition from the Middle Palaeolithic is still not clearly defined. Extensive research in the northern part of the Korean Peninsula in future would be very helpful if we are to understand how the Upper Palaeolithic moved in the peninsula.

Morphology and technical traits

In general, as in other parts of the world, appearance of blade technology is the most distinctive feature in the Upper Palaeolithic stone industries of the Korean Peninsula. However, some stone industries without any sign of blade technique are often called Upper Palaeolithic on the basis of the presence of new tool types of extensive secondary modification. Some industries have been classified as Upper Palaeolithic just because of the stratigraphic position of cultural material bearing layers. The irregular and less-refined tool-making tradition of the early and middle Palaeolithic persisted until the late part of the Upper Palaeolithic in the peninsula. Archaic tool types of coarse raw material like quartz vein, often described as 'pebble-tool tradition' occurring in Upper Palaeolithic industries are thought to be evidence for technological continuity to the Upper Palaeolithic age from the previous stage (Lee 2005; Seong 2006).

In the Upper Palaeolithic, more fine-grained, typically siliceous rocks were preferred for tool making. Shale, chert, obsidian, volcanic tuffs etc. were used for making blades. During the Upper Palaeolithic, raw materials for tool making were often exploited even in remote areas. For example some obsidian was evidently transported several hundreds of kilometres from its origin. In particular, obsidian from Mt Baekdu (Mt Jangbai in Chinese) is found at some sites in central part of the peninsula (e.g. Sohn 1989). More refined secondary retouch is often observed on some tool types such as tanged points, end-scrapers on blades, thumb-nail scrapers and various types of burin. Small becs, points, denticulates etc. became more common in the Upper Palaeolithic industries. Wedge-shaped prismatic cores are common, while several blades were detached from a circular single striking platform. In spite of some claims for works of art, few have been reported from Upper Palaeolithic sites in Korea and survival of organic objects at open-air Palaeolithic sites should not be expected.

Chronology

Several attempts have been made to construct a chronology for the Palaeolithic of the Korean Peninsula (Chang 2005; Lee 2002; Yi 2001). It is common for Korean Palaeolithic archaeologists to use the three sub-age system, Lower, Middle and Upper Palaeolithic for chronological description. However, it is not based on the technological definition of each stage. In particular, 'Middle Palaeolithic' often refers simply to some stone industries earlier than Upper Palaeolithic industries but without any technological definition. It is hard to recognize at present any technological difference between Lower and Middle Palaeolithic industries in the Korean Peninsula. They are all crude and of 'expedient' style, mostly on quartz vein or quartzite rocks without any extensive secondary modification except for some typical tool types, for example handaxes, choppers and polyhedrals. No real Levallois technique has been observed in industries which are claimed to be 'Middle Palaeolithic'. For stone industries earlier than the Upper Palaeolithic in Korea, few dates are available on which to base a chronology. In spite of a few ^{14}C dates for 'Middle Palaeolithic' stone industries, the chronology of Lower and Middle Palaeolithic sites in the Korean Peninsula is still far from reliable.

In comparison, the Upper Palaeolithic chronology is considered to be well established. This is because relatively reliable ^{14}C dates have been accumulated, obtained relatively systematically for this period (Bae 2002). In addition to ^{14}C dating, some other absolute dating methods have been applied in order to construct a chronology. TL, OSL and tephra are used for dating Upper Palaeolithic industries. In addition to such absolute dating methods, so-called 'soil wedges' are often referred to in many articles (e.g. Han 2003) as an important clue for marking chronological boundaries in Palaeolithic cultural evolution, even when not always relevant. These 'soil wedges' are cracks which appear as long vertical lines on walls of excavation pits. They are often considered to be periglacial phenomena formed during the Pleistocene, in particular the Upper Pleistocene. The uppermost wedges of Palaeolithic sites are considered to have formed around 15,000 BP and the second from the surface around 65,000 BP by some geological scientists (Lee 1995; Kim *et al.* 2004). They believe that the wedges are the ice wedges formed under glacial maximum climatic conditions.

However, micro-morphological evidence from preliminary analysis indicates they were formed under wet climate rather than cold and dry conditions and under different processes to ice-wedges (Perrenoud 2003). A comparative morphological analysis shows they were formed under dry climatic conditions (Cho 2006), and their structure looks quite different from those ice wedges found in periglacial environments. It is simply premature to use them as indicators of a certain period in a Palaeolithic chronology without any proof of the climatic conditions under which they were formed. At any rate, actual observations of the formation processes causing such cracks still need to be made.

Tephra chronology is considered a very reliable method for dating Upper Palaeolithic sites in the Korean Peninsula. In particular, AT tephra provides valuable evidence for estimating ages of Upper Palaeolithic archaeological layers (Machida & Siroi 2003). The age of the AT tephra is often thought to be *c.* 24,000 BP, while some scientists estimate an age of 29,000 BP for the earliest example. At any rate, it is a very useful method for building up a chronology of Upper Palaeolithic industries because the tephra is found at many Palaeolithic sites.

In spite of some claims, it is not clear yet whether the earliest date for the Upper Palaeolithic industry goes back any earlier than 40,000 BP (Fig. 10.2). Some dates for early blade stone industries are older than 30,000 BP, but only one is around 40,000 BP. More reliable absolute dates are necessary if we are to consider these early dates to represent the time of introduction of blade stone technology to the Korean Peninsula. According to dates obtained from the recent excavation of the Hopyeong-dong site in Gyeonggi province, central Korea, blade technology clearly appeared around 30,000 BP (Hong & Kononenko 2005). Considering that ^{14}C dates back to nearly 40,000 BP from the Yongho-dong site (Han 2002), from which a tanged point made on blade was reported, the introduction of blade technology could be much earlier than the suggested age of 30,000 BP in Korea. An age of 35,000 BP has been suggested for the beginning of the Upper Palaeolithic (Jang 2005).

However, there are several different opinions when it comes to defining the lower boundary of the Upper Palaeolithic. A transient period has been suggested before the introduction of blade technology in the Korean Peninsula when, it is claimed, an increasing number of small tools and elongated flakes are observed compared to the previous stage. Elongated flakes and cores with several parallel flake scars are often present in this stage. Some archaeologists believe that these artefacts represent an initial Upper Palaeolithic development stage and that the technique involved in making elongated flakes eventually evolved into blade technology (Lee 2005; Seong 2006). It still remains to be resolved as to what a clear definition of Upper Palaeolithic culture would be.

Important technological developments occur in the Upper Palaeolithic around 20,000 BP. Blade cores and blades get smaller and eventually micro-cores appear around that time. The earliest age of micro-cores is the radiocarbon-dated occurrence from the Janghung-ri site in Gangwon province (Choi *et al.* 2001), in the upper reach of the Hantan river basin. At this stage, most tools became much more elaborate with extensive secondary retouch on fine-grained raw materials. It is apparent that most of the peninsula was occupied by humans at the peak of the Ice Age on the basis of the distribution of sites. A somewhat later age than that from the Janghung-ri has been suggested for the appearance of microlithic technology. It is a confusing situation for our understanding of the age of the introduction of microlithic technology, because much earlier dates have been reported at some localities. It is still uncertain how long the microlithic culture persisted, but probably at least until roughly sometime around the Pleistocene/Holocene transition or the early part of the Holocene.

Along with technological developments during the Upper Pleistocene, it should be noted that older types of stone industries were also present in most Upper Palaeolithic stone industries. Some of the quartz vein or quartzite stone artefacts appeared with more evolved stone industries, for example blade stone industries, while others occur quite independently of these. And therefore, one topic which needs serious consideration in the near future is whether both stone industries appeared at the same time or represent a mixture of stone artefacts of different cultural stages — although this is thought to be less likely. Although different cultural groups or traditions have been suggested for industries made of different raw materials, it is quite likely that Upper Palaeolithic people utilized various convenient locally available resources for their immediate purposes.

The tanged point is regarded as a type fossil of the Upper Palaeolithic in Korea, despite possibly being present in the late Middle Palaeolithic. This tool type is very common at Upper Palaeolithic sites and appeared throughout the Upper Palaeolithic in the Korean Peninsula. However, in Japan, where this tool type was introduced from the Korean Peninsula and is often called Hyakuhensentoki, it appeared at an earlier stage of the Upper Palaeolithic and disappeared when microlithic stone industries appeared. It would be interesting to know why tanged points

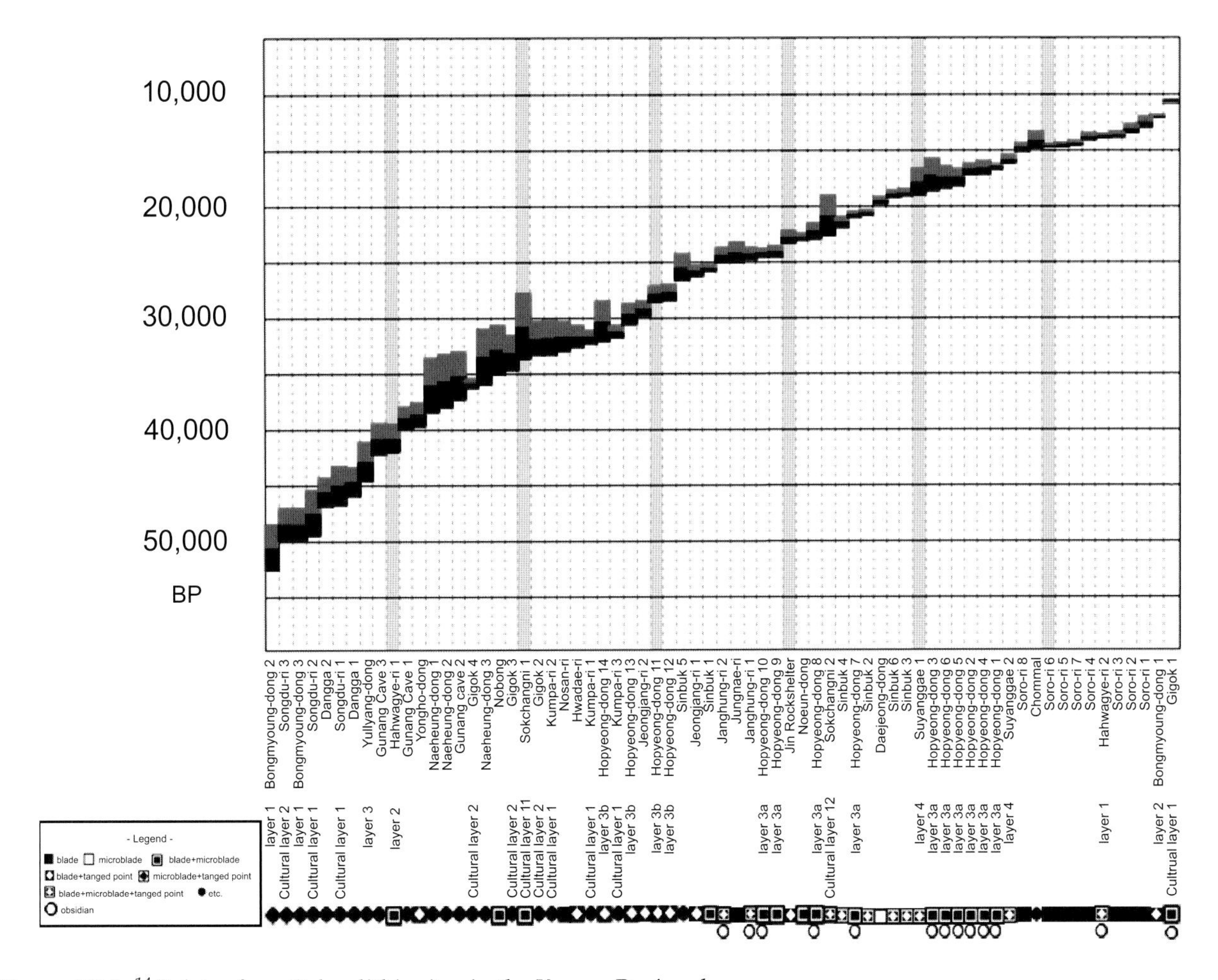

Figure 10.2. *^{14}C dates from Palaeolithic sites in the Korean Peninsula.*

continued to appear until the final stage of the Upper Palaeolithic Age in Korea.

Origin of Upper Palaeolithic stone industries

It is generally believed that Upper Palaeolithic stone industries were introduced by anatomically modern humans, *Homo sapiens sapiens*. In Korea, we have several fossil remains of *Homo sapiens sapiens*. In particular, two human fossils from cave sites in North Korea, Mandal man and Seungnisan man, were preserved in a relatively good condition, but these are now assigned to the final stage of the Upper Pleistocene. Micro-cores were found in association with the Mandal fossil, which would be quite late. Some earlier ones from Ryokpo Cave and Yonggok Cave are believed to be associated with *Homo sapiens*. No stone artefact was reported from Ryokpo while crude artefacts on quartz vein were recovered at Yonggok. The human fossil from Yonggok Cave was dated 43,000–48,000 BP by U-Th method, which seems reasonable considering its modern traits. However, the quartz-vein industry looks very primitive and no sign of laminar flaking was observed at the stone industries from the cave. Although North Korean anthropologists claim that these two fossils could represent the origin of anatomically modern humans in this area (Chang 2002), current human fossil evidence is not relevant when discussing the introduction of the Upper Palaeolithic in the peninsula.

It is not clear at present how blade technology moved into the Korean Peninsula. In northeast China, the blade industry from the Shuidonggu site in Inner Mongolia is thought to be one of the oldest sites, although the previous collection turned out to be a mixture of different stages of lithic development. No absolute date is yet available, but it is thought to be older than 30,000 BP, probably 35,000 BP (Jia & Huang 1985; Matsufuji 2003; Wang 2004). Blade cores include prismatic cores from which several large blades were taken from striking platforms.

It is still the prevailing view that lithic blade

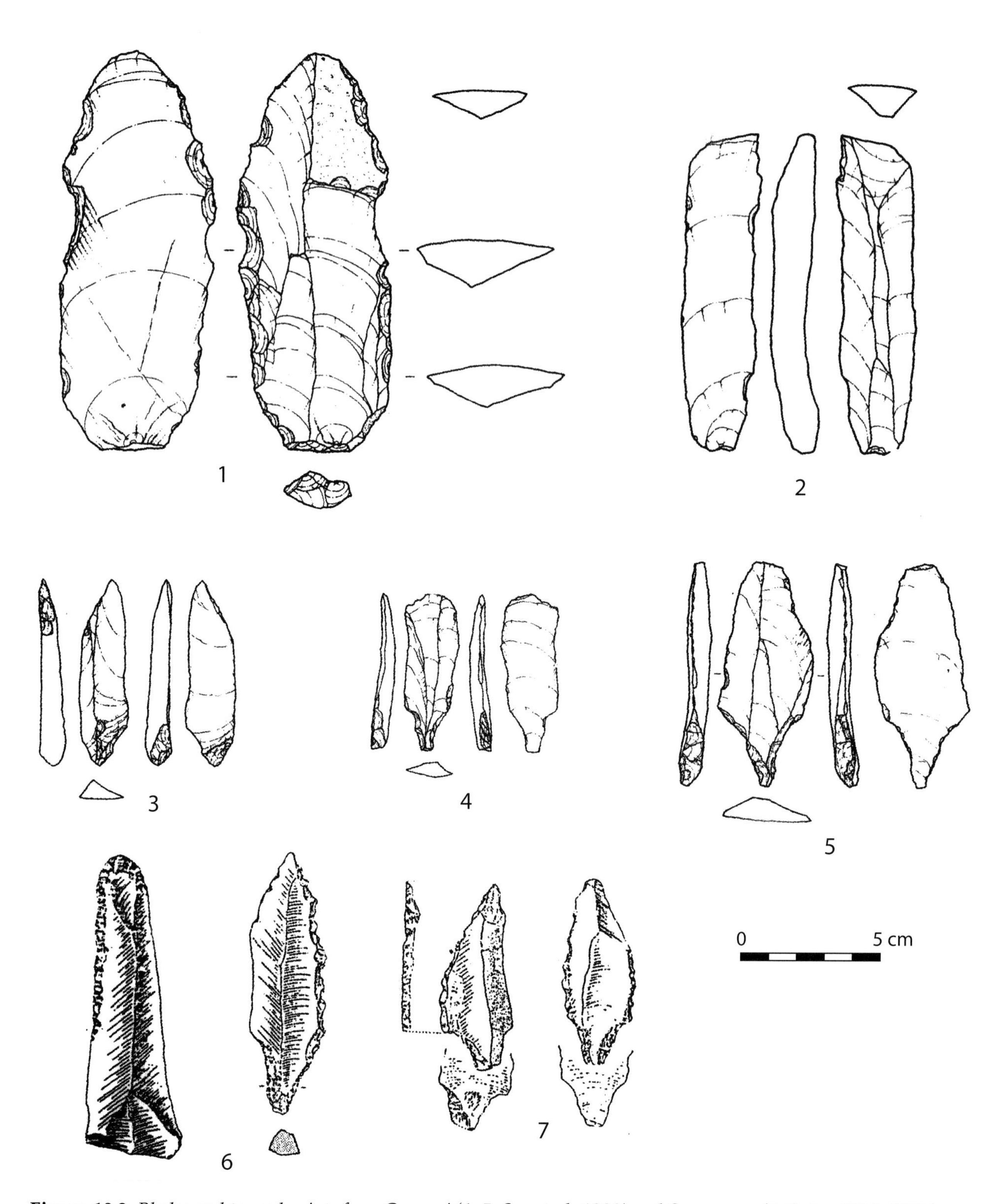

Figure 10.3. *Blades and tanged points from Gorye-ri (1–5: Seo* et al. *1999) and Suyanggae (6–7: Lee 1984; 1985).*

industries in Inner Mongolia could be the origin of the Upper Palaeolithic culture in the northeastern part of Asia. However, at present it is hard to understand how Upper Palaeolithic or blade technology moved down to China and the Korean Peninsula. Few chronometric age dates are available for the early industries of the Upper Palaeolithic in northern China. Considering that few blade industries have been reported in southern China, it is quite likely that blade technology may have originated from Siberia. Blade industries in Siberia have been dated back to earlier than 34,000 BP (Derevianko & Rybin 2005).

Local development of blade technology has been claimed by some Korean Palaeolithic archaeologists (Lee 2005; Seong 2006). The technology of making elongated flakes in the previous Palaeolithic stage might have developed into real blade technology. Elongated flakes and cores with parallel flaking from single striking platforms are observed in some stone industries presumably slightly earlier than the Upper Palaeolithic. It has been argued that elongated flake industries represent a transient stage towards blade technology during the Upper Palaeolithic. It is a more evolved débitage technique. However debates remain as to whether it could belong to the initial stage of Upper Palaeolithic because it appears in some layers which are older than the Upper Palaeolithic.

However, the technology producing elongated flakes is believed to be different to real blade technology in some significant ways. Much more sophisticated flaking techniques including controlled flake angle, more deliberate use of cores and relevant raw materials, are required for blade technology of the Upper Palaeolithic. Considering the earlier dates of some Upper Palaeolithic sites in the Korean Peninsula in comparison to northern China, local development of blade industry in the Korean Peninsula would have some support from local archaeologists. It is unclear whether elongated flake technology was an initial or transient stage in the development of blade technology. It is not clear how much the two different technological stages related each other. A period transient to the real blade stone-tool technology has been identified in Siberia, China and Japan. However, it should be noted that no Levallois flake has been observed in Korea and this is a significant difference to the so called 'transient' period between the Middle and Upper Palaeolithic in the Altai region (Derevianko & Shunkov 2004).

Elongated flake technology in Korea could be considered as an innovation in Palaeolithic technological development. However, the introduction of real blade technology was probably a much more serious innovation than this if we consider that other technological evolution appeared along with blade technology. We should be careful about interpreting currently available data and should leave some room for future research when we investigate the development of Upper Palaeolithic culture and contemplate reasons why some data do not fit a more general explanation of modern human dispersal.

Concluding remarks

As most Palaeolithic archaeologists in Asia believe, Upper Palaeolithic blade technology in the Korean Peninsula was quite likely introduced sometime between 40,000 BP and 30,000 BP — possibly from Inner Mongolia via northern China. Current archaeological evidence in northern China and Manchuria may not be sufficient to understand the process of introduction of blade technology to the peninsula. More extensive research needs to be carried out in this area.

In the meantime, it is an interesting proposition that blade technology evolved from previously existing elongate flake technology in the peninsula, in spite of some significant differences in technological elements between the two stages. The 'transient' phase of Palaeolithic evolution has been observed in the Altai and in northern China. As indicated above, there are some serious differences between the transient elements in Altai and elongated flake technology in Korea. Whatever technological elements of 'transient' are, if we consider the transient technology as an initial stage of blade technology, serious questions would be who developed it and how. Did *Homo sapiens* develop the initial stage of blade technology? Or *Homo sapiens sapiens*? If *Homo sapiens sapiens*, did they lose their fully developed blade technology when they came to Korea or other areas where these transient stages are observed? Another question to be answered could be why archaic types of tools persisted until quite late period in the Upper Palaeolithic in Korea even though Upper Palaeolithic humans had far more sophisticated technology for resource exploitation and making tools. We have to wait until more concrete archaeological evidence is available and at the same time need to carry out extensive comparative analysis of technological traits in industries belonging to the initial stages of and the later part of the Upper Palaeolithic that have been found in northern China including Manchuria and Mongolia as well as in the Korean Peninsula.

References

Bae, K.D., 2002. Radiocarbon dates from Paleolithic sites in Korea. *Radiocarbon* 44(2), 473–6.

Chang, W.J., 2002. *Origin of Korean People*. Pyoungyang: Baekgwasajeonchulpansa. [In Korean.]

Chang, Y.J., 2005. Development of blade technology of vein-quartz industries in the Korean Peninsula. *Paleolithic Archaeology* 12, 57–91.

Cho, C.J., 2006. *Comparative Study of Formation of Dry Cracks and Ice Wedges of Upper Pleistocene at Paleolithic Sites*. Osaka: Association of Osaka Cultural Heritage.

Choi, B.K., S.Y. Choi, H.Y. Lee & J.H. Cha, 2001. *Jangheungni [Janghung-ri] Paleolithic Site*. Chuncheon: Institute of Archaeology, Kangwon University.

Derevianko, A.P. & E.P. Rybin, 2005. The earliest representations of symbolic behavior by Paleolithic humans in the Altai Mountains, in *The Middle to Upper Paleolithic Transition in Eurasia: Hypothesis and Facts*, ed. A.P. Derevianko. Novosibirsk: Russian Academy of Sciences Siberian Branch, Institute of Archaeology and Ethnography, 232–55.

Derevianko, A.P. & M.V. Shunkov, 2004. Formation of the Upper Paleolithic transition in the Altai. *Archaeology, Ethnology and Anthropology of Eurasia* 3(19), 12–40.

Han, C.K., 2002. Yonghodong Paleolithic site, Taejeon, Korea, in *Paleolithic Archaeology in Northeast Asia*. Ansan: Institute of Cultural Properties, Hanyang University and Yeoncheon County, 163–72.

Han, C.K., 2003. Chronological problems of the Korean Paleolithic sites. *Journal of the Korean Paleolithic Society* 7, 1–39.

Hong, M.Y. & N. Kononenko, 2005. Obsidian tools and their use excavated from the Hopyeong-dong Upper Paleolithic site, Korea – Preliminary progressive report 1. *Journal of the Korean Paleolithic Society* 12, 1–20.

Jang, Y.J., 2005. A critical review for regional independent of quartz artefacts. *Journal of the Korean Paleolithic Society* 12, 57–90.

Jia, L. & W.W. Huang, 1985. The late Paleolithic of China, in *Paleoanthropology and Paleolithic Archaeology in the People's Republic of China*, eds. R. Wu & J. Olsen. New York (NY): Academic Press, 211–23.

Kim, J.Y., H.J. Lee & D.Y. Yang, 2002. Environment and chronology of Quaternary sediments of Paleolithic sites in Korea. *Journal of Korean Paleolithic Society* 6, 165–80.

Kim, J.Y., G.K. Lee, D.Y. Yang, S.S. Hong, U.H. Nam & J.Y. Lee, 2004. Distribution and formation of Quaternary deposits in South Korea. *Journal of Korean Paleolithic Society* 10, 1–23.

Lee, D.Y., 1995. Geological studies of prehistoric layers for reconstruction of environment and chronology. *Journal of Korean Ancient Historical Society* 20, 521–46.

Lee, H.J., 2002. Evolution of Paleolithic technology in the Korean Peninsula, in *Paleolithic Culture in Korea*, eds. Yonsei University Museum. Seoul: Yonsei University Press, 85–106.

Lee, H.J., 2005. The Middle Paleolithic to Upper Paleolithic transition and the tradition of flake tool manufacturing on the Korean Peninsula. in *The Middle to Upper Paleolithic Transition in Eurasia: Hypothesis and Facts*, ed. A.P. Derevianko. Novosibirsk: Russian Academy of Sciences Siberian Branch Institute of Archaeology and Ethnography, 486–500.

Lee, K.G., 2004. Sinbuk Upper Paleolithic site, Jangheung County, Jollanam Province, Korea, in *Evaluating the Cultural Features of the Sinbuk Upper Paleolithic Site in the Northeastern Asia*, ed. Gi-kil Lee. Kwangju: Jangheung County and Chosun University Museum, 31–8.

Lee, Y.J., 1983. Excavation of the Suyanggae site, Danyang, Korea, in *Report of Excavation of Archaeological Sites in Submerged Area by the Chungju Dam*. Cheongju: Chungbuk University Museum, 45–66.

Machida, I. & P. Siroi, 2003. *Tephra Atlas*. Tokyo: Tokyo University.

Lee, Y.J., 1984. Excavation report on the Suyanggae Paleolithic site, in *Report on Excavation of Archeological Sites in the Submergence Area of Chungju Dam Construction*. Cheongju: Chungbuk National University Museum, 101–86.

Lee, Y.J., 1985. 1985 excavation of Suyanggae site, in *Report on Extended Excavation of Archeological Sites in the Submergence Area of Chungju Dam Construction*. Cheongju: Chungbuk National University Museum, 101–253.

Matsufuji, K., 2003. Origin of the Upper Paleolithic in Northeast Asia. Unpublished paper presented at the International Seminar for Commemorating the Chongokni site. Yeonchen: Yeonchen County, the Institute of Cultural Properties, Hanyang University and the Korean Paleolithic Society.

Park, Y.C., 2001. The chronology and cultural traditions of the Upper Paleolithic in southern Korea, in *Seonggoknonchong*. (Collection of Papers of the Seonggok Foundation.) Seoul: Seonggok Cultural Foundation, 327–88.

Perrenoud, C., 2003. Preliminary micromorphological study of Chongokni Paleolithic site (Korea), in *Geological Formation of the Chongokni Paleolithic Site and Paleolithic Archaeology in East Asia*, eds. K.D. Bae & J.C.Lee. Ansan: The Institute of Cultural Properties, Hanyang University and Yeonchon County, 171.

Seo, Y.N., H.J. Kim & Y.J. Jang, 1999. Upper Paleolithic culture from the Goryeri site, Milyang in Gyeongnam province, in *Paleolithic Culture in Youngnam Area, The 8th Symposium Proceedings of the Youngnam Archaeological Society*. Daegu: The Youngnam Archaeological Society, 65–82.

Seong, C.T., 2006. A comparative and evolutionary approach to the Korean Paleolithic assemblages. *Journal of Korean Ancient Historical Society* 51, 5–42.

Sohn, P.K., 1967. Stratified Sokchangni Paleolithic site. *Ryoksahakbo* (Korean Historical Review) 35/36, 1–25.

Sohn, P.K., 1989. Origin of obsidian artefacts from the Sangmuryongni site, in *Sangmuryongni*, eds. B.K. Choi & Y.H. Hwang. Chunchon: Kangwon University Museum, 781–96.

Wang, Y.P., 2004. The Upper Paleolithic in China, in *Evaluating the Cultural Features of the Sinbuk Upper Paleolithic Site in the Northeastern Asia*, ed. G.G. Lee. Kwangju: Chosun University Museum and Jangheung County, 101–7.

Yi, S.B., 2001. Middle and Upper Paleolithic transition in Korea: a brief review. *Journal of the Korean Paleolithic Society* 4, 17–24.

Chapter 11

The Middle to Upper Palaeolithic Transition North of the Continental Divide: Between England and the Russian Plain

Janusz K. Kozłowski

The beginnings of the Upper Palaeolithic in Europe have been discussed primarily in terms of the Balkan–Danube axis as a migration route of anatomically modern humans, and with reference to the northern coast of the Mediterranean Sea and the Franco-Cantabrian provinces as regions of autochthonous formation of leptolithic (i.e. blade-dominated) cultures. The territories north of the central European divide, notably the European Lowland (Fig. 11.1), have been regarded as a zone of marginal and sporadic influences from the central European plateaux or upland zone that was uninhabited between OIS-4 and OIS-2 and re-settled only at the end of the glacial period. This viewpoint has been emphasized particularly with reference to the western and central part of the European Lowland, which was almost entirely covered by ice during the Last Glacial Maximum (LGM, OIS-2). Analysis of finds from the southern borders of this part of the European Lowland and from the northern boundaries of the central–western European plateaux suggests, however, that there were Early Upper Palaeolithic units specific to this territory which were different from the units of the central and southern belt of Europe (see e.g. Kozłowski & Kozłowski 1981; Kozłowski 2002; Otte 1981).

Recent years have seen important discoveries of sites from the early phase of the Upper Palaeolithic in northeastern Europe. These sites are situated far beyond the range of the well-known centres in the Middle Don basin or even in the Upper Volga basin. Some of these recently discovered sites represent taxonomic units known from the central part of the eastern European Lowland; but there are also idiosyncratic assemblages which show no similarity to units from the central part of the Russian Plain. The presence of sites dated to the Interpleniglacial (late phase of OIS-3) in high latitudes (up to 66°N), such as a group of sites discovered in the Upper Kama and the Pechora basins, confirm that the extreme north of the European Lowland could have been penetrated by the hunters of the Early Upper Palaeolithic.

Palaeogeography of northern Europe in the Interpleniglacial (Fig. 11.2)

The Interpleniglacial saw an acceleration in the rhythm of palaeotemperature oscillations: from relatively low values (corresponding to Dansgaard/Oeschger cold events) to relatively high values especially in summer (Dansgaard/Oeschger warm events). During the entire OIS-3 the ice-sheet cover was confined to the mountainous part of Scandinavia, whereas the coast line approximated –70 m (Arnold *et al.* 2002; Cacho *et al.* 1999; Coope 2002; Lambeck 1995).

Palaeoenvironmental reconstructions for Dansgaard/Oeschger cold events during OIS-3 have shown that nearly the whole of the European Lowland, as far as Belgium in the west, was a polar desert. The northern boundary of the Lowland was covered by steppe-tundra which also embraced the northwest part of the Lowland, the plateaux and mountains of central Europe (Huntley & Allen 2003).

The palaeogeographical reconstructions by Van Andel & Tzedakis (1996) suggest that during Dansgaard/Oeschger warm events almost the entire European Lowland was covered by shrub tundra. Only some southern areas, adjacent to the belt of plateaux and the northwest part of the Lowland, were open coniferous parkland. The northern border between the shrub tundra and the polar desert ran across the Baltic and the North Sea basin. A map published by Huntley & Allen (2003) for Dansgaard/Oeschger warm events is, however, more complex. The European Lowland is divided into two parts: western (from the British Isles to the Vistula basin), with a landscape of patchy woodland), and eastern (in the Dnepr and the Dnestr basins together with the upper and middle Danube), which is assumed to have been predominantly a park-

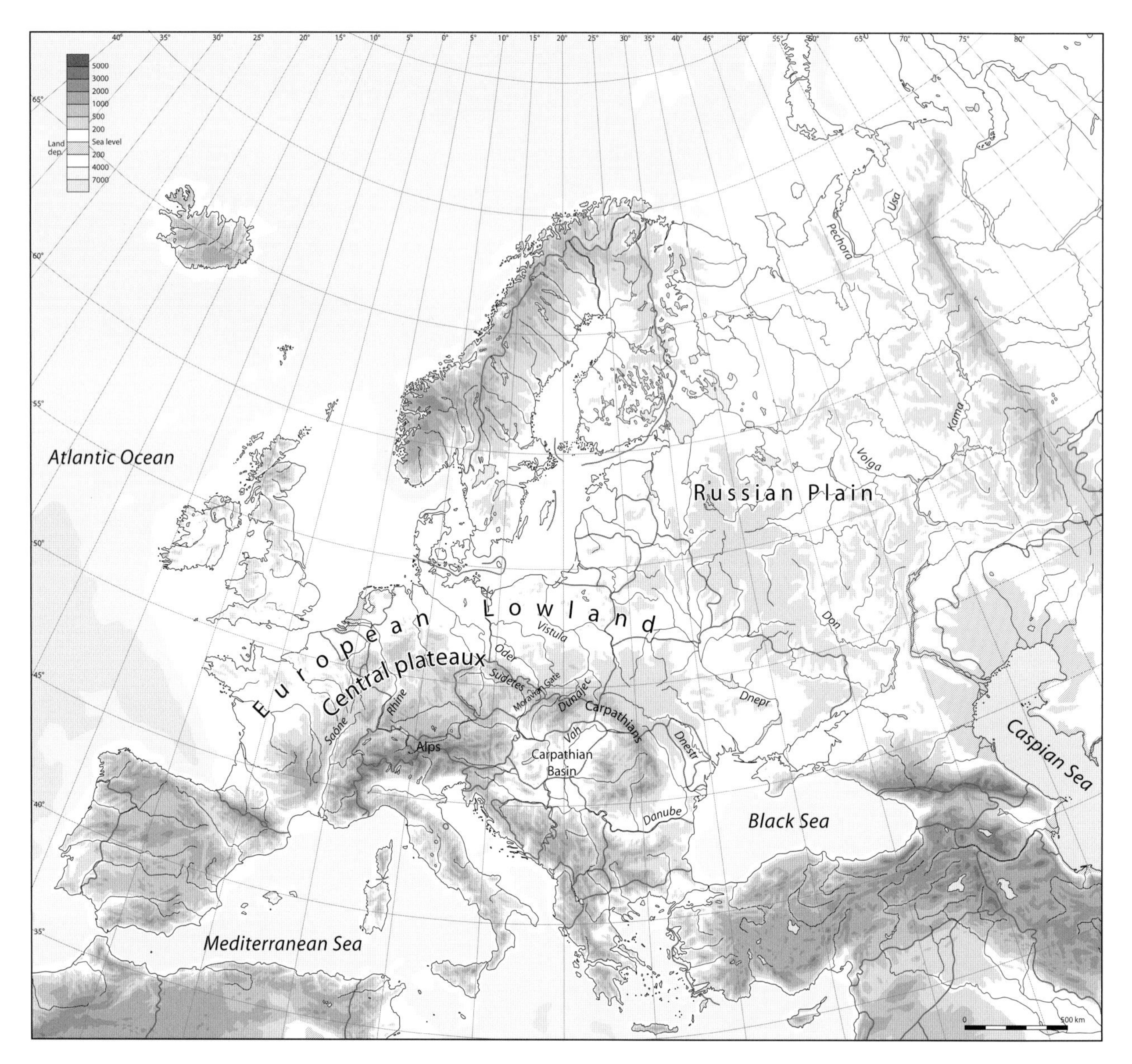

Figure 11.1. *Topographic map of Europe and western Russia showing areas mentioned in the text (base map after Collins atlas 1:20,000,000).*

land or savannah-like landscape. This reconstruction shows the northeastern part of the European Lowland as especially privileged in its vegetational cover. The Baltic countries are assumed to have been covered by temperate grassland, the upper Volga basin by taiga. Typical tundra is claimed to have covered only central and northern Finland.

In central Europe, at the border between the plateaux with the patchy woodland and the European Lowland with parkland, a mosaic landscape is expected. In the Upper Vistula basin for example, parallel to the northern boundary of the Carpathians, pollen diagrams for the period 32 kyr BP show low frequencies of arboreal pollen (Kraków-Kryspinów – up to 15 per cent maximum: Mamakowa & Rutkowski 1989). For the same period in the Carpathian Foreland tree-pollen frequencies increase to more than 50 per cent (Dobra palynological site, southern Poland: Środon 1969). It should be emphasized that in the earliest Dansgaard/Oeschger warm event (*c.* 39–38 kyr BP) the proportion of arboreal pollen in the Vistula basin had been higher (e.g. at Ściejowice near Kraków: Mamakowa & Rutkowski 1989).

Archaeobotanical and archaeozoological data from the northeast part of the European Lowland, particularly the northeastern part of the Russian Lowland and the Pechora basin, indicate, however, that in the Dansgaard/Oeschger warm events during the

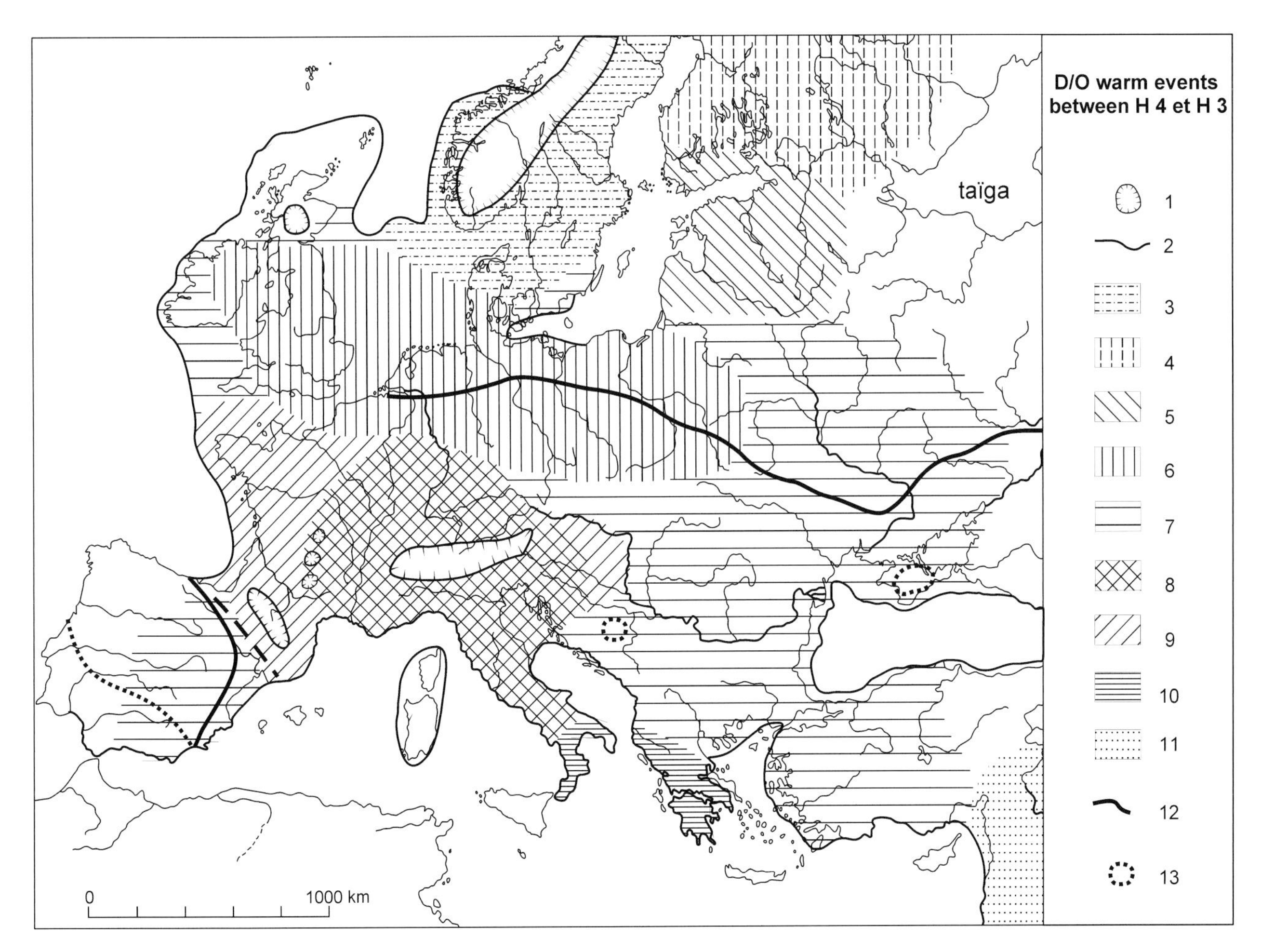

Figure 11.2. *Map of the environment during Dansgaard/Oeschger warm events (Huntley & Allen 2003, modified): 1) ice-sheet; 2) shorelines; 3) periglacial/polar desert; 4) tundra/deciduous taiga mosaics; 5) temperate grassland with cool coniferous woodland; 6) cool steppe/patchy woodland; 7) parkland; 8) temperate grassland; 9) mixed-forest mosaics; 10) temperate woodlands; 11) warm steppe; 12) maximum extension of the Aurignacian (36–32 kyr* BP*); 13) Neanderthal refuges (32–28 kyr* BP*).*

younger part of OIS-3 (about 40–36 kyr BP and about 32–30 kyr BP) this territory was treeless steppe-tundra with herbs and grasses; the presence of trees, mainly willow shrubs, is recorded only in river valleys. This landscape is evidenced by faunal assemblage composition at the northernmost sites of eastern Europe where we see typical steppe-tundra species such as mammoth, woolly rhinoceros, reindeer, bison, horse, polar fox, musk ox and lemming.

A settlement hiatus at the border between the plateaux and the Lowland?

Until fairly recently it was thought that — unlike in the Carpathian basin and the Upper Danube basin — a settlement hiatus occurred north of the Carpathians and the Sudetes, starting from before OIS-4, between the Middle and Upper Palaeolithic, dated to the second half of OIS-3. Only the discoveries in the Vistula basin near Kraków at the sites of Piekary IIa and Kraków-Księcia Józefa Street have allowed us to document, with any certainty, the existence of settlement between 61.3±8.7 and 47.8±8.9 kyr TL BP (Piekary IIa, layer 7c: Valladas *et al.* 2003), 44.4±1.4 and 40.38±0.94 kyr uncal. bp (Kraków-Księcia Józefa Str. lower and middle layers: Sitlivy *et al.* 2004) and 35.6±2.6 and 33.0±2.8 kyr TL BP (Piekary IIa layer 7a: Valladas *et al.* 2003). Thus, assemblages from these layers fill the gap spanning the end of OIS-4 to the second half of OIS-3.

In terms of taxonomy these assemblages should be ascribed to the Middle Palaeolithic when — besides Levallois linear, convergent and recurrent methods — we first see volumetric blade technology with advanced preparation based on crests, installation

Figure 11.3. *Refitted bidirectional volumetric blade cores from Kraków-Księcia Józefa Street.*

and rejuvenation of neo-crests. Blade débitage started from a single platform, however the opposed platform bidirectional method is well represented, particularly at Piekary IIa in layers 7c and 7a. Besides the advanced blade technique the assemblages contain fine flakes detached from centripetal cores. Among tools there are Middle Palaeolithic (side-scrapers) as well as Upper Palaeolithic forms (e.g. a burin on a retouched truncation and truncated backed blades in the assemblage from layer 7c from Piekary IIa: Valladas *et al.* 2003, fig. 9).

The highly advanced volumetric blade technique based on double-platform cores (Fig. 11.3) is recorded in the middle level at the site of Kraków-Księcia Józefa Street (Sitlivy *et al.* 1999; 2004).

The presence of the volumetric blade technique in assemblages from the Vistula basin makes them somewhat similar to the Middle Palaeolithic blade industries of northwestern Europe, especially France and Belgium (comp. e.g. Revillon & Tuffreau 1994), but in these countries they are probably earlier than OIS-3 or even OIS-4. Although in southern Poland the volumetric blade technique could have persisted until the younger part of OIS-3 (it is represented as late as layer 3 at Piekary II and layer 6 at Piekary IIa – the so-called Naskalanski industry: Krukowski 1939), the fully developed Upper Palaeolithic Aurignacian at Piekary IIa and Piekary II (Sachse-Kozłowska & Kozłowski 2004) shows no technological continuation from earlier assemblages.

Piekary-type assemblages show us that at the end of the Middle Palaeolithic, at the border between the Uplands and the Lowland, taxonomic units which could have persisted until the end of OIS-3 existed that were specific to this zone of western and central Europe but unknown in the Middle and Upper Danube basin. This double-platform blade core technique is not seen in other Early Upper Palaeolithic industries (Szeletian, Aurignacian) except in Lincombian-Ranisian-Jerzmanowician (L-R-J) type assemblages.

The early Upper Palaeolithic at the upland–lowland border, central and western Europe

Penetration of culture groups from the plateaux into the southern part of the European Lowland

The Early Upper Palaeolithic units of younger OIS-3 date in the middle and upper Danube basin crossed the mountain ranges of the Carpathians and the Sudetes, the borders of the western European loess uplands, to appear in the southern borderland of the European Lowland. The two units were the Szeletian in southern Poland, and the Aurignacian at the northern borders of the uplands, ranging from southern England as far as Poland.

Penetration of Szeletian groups into southern Poland (Fig. 11.4)

Szeletian groups entered southern Poland from the middle Danube basin seasonally, from approximately 36–34 kyr BP. This incursion took place in the foreland of three Transcarpathian passes:

1. The Moravian Gate: sites in Upper Silesia in the Upper Oder basin (Kozłowski 1964: Foltyn 2003);
2. The pass between the Upper Vah basin and the Poprad and Dunajec rivers, this route being proven by a Szeletian camp in layer XI in the Obłazowa Cave in the very heart of the western Carpathians (Valde-Nowak *et al.* 2003); along this route Szeletian groups penetrated the Upper Vistula basin (sites near Kraków: Kozłowski 1969; Chmielewski 1975; Sachse-Kozłowska & Kozłowski 1975); and
3. passes in the eastern part of the Polish Carpathians: single finds of leaf points in the Dunajec and the Raba valleys.

These routes are documented by three types of site:

1. occasional base camps evidencing longer sojourns of Szeletian hunters north of the Carpathians and the Sudetes. Among such camps are Dzierżysław with an oval dwelling feature ringed by erratic

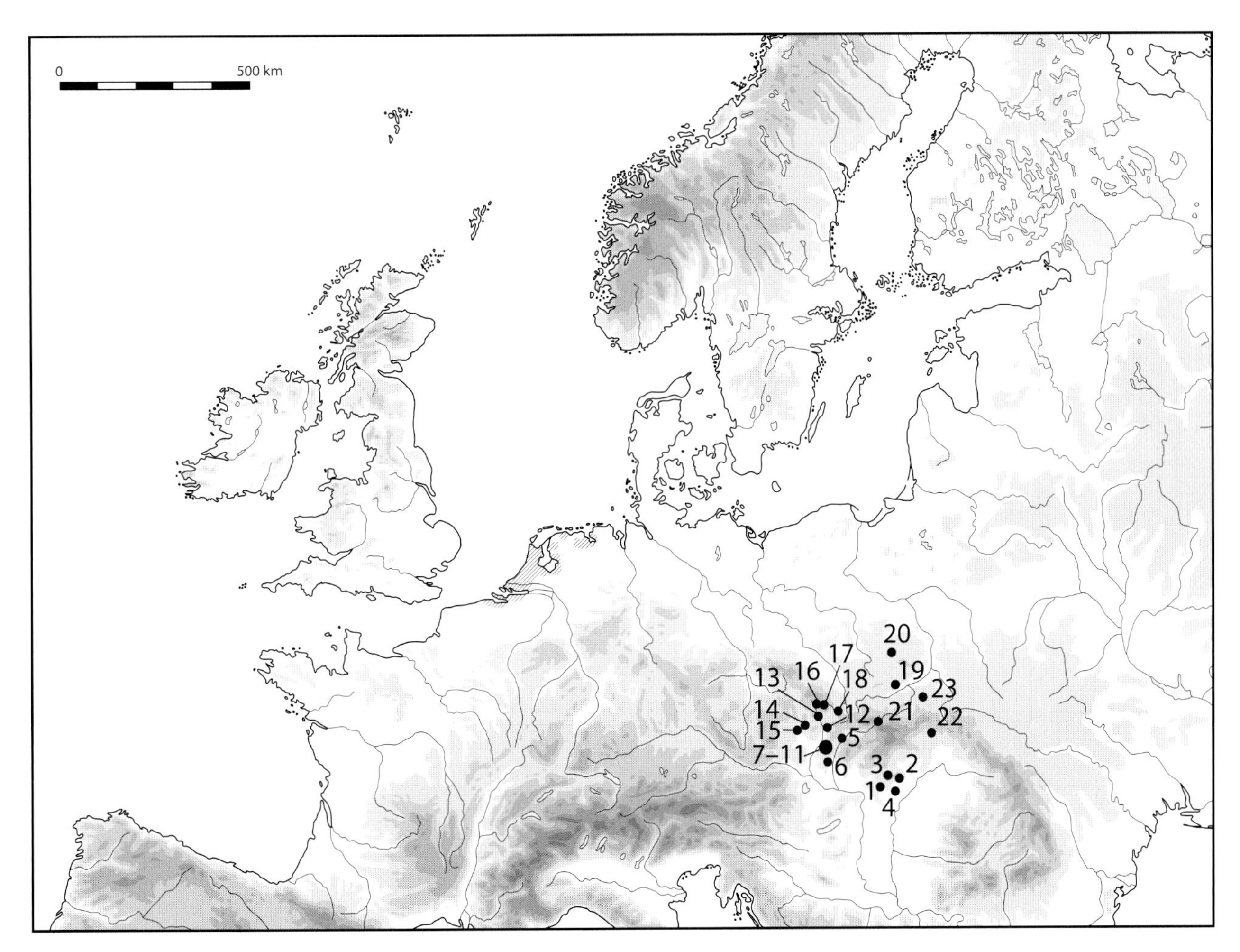

Figure 11.4. *Szeletian in northern central Europe. Hungary: 1) Szeleta; 2) Puskaporos; 3) Balla; 4) Eger (open-air sites near Eger). Czech Republic: 5) Ivanovce; 6) Vedrovice V; 7) Rozdrojovice; 8) Modrice; 9) Neslovice; 10) Orechov; 11) Moravsky Krumlov IV; 12) Ostrava; 13) Otice; 14) Ondratice; 15) Vicencov; 16) Trebom. Poland: 17) Dzierżysław 1 and 8; 18) Cieszyn; 19) Kraków-Zwierzyniec I; 20) Strzegowa – Jasna Cave; 21) Obłazowa Cave; 22) Gostwica; 23) Dobczyce.*

stones (Kozłowski 1964), and those in sectors 1, 2, 3 and 4a at Kraków-Zwierzyniec I (Sawicki 1957; Chmielewski 1975; Allsworth-Jones 1986; Manka 2006);
2. erratic flint workshops in the Upper Oder basin (e.g. Dzierżysław 8); and
3. ephemeral camps in caves (Mamutowa Cave layer 6: Kowalski 1967; 1969; Obłazowa Cave layer XI: Valde-Nowak *et al.* 2003) and numerous single finds of leaf points at open-air sites (Cieszyn I, Gostwica, Dobczyce and others: Foltyn 2003).

Transcarpathian contacts during the Szeletian are illustrated by imports of raw materials such as:
1. radiolarite from the Vah valley and quartzite from the Drahany Plateau in Moravia and, recently, even quartz-porphyry from the Bükk Mountains found at sites in southern Poland; and
2. erratic Silesian and Jurassic flint from the Kraków region found at Moravian sites (Oliva 2002; Kozłowski 1965), as well as flint from the middle Vistula basin (Swieciechów flint and so-called chocolate flint) found at sites in northeastern Hungary (Kozłowski 1963; Vértes 1965).

Unfortunately, faunal remains are preserved at neither southern Polish nor Moravian sites so nothing can be said about hunting strategies and seasonality of occupation by Szeletian groups at the northern edge of the uplands of central Europe.

The northern peripheries of Aurignacian settlement (Fig. 11.5)

Aurignacian settlement is recorded from about 32–28 kyr BP along the entire border of the European Lowland and the northern borders of the uplands, from England as far as Poland. Aurignacian camps — both ephemeral and residential — occur in

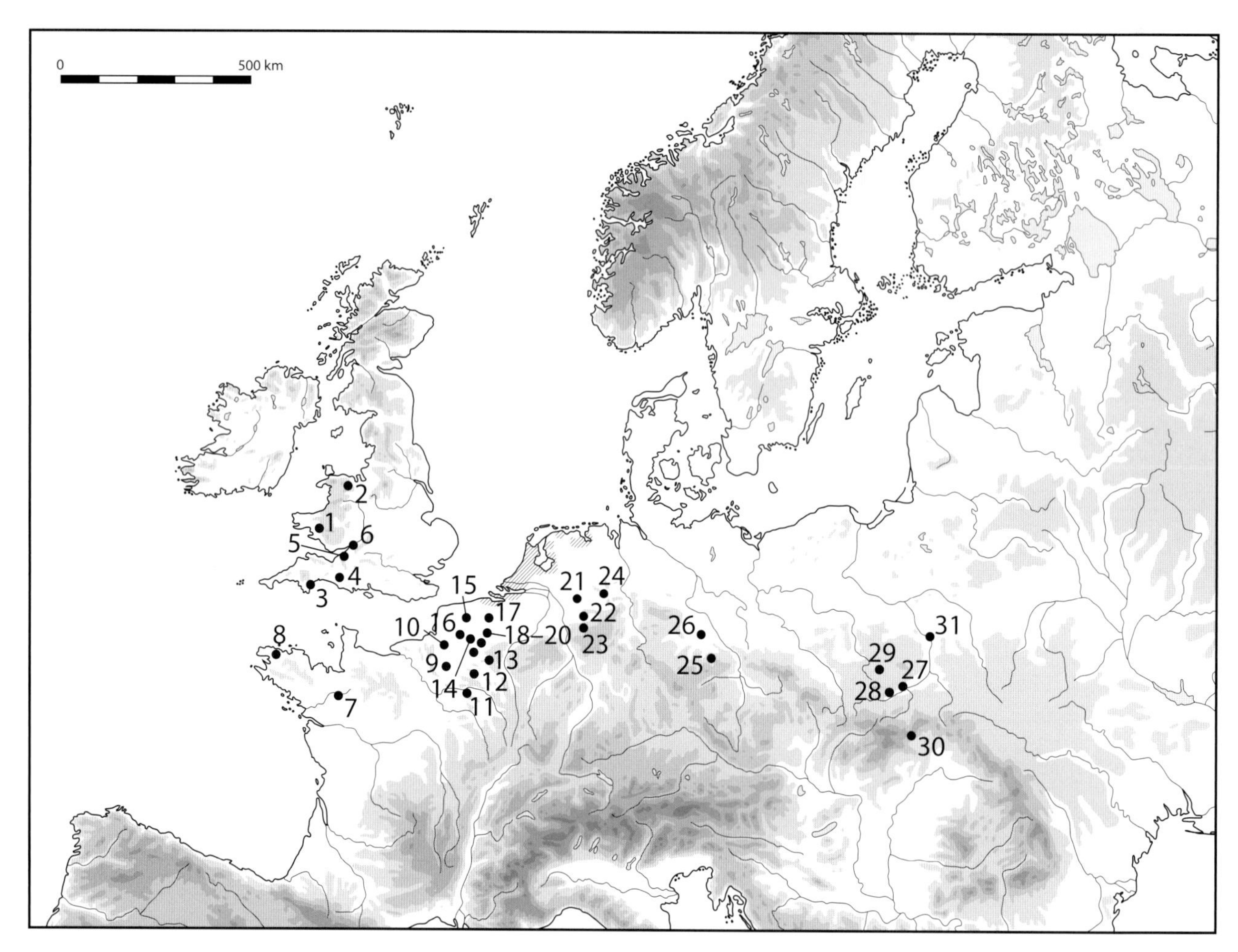

Figure 11.5. *Northernmost Aurignacian sites in western and central Europe. United Kingdom: 1) Paviland Cave; 2) Pfynnon Beuno Cave; 3) Kent's Cavern; 4) Aston Mills; 5) Uphill Quarry; 6) Hyaena Den. France: 7) Saulges Cave; 8) Belloy-en-Santerre; 9) Attilly; 10) Rouvroy; 11) Auboue; 12) Havange. Belgium: 13) Altwies-Laagen Aker; 14) Maisieres; 15) Spy Cave; 16) Princesse Pauline Cave; 17) Du Prince Cave; 18) Trou Magrite Cave; 19) Goyet Cave; 20) du Docteur Cave. Germany: 21) Lommersum; 22) Wildschauer Cave; 23) Wildhaus Cave; 24) Balve Cave; 25) Breitenbach; 26) Ranis layer 3 (?). Poland: 27) Kraków-Zwierzyniec I; 28) Piekary I, II, IIa, IV; 29) Mamutowa Cave; 30) Obłazowa Cave; 31) Góra Puławska II.*

Wales and southwest England (e.g. caves such as Paviland Cave and Kent's Cavern: Campbell 1986), in northern France (Normandy to Lorraine), in Belgium (Spy, Goyet, Maisières and others: Otte 1979), Germany (from the middle Rhineland, e.g. Lommersum, across Hesse as far as the Lower Elbe e.g. Breitenbach: Hahn 1977; 1989) and Poland (sites near Kraków: Sachse-Kozłowska 1978; Sachse-Kozłowska & Kozłowski 2004). Lithic inventories at these sites are fairly varied longitudinally, e.g. west of the Rhine there are *busqué*-type burins, whereas east of the Rhine carenoidal end-scrapers seem to be more frequent. At the same time, links with Aurignacian groups further to the south are recorded. For example, sites in northern France maintained contacts with upland sites at the edge of the Massif Central and in the Saône basin, sites on the Rhine with those in the Upper Danube basin, and sites in the Upper Vistula basin with sites in Moravia or — more broadly — in the Middle Danube basin. Danubian links have been confirmed by imports of Transcarpathian raw material and raw-material exchange via the Moravian Gate.

The Aurignacian settlement along the southern edges of the European Lowland displays no specific characteristics, differing from those of Aurignacian assemblages in the upland zone. In fact this settlement is the result of expansion from upland territories, most probably in the warmer episodes at the end of OIS-3. An exception is the site of Góra Puławska — the east-

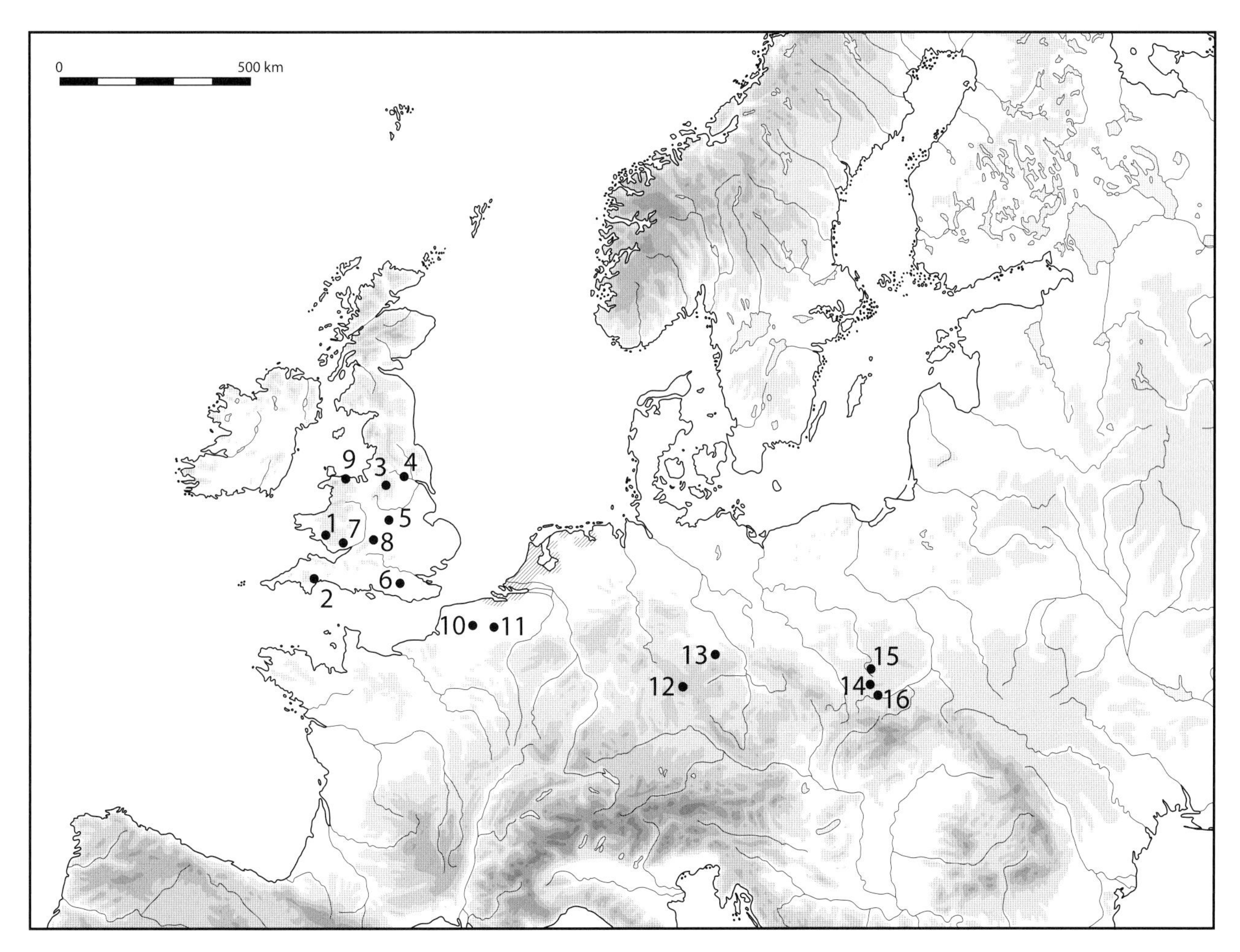

Figure 11.6. *Most important Lincombian-Ranisian-Jerzmanowician sites. United Kingdom: 1) Paviland Cave; 2) Kent's Cavern; 3) Pin Hole; 4) Robin Hood Cave; 5) Glaston, Grange Farm; 6) Beedings; 7) King Arthur's Cave; 8) Badger Hole; 9) Pfynnon Beuno Cave. Belgium: 10) Spy Cave; 11) Goyet Cave. Germany: 12) Zwergloch; 13) Ranis Cave, layer 2. Poland: 14) Nietoperzowa Cave; 15) Koziarnia Cave; 16) Puchacza skala Cave.*

ernmost site in the Vistula basin, the assemblage from which is closest to the eastern Aurignacian Muralovka type (Krukowski 1939).

The hypothesis that the Aurignacian could form a separate 'Lowland group' was based, on one hand, on the different technology represented by materials from the northern French site of Herbeville-le-Murger (Gouedo *et al.* 1996) and, on the other hand, a tentative assumption that in the 'northern Aurignacian' we are dealing with the co-occurrence of Jerzmanowice-type points and carenoidal forms. The grounds for the latter hypothesis were provided by mixed collections containing both Aurignacian points and points representing the L-R-J complex in Britain and in Belgium. In its extreme form this hypothesis was intended to show that the L-R-J complex is merely a 'hunting facies' of the Aurignacian (see Flas 2006).

Early Upper Palaeolithic of the northwestern European Lowland: the Lincombian-Ranisian-Jerzmanowician (L-R-J) complex (Fig. 11.6)

Unlike the central European units (Szeletian and Aurignacian) the L-R-J complex is not known in the middle and upper Danube basin. Its distribution, in the period from 38–30 kyr BP, is confined to the belt between the European Lowland and the uplands, from England to Poland. Our knowledge of this unit is incomplete as L-R-J sites are, in practice, only short-term hunting camps (mainly in caves) with, almost exclusively, leaf points and no local lithic production. Only in England are open-air camps recorded with greater variety in tools and even traces of local stone working (Beedings: Jacobi 1980; 1986; Glaston: Thomas & Jacobi 2001). L-R-J settlement occurs in clusters which form a belt, isolated from one another,

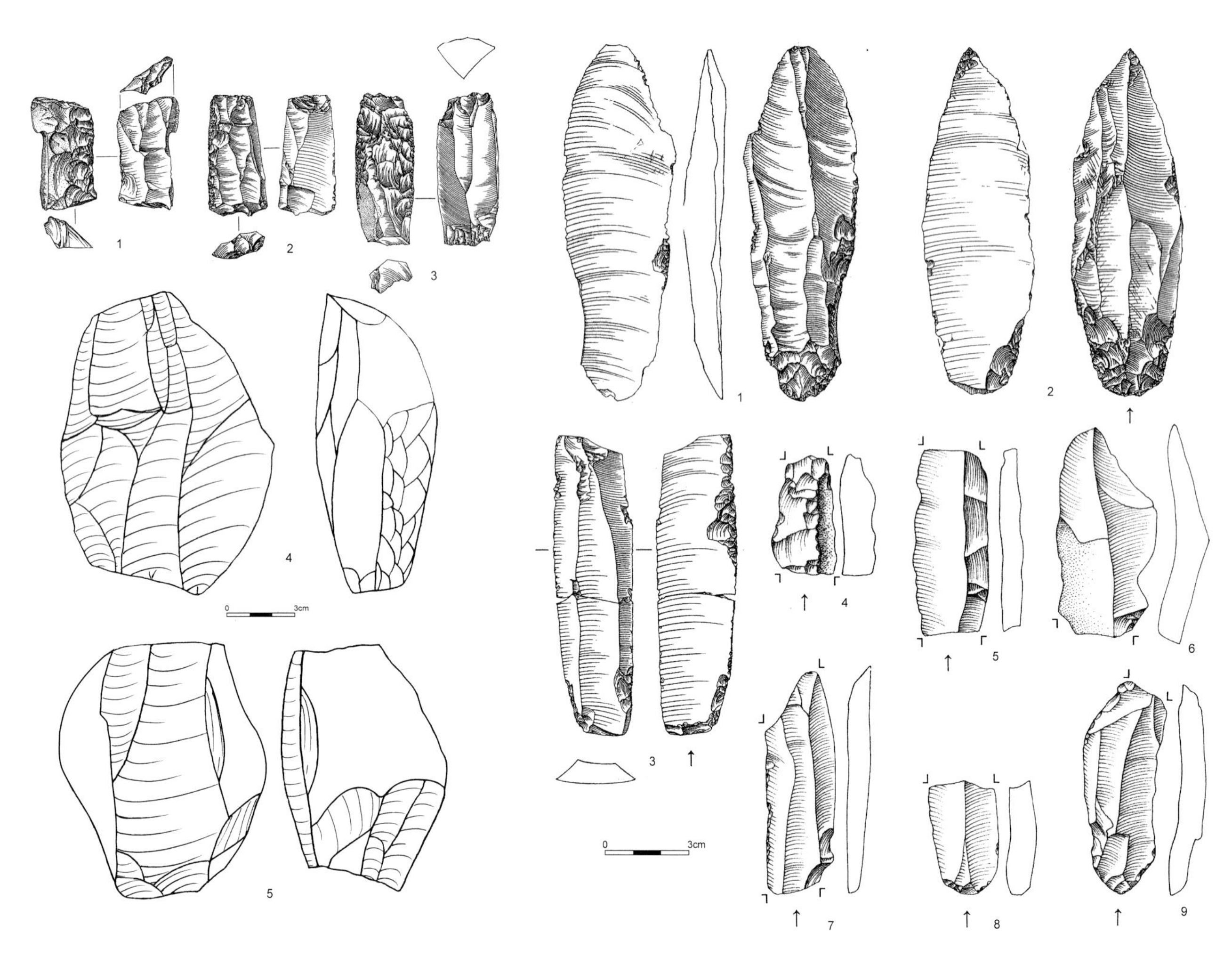

Figure 11.7. *Lincombian-Ranisian-Jerzmanowician cores: 1–3) Beedings (England) (according to R. Jacobi); 4–5) Jerzmanowice, Nietoperzowa Cave.*

Figure 11.8. *Lincombian-Ranisian-Jerzmanowician blades: 1–3) Beedings (according to R. Jacobi); 4–9) Jerzmanowice, Nietoperzowa Cave (according to D. Flas).*

in England, Belgium and Thuringia, southern Poland. This structure of L-R-J settlement indicates latitudinal contacts between clusters, in all likelihood via the southern edges of the European Lowland.

As regards technology and typology, L-R-J assemblages are unique in comparison, although contemporaneous, with Szeletian and Aurignacian assemblages. For this reason the claim about the identity of the L-R-J with the northern facies of the Szeletian culture (Allsworth-Jones 1986; Valoch 1972), or the 'hunting' facies of the Szeletian or the Aurignacian is ungrounded.

Some of the distinctive features of the L-R-J complex, provided below, confirm its separate character, which differs from that of the Szeletian or the Aurignacian:

1. although among leaf points there are wholly bifacial specimens, the majority are made on blades and have partial bifacial retouch. Such points do not occur in other industries of the early phase of the Upper Palaeolithic, particularly the Szeletian. Single occurrences of blade points appear only in some Bohunician assemblages (Kozłowski 1990; Oliva 1981; 1985);
2. the presence of bilaterally retouched blades and *lames appointées*;
3. except for points the most common tools are burins and *couteaux de Kostenki*, whereas end-scrapers and retouched flakes are rare;
4. blade-production technique is based on double-platform volumetric cores with central or postero-central crests (Fig. 11.7). This facilitated production of relatively long blades (9–15 cm) with a straight profile which were suited for the production of blade points (Fig. 11.8). To detach blades a soft punch was employed which was not used in the Szeletian or other leaf-point industries;

5. a very small flake production component: at Jerzmanowician sites (e.g. at Nietoperzowa Cave in Poland) only 26 out of about 280 stone artefacts are flakes, and at the British site of Beedings (where lithic production took place on-site) only 13 out of 142 artefacts are flakes. In phase 2 at Ilsenhöhle-Ranis Cave only one flake was found. This is a vital difference in comparison with the Szeletian where flake blanks dominate over blade blanks.

The genesis of the L-R-J complex is hard to explain. None of the Middle Palaeolithic units in the northern European belt combines the use of leaf points and blade technique, particularly the technique based on bidirectional volumetric cores. Moreover, in the south of England a large settlement hiatus still exists between OIS-4 and the younger OIS-3 (64–38 kyr BP). In northern France, at the beginning of OIS-3, there are MTA-type Mousterian industries, La Quina-type Charentian and weakly defined assemblages with Levallois technology. In Belgium, too, the La Quina-type Charentian appears at the beginning of OIS-3, and in Germany the Micoquian. As has been stated earlier, it is only in southern Poland that, during OIS-3, we see assemblages with volumetric blade technique (Piekary IIa, Kraków-Księcia Józefa Str.) where the most common technique is the production of straight blades from double-platform cores, similar to those in the L-R-J (Fig. 11.8). Thus, these assemblages could be regarded as the technological ancestors of the L-R-J, although by themselves they do not account for the origin of the complex. The concept of points — at first wholly bifacial, subsequently blade points — must have developed from one of the Middle Palaeolithic units with leaf points. The technique of production of wholly bifacial L-R-J points (the sequence of shaping the edges and surfaces of points) suggests that these points do not derive from the Micoquian (the sequence of retouches of some of the Szeletian points is closer to Micoquian points) but, as was argued earlier, from the Altmühlian in the Upper Danube basin (Kozłowski 1990). Genetically the L-R-J complex can be assigned to the 'transitional industries' (as can the Szeletian) but, at the same time, we must be aware that between 30–25 kyr BP, i.e. in the middle phase of the Upper Palaeolithic, the L-R-J complex continues to evolve stimulating the formation of the Maiserian and, possibly, northern facies of the Gravettian of Font Robert type (Kozłowski 1974; Flas 2000–2001).

We can therefore suggest that, between 40 and 30 kyr BP, the L-R-J complex was a specific adaptation of two Middle Palaeolithic traditions which are territorially close to the European Lowland, namely the blade and the leaf-point traditions. This adaptation took place in the border zone of the upland and Lowland. In the cool Dansgaard/Oeschger episodes these territories were covered by steppe-tundra with rare trees, whereas in the warm episodes this was the boundary between the parkland with coniferous trees and patchy woodland.

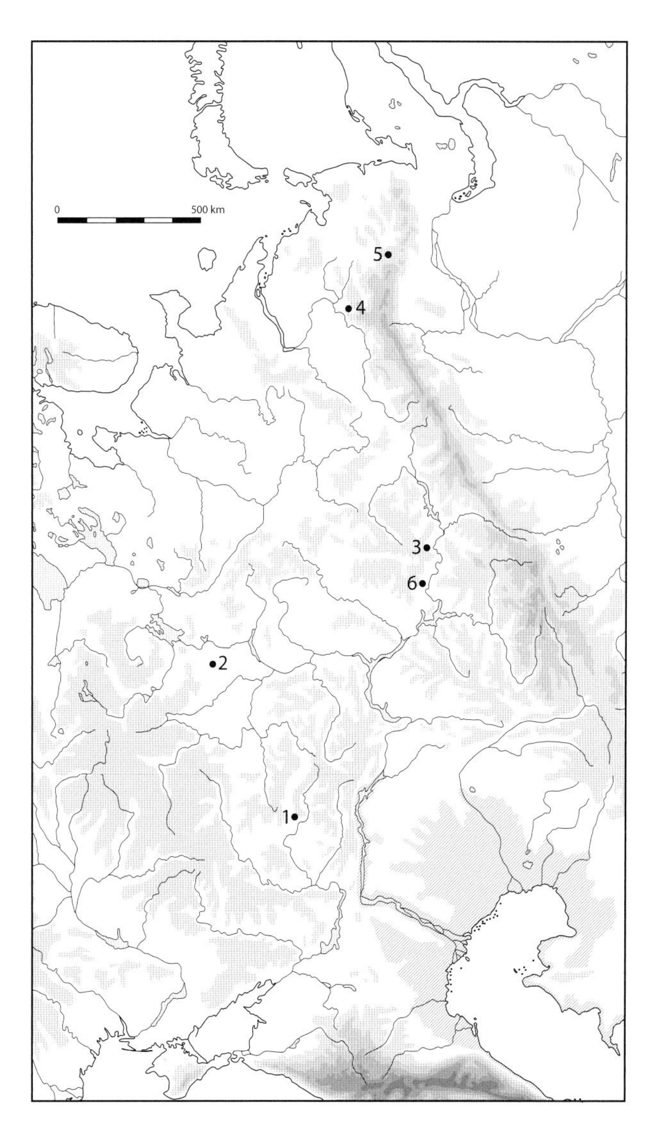

Figure 11.9. *Early Upper Palaeolithic sites in Russia: 1) Kostenki; 2) Sungir; 3) Garchi; 4) Byzovaya; 5) Mamontovaya Kuria; 6) Zaozerie.*

The Early Upper Palaeolithic in the northeast European Lowland (Fig. 11.9)

The settlement of the northeast part of the East European Lowland in the early Upper Palaeolithic is quite different. First of all, settlement embraced the northernmost part of the Lowland as far as the Pechora river basin. In these territories there are culture units originating from the south — mainly from the middle Don basin — and units that are specific for the Lowland of northeast Europe.

The expansion of the Streletskian-Sungirian to the northeast confines of eastern Europe

Discoveries in the Kama river basin, which belongs to the Caspian Sea catchment basin, and in the basin of the Pechora, which flows into the Arctic Sea, confirm penetration of the area between 59°N and 66°N between 36/35 and 28 kyr BP (Pavlov & Inderlid 2000; Svensen & Pavlov 2003). Among the sites discovered, the inventory from Garchi I is the most distinctive. It is situated in the Kama valley at 59°N. It is dated to 28.7±0.8 kyr BP and has yielded typical Sungirian leaf points — triangular, with a slightly concave or straight base. Some of the leaf points were made on flakes, as were some side-scrapers and retouched flakes.

Further north still there are sites in the Pechora basin. However, the leaf points discovered at these sites do not represent forms that are diagnostic of the Sungirian. For example, the site of Byzovaya, situated at 65°N (Guslitser & Liiva 1972), provided a rich inventory of wholly bifacial laurel-leaf points, accompanied by a large number of side-scrapers which are mainly lateral but also trapezoidal and convergent; some of the side-scrapers have thinned bases. End-scrapers are few in number, but particularly interesting is a carenoidal specimen. The site provided a large series of radiocarbon determinations spanning the period from 33.2 to 25.5 kyr BP, although most dates group at about 28/27 kyr BP. However, the taxonomic position of Byzovaya is controversial. Some researchers tend to assign the assemblage to the early Streletskian phase of the Sungiarian (Kanivets 1976), while others emphasize its links with the classical phase of the Sungirian (Bader 1978; Anikovich 1986). In either case, Byzovaya demarcates the range of penetration of the population from the central part of the Russian Lowland to the north. This range could be pushed even further north if we assign to the same tradition the site of Mamontovaya Kuria on the Usa river, a tributary of the Pechora, lying at 66°N (Fig. 11.1). This site is even earlier than Garchi and Byzovaya: the radiocarbon dates obtained on a mammoth tusk from three laboratories fall between 36.6 and 34.6 kyr BP. Apart from the mammoth tusk, which has intentional incisions, the site yielded only one fragment of a bifacial point transformed into a side-scraper. Although this is not a diagnostic form, the point belongs, in all likelihood, to a taxonomic unit similar to that seen at Byzovaya and, thus, supports a claim that the expansion towards the northeast had already begun during the older, Streletskian phase of the Sungirian. It should be added that the hunters' objective when entering these territories was, without doubt, the hunting of mammoth which at Byzovaya and Mamontovaya Kuria accounts for more than 98 per cent of the fauna (Svensen & Pavlov 2003).

Did the typical Lowland units of the Early Upper Palaeolithic form in northeastern Europe?

To answer this question — whether specific cultural traditions or adaptations, independent of those in the central part of the Russian Lowland, formed at the northern confines of the eastern European Lowland — is not easy. In any case, new analysis of the assemblage from the upper layer of the site of Kostenki 8 (Telmanskaya) does not attribute this inventory to the L-R-J complex (Flas 2006).

Recently, the site of Zaozerie which is part of the early phase of the Upper Palaeolithic was discovered in northeastern Europe. Its inventory exhibits highly original technological and typological features. The site, investigated by P. Pavlov, is situated on the Chusovaya river in the Kama basin at 58°N. A number of radiocarbon dates of around 31 kyr BP have been obtained on bones from the cultural layer, with an AMS date of 34.2±0.4 kyr BP being obtained on charcoals in the fossil soil.

A distinctive feature of the industry at Zaozerie is the use of the volumetric blade technique with double-platform cores from which fairly long blades with straight profiles were obtained. Besides the blade technique a flake technique was also used; indeed seven times as many flakes as blades are recorded at the site. The largest tool group consists of all kinds of end-scrapers, usually on flakes, and sometimes fairly thick with high fronts; only two specimens are carenoidal. Ogival and discoidal end-scrapers also occur. Moreover, there were side-scrapers (including partially bifacial, oval specimens), blades with fairly steep marginal retouch, burins-on-snap and lateral truncation burins, but the most diagnostic items are backed tools: a convex backed piece with retouch on the opposite edge, a mildly angulated backed piece, a robust convex truncation, a transversal truncation with steep, partially lateral retouch (Fig. 11.10).

The co-occurrence of the double-platform blade core technique and specific backed tools is not known in any eastern European industry of the early phase of the Upper Palaeolithic. This suggests that the industry from Zaozerie is a highly specific expression of the evolution of a separate cultural tradition in the northeast of Europe. The presence of personal adornments at Zaozerie is noteworthy, such as for example an unfinished shell pendant. The fauna from the site is that of steppe-tundra or 'mammoth steppe', and horse, rhinoceros, hare and mammoth were hunted.

So is this unit from Zaozerie 'transitional' between the Middle and Upper Palaeolithic, or is it fully Upper Palaeolithic? The problem is difficult to resolve as the assemblage at Zaozerie exhibits — so to speak — distant echoes of 'transitional' traditions (bifacial retouches, backed tools) as well as Upper Palaeolithic components (developed blade technology, high end-scrapers — echoes of an 'Aurignacoidal' tradition). Altogether, the assemblage is idiosyncratic.

Conclusions

During Interpleniglacial OIS-3, especially its younger part, the European Lowland was a territory of seasonal penetration and settlement from the middle in the central belt of Europe, but also a territory where specific cultural adaptations evolved.

By the end of OIS-4, and during the early phase of OIS-3, Middle Palaeolithic industries using blade technology appear in northwestern Europe at the southern border of the Lowland (except England). In northeastern Europe during OIS-4 and early OIS-3 settlement is absent.

In the early part of OIS-3 the territories adjacent to the European Lowland were penetrated first (36–34 kyr BP) by 'transitional' cultures: the Szeletian in central Europe and the Streletskian-Sungirian in eastern Europe, and later (34–32 kyr BP) by Aurignacian groups in western and central Europe. Behind these penetrations were settlement centres in the middle belt of Europe.

At the same time, in the territories neighbouring the Lowland, new cultural adaptations emerge and specific taxonomic units develop that are 'transitional' in their tool morphology although at the same time exhibiting Upper Palaeolithic technology. In the west-central part of the Lowland this is the L-R-J complex, whereas in eastern Europe we see the industry at Zaozerie. The occurrence of settlement at high latitudes in eastern Europe in the Early Upper Palaeolithic indicates that in northwestern and central Europe human penetration could have extended far into the Lowland region. However the ice-sheet transgression of the LGM caused Lowland sites older than the LGM to be buried beneath glacial and fluvioglacial sediments.

Unfortunately we have no fossil human remains from territories north of the central European divide, a fact which precludes any conclusions as to whether the occupation of the Lowland was related to the appearance of anatomically modern humans. One fact is, nonetheless, certain: culture complexes that developed in the northwest European Lowland, such as the L-R-J, or penetrated far into the East European Lowland (the Streletskian-Sungirian) continued to evolve until the middle phase of the Upper Palaeolithic when their makers were identified as anatomically modern humans, although some Neanderthal heritage may have been involved (see fossil human remains from the graves of Sungir).

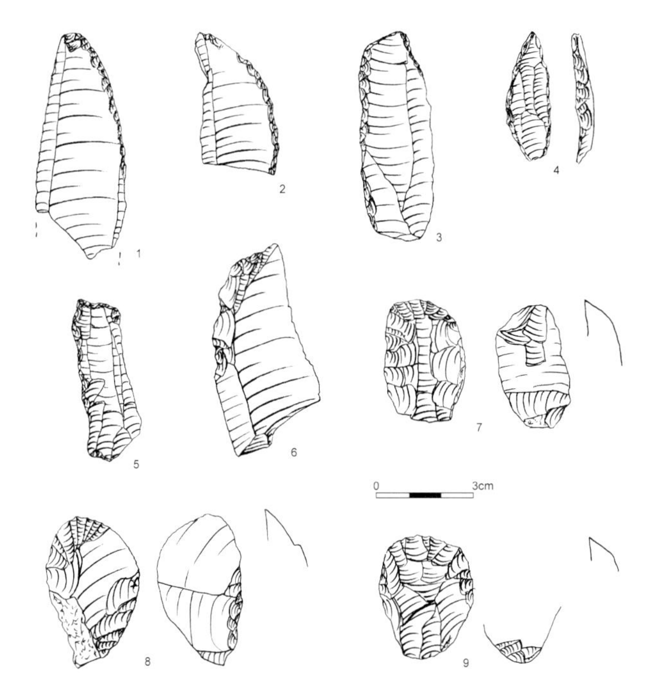

Figure 11.10. *Lithic artefacts from Zaozerie (excavations by P. Pavlov): 1, 2) retouched blades; 3–5) backed implements; 6) convex truncation; 7–9) thick end-scrapers.*

Acknowledgements

I would like to express my gratitude to Dr Pavel Pavlov for giving me the opportunity to study unpublished lithic finds from Zaozerie.

References

Allsworth-Jones, P., 1986. *The Szeletian and the Transition from Middle to Upper Palaeolithic in Central Europe.* Oxford: Clarendon Press.

Van Andel, T.H. & P.C. Tzedakis, 1996. Palaeolithic landscapes of Europe and environs, 150,000–25,000 years ago: an overview. *Quaternary Science Reviews* 15(5–6), 481–500.

Anikovich, M.V., 1986. O niekotorych spornikh problemach paleolita baseyna Petchory, in *Paleolit i Neolit.* Moscow: Nauka, 36–47.

Arnold, N.S., T.H. van Andel & V. Valen, 2002. Extent and dynamics of the Scandinavian ice sheet during Oxygen Isotope Stage 3 (65,000–25,000 yr B.P.). *Quaternary Research* 57(1), 38–48.

Bader, O.N., 1978. *Sungir - Verkhnepaleoliticheskaya stoyanka.* Moscow: Nauka.

Cacho, I., J.O. Grimalt, C. Pelejero *et al.*, 1999. Dansgaard-Oeschger and Heinrich event imprints in Alboran Sea paleotemperatures. *Paleoceanography* 14(6), 698–705.

Campbell, J.B., 1986. Hiatus and continuity in the British Upper Palaeolithic: a view from the Antipodes, in *Studies in the Upper Palaeolithic of Britain and Northwest Europe,* ed. D Roe. (British Archaeological Reports, International Series 296.) Oxford: BAR, 7–42.

Chmielewski, W., 1975. The Upper Pleistocene archaeological site Zwierzyniec I in Cracow. *Swiatowit* 34, 7–59.

Coope, G.R., 2002. Changes in the thermal climate in northwestern Europe during Marine Oxygen Isotope Stage 3, estimated from fossil insect assemblages. *Quaternary Research* 57(3), 401–8.

Flas, D., 2000–2001. Étude de la continuité entre le Lincombien-Ranisien-Jerzmanowicien et le Gravettien aux pointes pédonculées septentrional. *Préhistoire Européenne* 16–17, 163–89.

Flas, D., 2006. La transition du Paléolithique moyen au superieur dans la Plaine septentrionale de l'Europe. Unpublished PhD dissertation, University of Liège.

Foltyn, E., 2003. Uwagi o osadnictwie kultur z ostrzami lisciowatymi na polnoc od luku Karpat. *Przeglad Archeologiczny* 51, 5–48.

Gouedo, J.M., F. Lecolle & G. Drwila, 1996. Le gisement aurignacien de plein-air d'Herbeville-le-Murger (Yvelines). Bilan des fouilles 1991–1992. *L'Anthropologie* 100(1), 15–41.

Guslitser, B.I. & A. Liiva, 1972. O vozrastie mestonakhozhdenia pleystotsenovoy fauny i paleoliticheskoy stayanki Byzovaya na sredniey Petchorie. *Izvestia AN ESSR, Biologia* 21(3), 250–54.

Hahn, J., 1977. *Aurignacien: das ältere Jungpalaolithikum in Mittel- und Osteuropa.* (Fundamenta.) Cologne: Bohlau Verlag.

Hahn, J., 1989. *Genese und Funktion einer jungpaläolithischen Freilandstation: Lommersum im Rheinland.* (Rheinische Ausgrabungen 29.) Cologne: Rheinland-Verlag.

Huntley, B. & J.R.M. Allen, 2003. Glacial environments III: Palaeo-vegetation patterns in Late Glacial Europe, in *Neanderthals and Modern Humans in the European Landscape during the Last Glaciation: Archaeological Results of the Stage 3 Project*, eds. T.H. van Andel & W. Davies. (McDonald Institute Monographs.) Cambridge: McDonald Institute for Archaeological Research, 79–102.

Jacobi, R.M., 1980. The Upper Palaeolithic of Britain with special reference to Wales, in *Culture and Environment in Prehistoric Wales*, ed. J.A. Taylor. (British Archaeological Reports 76.) Oxford: BAR, 15–100.

Jacobi, R.M., 1986. The contents of Dr Harley's showcase, in *The Palaeolithic of Britain and its Nearest Neighbours: Recent Trends,* ed. S.N. Collcutt. Sheffield: Sheffield University Press, 62–8.

Kanivets, V.I., 1976. *Paleolit krayniego severo-vostoka Evropy.* Moscow: Nauka.

Kowałski, S., 1967. Wstepne wyniki badan wykopaliskowych w jaskini Mamutowej prowadzonych w latach 1957–1964. *Materialy Archeologiczne* 8, 47–60.

Kowałski, S., 1969. Nowe dane do poznania kultury jerzmanowickiej w Polsce. *Swiatowit* 30, 177–88.

Kozłowski, J.K., 1963. Nowe znalezsko importu z krzemienia swieciechowskiego na terenie Wegier. *Archeologia Polski* 7, 31–5.

Kozłowski, J.K., 1964. *Paleolit na Gornym Slasku.* Wroclaw: Ossolineum.

Kozłowski, J.K., 1965. *Studia nad roznicowaniem kulturowym w paleolicie gornym Europy srodkowej.* Kraków: PWN.

Kozłowski, J.K., 1969. *Problemy geochrologii paleolitu w dolinie Wisly pod Krakowem.* (Folia Quaternaria 31.) Kraków: PAN.

Kozłowski, J.K., 1974. Review of J. de Heinzelin, *L'industrie du site paléolithique de Maisière-Canal. Helinium* 14, 274–6.

Kozłowski, J.K., 1990. Certains aspects techno-morphologiques des pointes foliacées de la fin du Paléolithique moyen et du début du Paléolithique supérieur en Europe. *Mémoire du Musée de Préhistoire d'Ile de France, Nemours*, 125–33.

Kozłowski, J.K., 2002. *La Grande Plaine de l'Europe avant le Tardiglaciaire.* (ERAUL 99.) Liège: ERAUL, 53–65.

Kozłowski, J.K. & S.K. Kozłowski, 1981. Paléohistoire de la Grande Plaine européenne. *Archaeologia Interregionalis* 1, 143–62.

Krukowski, S., 1939. *Palolit Polski.* Kraków: PAU.

Lambeck, K., 1995. Late Devensian and Holocene shorelines of the British Isles and North Sea from models of glacio-hydro-isostatic rebound. *Journal of the Geological Society London* 151, 437–48.

Mamakowa, K. & K. Rutkowski, 1989. Wstepne wyniki badan paleobotanicznych profilu ze Sciejowic, in *Przewodnik LX Zjazdu Polskiego Towarzystwa Geologicznego.* Kraków: PTG, 113–17.

Manka, D., 2006. The assemblage of the Jerzmanowicie Culture from Kraków-Zwierzyniec I. *Sprawozdania Archeologiczne* 57, 371–420.

Oliva, M., 1981. Die Bohunicien-Station bei Podolí (Bez. Brno-Land) und ihre Stellung im beginnenden Jungpaläolithikum. *Casopis Moravskeho Musea* 66, 7–45.

Oliva, M., 1985. La signification culturelle des industries paléolithiques: l'approche psychosociale, in *La Signification Culturelle des Industries Lithiques*, ed. M. Otte. (British Archaeological Reports 239.) Oxford: BAR, 92–114.

Oliva, M., 2002. Vyuzivani krajiny a zdroju kamennych surovin v mladem paelolitu ceskych zemi. *Archeologicke Rozhledy* 54, 555–81.

Otte, M., 1979. *Le Paléolithique superieur ancien en Belgique.* Bruxelles: MRAH.

Otte, M., 1981. Les industries a pointes foliacees du Nord-Ouest européen. *Archaeologia Interregionalis* 1, 95–116.

Pavlov, P. & S. Inderlid, 2000. Human occupation in northeastern Europe during the period 35,000–18,000 bp, in *Hunters of the Golden Age: the Mid-Upper Palaeolithic of Eurasia 30,000–20,000* BP, eds. W. Roebroeks, M. Mussi, J. Svoboda & K. Fennema. Leiden: University

of Leiden, 165–72.
Revillon, S. & A. Tuffreau (eds.), 1994. *Les Industries Laminaires au Paléolithique Moyen*. Paris: Éditions CNRS.
Sachse-Kozłowska, E., 1978. Polish Aurignacian assemblages. *Folia Quaternaria* 50, 1–37.
Sachse-Kozłowska, E. & S.K. Kozłowski, 1975. Nowa kultura gornopaleolityczna w Europie srodkowej. Ze studiow nad materialami ze stanowiska Zwierzyniec I. *Archeologia Polski* 20, 275–86.
Sachse-Kozłowska, E. & S.K. Kozłowski (eds.), 2004. *Piekary pres de Cracovie (Polgne): Complexe des sites paleolithiques*. Kraków: PAU.
Sawicki, L., 1957. Sprawozdanie z badan stanowisk paleolitycznych Zwierzyniec I i Piekary II przeprowadzonych w roku 1955. *Sprawozdania Archeolgiczne* 4, 11–19.
Sitlivy V., K. Sobczyk, T. Kalicki, C. Escutenaire, A. Zieba & K. Kaczor, 1999. The new Paleolithic site of Ksiecia Jozefa (Cracow, Poland) with blade and flake reduction. *Préhistoire Européenne* 15, 87–111.
Sitlivy V., K. Sobczyk, C. Escutenaire *et al.*, 2004. Late Middle and Early Upper Palaeolithic complexes of the Cracow region, Poland, in *Acts of the XIVth UISPP Congress, University of Liège, Belgium, 2–8 September 2001*, ed. Le Secrétariat du Congrès. (British Archaeological Reports S1240.) Oxford: BAR, 305–17.
Środon, A., 1969. O roslinnosci interstadialu Paudorf w Karpatach Zachodnich. *Acta Palaeobotanica* 15, 17–41.
Svensen, J.I. & P. Pavlov, 2003. Mamontovaya Kurya: an enigmatic nearly 40 000 years old Paleolithic site in the Russian Arctic, in *The Chronology of the Aurignacian and of the Transitional Technocomplexes*, eds. J. Zilhão & F. d'Errico. Lisbon: Instituto Portuguès de Arqueologia, 109–20.
Thomas, J. & R. Jacobi, 2001. Glaston. *Current Archaeology* 173, 180–83.
Valde-Nowak, P., A. Nadachowski & T. Madeyska (eds.), 2003. *Obłazowa Cave: Human Activity, Stratigraphy and Palaeoenvironment*. Kraków: PAN.
Valladas, H., N. Mercier, C. Escutenaire *et al.*, 2003. The late Middle Palaeolithic blade technologies and the transition to the Upper Palaeolithic in southern Poland: TL dating contribution. *Eurasian Prehistory* 1(1), 57–82.
Valoch, K., 1972. Rapports entre le Paléolithique moyen et le Paléolithique supérieur en Europe centrale, in *Origine de l'Homme Moderne*, ed. F. Bordes. Paris: UNESCO, 161–71.
Vértes, L., 1965. *Az oskekor es az atmeneti kokor emlekei Magyarorszagon*. Budapest: Akademiai Kiado.

Chapter 12

Rethinking the 'Ecological Basis of Social Complexity'

Katherine V. Boyle

In 1985 Paul Mellars wrote of the ecological basis of the social complexity which he viewed as characterizing the Upper Palaeolithic. This 'ecological basis' was based largely on the importance of reindeer and the rarity of fish (mainly salmon) in the classic Périgord area of southwestern France. This chapter reviews the evidence for an ecological basis for evolving complexity during the period associated with the Middle–Upper Palaeolithic transition rather than the later phases of the Upper Palaeolithic which were of more concern in the 1985 publication. It looks mainly at Franco-Cantabria (the Châtelperronian) but also at the equivalent period in Italy, the Uluzzian.

> Almost everything I have done ... is cultural ecology, within a strongly ecological, adaptive framework, looking at the relationships between humans and their environment. I liked ecological thinking because it put the emphasis on the kinds of things that I felt archaeology can get at, technology, environments, subsistence, seasonality; this approach did then and still does lead to understanding long-term change. (p. 14)

So says Paul Mellars in Pamela Jane Smith's 'microbiography' in Chapter 2 of this volume. It, cultural ecology, is an approach which has long characterized research on the Palaeolithic of western Europe, and continues to yield interesting and significant results which show just how complex the period is.

In 1985 Paul Mellars wrote, in 'The ecological basis of social complexity in the Upper Paleolithic of southwestern France', that the most striking aspects of the Upper Palaeolithic of Franco-Cantabria included the 'remarkably high density and concentration of sites within a relatively restricted area' and the 'depth, stratigraphic complexity, and archaeological wealth' seen at these sites, characteristics which have been reiterated more recently by Bocquet-Apel & Demars (2000, 553) when referring to the northern Aquitaine region as a permanent 'refuge zone' from the Aurignacian onwards. Mellars then went on to write of critical ecological features which supported Upper Palaeolithic behavioural patterns in southwest France and northern Spain including high density, diversity and concentration, of resources, their security and predictability (Mellars 1985, 273–4). Meanwhile, Rigaud (Chapter 14, this volume) describes Middle Palaeolithic Neanderthal populations as 'dispersed in small groups over a vast territory — there being a balance between human density and environmental resources' so that

> [t]he territory occupied and exploited by these groups is determined by the abundance of available resources, while the size of the group and their mobility is caused by exhaustion, or even collapse, and renewal of food resources as a function of season and the search for available raw materials. (p. 162)

The difference between resource security and predictability and 'exhaustion, or even collapse' is a significant one that can only be investigated if we have a sound knowledge of this resource base, its structure, distribution and changing nature.

In 2007 I looked at Châtelperronian biodiversity, considering faunal assemblages formed by both human (Neanderthal) and contemporary carnivore activity. In this paper the resource base and its diversity are viewed as the background against which we can set archaeological evidence for developing social complexity during the transition to, and the earliest phase of, the Upper Palaeolithic in Franco-Cantabria. An initial comparison with Italian Uluzzian faunal assemblages is also presented as this industry, although unrelated to the Châtelperronian (Gioia 1990, 244; Mussi 2001, 172; Palma di Cesnola 2001, 29), also falls within what Riel-Salvatore (2009, 387) calls the 'transitional interval' of 43,000 to 30,000 BP.

The faunal data with which we are concerned has been collected from a number of sources: monographs (e.g. Boyle 1990; Delpech 1983; Serangeli 2006), individual site reports, the key 2002 *Journal of Archaeological Science* paper by Grayson and Delpech, and reports of other large-scale on-going projects, e.g. Demars (2008) and the Stage Three Project (http://www.

esc.cam.ac.uk/research/research-groups/oistage3), Using the information available, I shall examine the resource-environment changes associated with the 'transition'. The time span covered actually appears to be that during which the Protoaurignacian of the west-central Mediterranean, identified on the basis of relative abundance of Dufour bladelets of Dufour sub-type (Teyssandier 2008) and dating as far back as at least 40,000 BP in Venetia (Italy).

The ecological setting of the transition

The demographic and ecological features of the Upper Palaeolithic which Mellars views as key to complex behaviour of which we have 'evidence' in southwest France include:

- high density and concentration of the human population based on the large, ever-growing number of sites recorded;
- relative stability and permanence of residential patterns over much of the year;
- large co-residential patterns exploiting periodic abundances of certain resources;
- relative stability of residential patterns (Mellars 1985, 273).

Ecological and geographic features which underlie these patterns include the relatively limited, well-defined ecological uniformity of the small geographic area of southwest France which has been the focus of so much attention. Expanding the region to cover a larger area means that this ecological uniformity disappears to be replaced by greater diversity of landscapes and resources. By comparing the resource-environments associated with the Transitional Industries of Italy and the Châtelperronian region we can place each in its wider context and go some way towards explaining the variability seen during this key period of apparent change.

Various forms of environmental data are available, although the scale of this record for the Châtelperronian and Uluzzian is significantly smaller than for the Aurignacian. Indeed, in 1990 Gioia made the point that there is little available in the way of environmental data for the Uluzzian, and that information had to come from fauna and sediments (Gioia 1990, 249). Today, however, the picture has improved somewhat.

The Châtelperronian, Uluzzian and earliest Aurignacian all fall within Marine Isotope Stage 3. Each is characterized by geographically and chronologically varying environmental conditions ranging from warm temperate environments similar to those seen today but usually with less tree cover to cold, continental conditions (Finlayson 2004, 138–9). Results of pollen analysis at sites across the region identify biotopes ranging from temperate forest, through cold mixed forest and taiga (evergreen and deciduous), to cold and temperate open grassland (d'Errico & Sánchez Goñi 2003) supporting large populations of both herbivores and carnivores (Pärtel *et al.* 2005). Of course, given the duration of the period under consideration (*c.* 45,000 to 30,000 BP), chronological variation is to be expected. Similarly, latitudinal extent of assemblages (*c.* 40°N to 47°N) means that a degree of difference in structure and complexity is likely; based on nearly universal patterns of decreasing diversity, richness and productivity with increasing latitude (Hutson 1994; Brown 1995; Lévêque & Mounolou 2001; Magurran 2004), it would be surprising if this were not the case.

As to the specific environments, in the north of our region conditions are open, almost steppic, e.g. at the Arcy-sur-Cure sites. Further south and west environments remain open but with an arboreal component of pine, birch and alder (Sémah & Renault-Miskovsky 2004, 92). At Grande Roche de la Plématrie à Quincay (Vienne) we see four temporally-distinct Châtelperronian pollen assemblages and environments (Leroyer 1990, 49): temperate (Egb, Egc, Egf and En), transitional (En, Em), cold and dry (Ejm) and finally temperate again (Ejo). At St Césaire (Charente-Maritime), after a relatively temperate and humid early Châtelperronian, conditions become colder and drier (Morin 2004, 115–16); the abundance of *Microtus gregalis* (narrow-headed vole) points to the occurrence of local tundra and wooded steppe conditions at both St Césaire and, further south in the Dordogne region, at Roc de Combe (Morin 2004; Grayson & Delpech 2008).

Further south still, in Cantabrian Spain, we see a wide variety of spatially diverse environments which owe as much to topography as changing climate. Owing to topographic extremes, from coastal belt to mountain, the area is one of mixed ecotones throughout the period and therefore highly variable resources (Dari & Renault-Miskovsky 2001, 138). For much of the period the flora includes a significant arboreal component of *Pinus sylvestris*, *Sorbus aria* and *Betula* (Uzquiano *et al.* 2008, 126–8). To this can be added a significant shrub layer of *Arbutus*, *Rhamnus*, *Prunus*, *Erica*, *Ilex*, etc. At sites such as El Castillo, the environment ranges from relatively dry through to humid cold temperate or cold conditions; landscapes of open grassland and a sometimes significant arboreal component restricted to the deep valley areas of the region are seen (Dari & Renault-Miskovsky 2001).

Further east, into Italy, we see a steppic environment prior to 39,000 BP replaced by a temperate, more humid, wooded, coniferous and mixed oak woodland

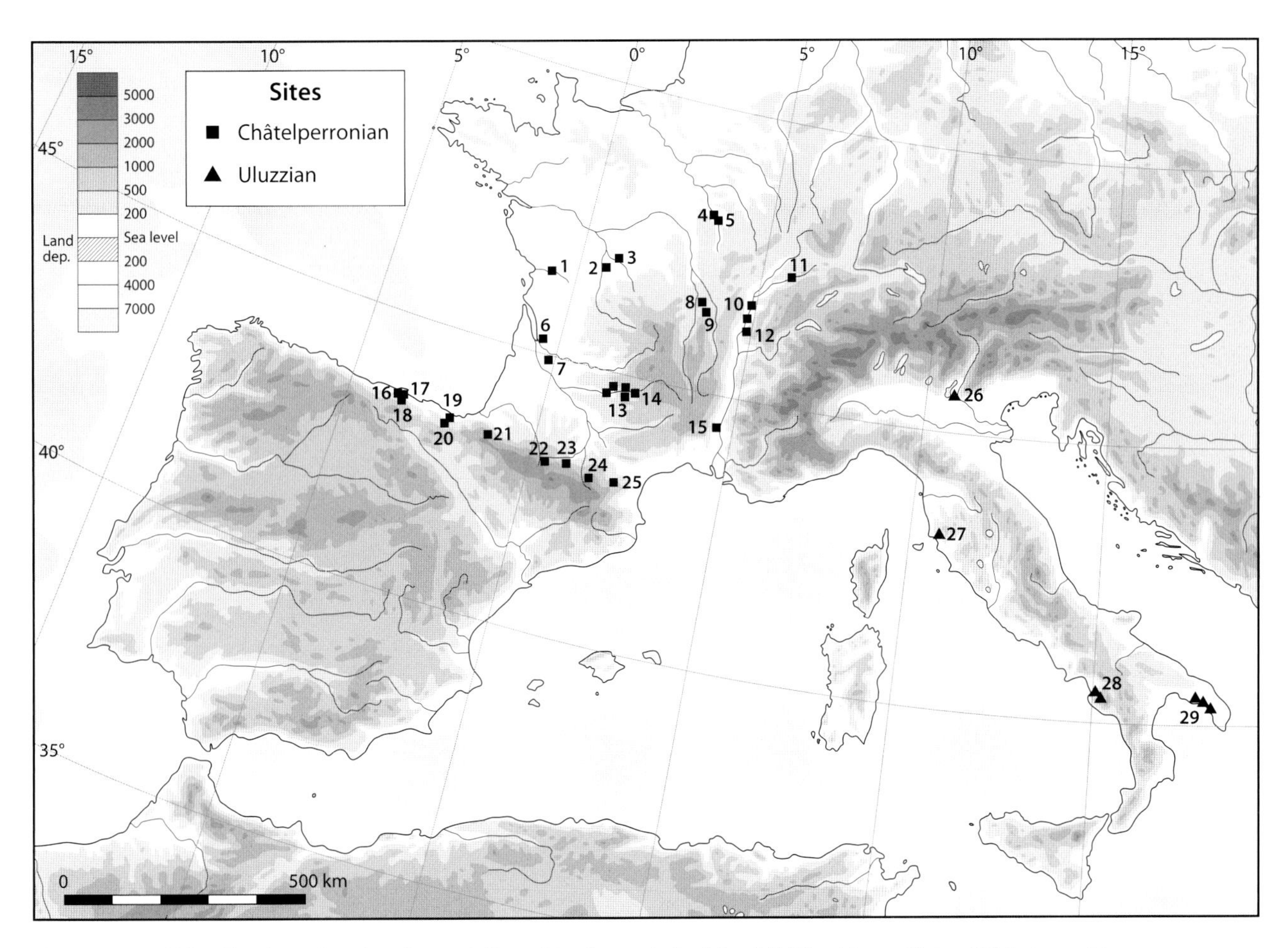

Figure 12.1. *Distribution of major sites attributed to the period of the Middle–Upper Palaeolithic transition: 1) La-Roche-à-Pierrot; 2) Fontenioux; Les Cottés; 3) Les Roches; 4) Grotte du Renne; 5) Roche au Loup; 6) Pair-non-Pair; 7) Camiac; 8) Grotte des Fées, Châtelperron; 9) Theillat; 10) Grotte de la Verpillière, Germolles; 11) Trou de la Mère Clochette; 12) Solutré; 13) Grotte XVI; 14) Roc de Combe; Le Piage; 15) Grotte Figuier; 16) El Pendo; 17) Cueva Morín; 18) El Castillo; 19) Ekain; 20) Labeko Koba; 21) Isturitz; Gatzarria; 22) Gargas; 23) Les Tambourets; 24) Le Portel; 25) Belvis; 26) Grotta Fumane; 27) La Fabbrica; 28) Castelcività; La Cala; 29) Grotta Bernardini; Grotta di Uluzzo; Grotta del Cavallo. (Base map after Collins atlas 1:20,000,000.)*

environment between 38,000 and 34,000 BP (Mussi 1990). Grotta di Paina level 9 (Berrici Hills, north Italy), dated to between 38,600+1400/–1800 BP (UtC-2695) and 37,900±800 BP (UtC-2042), sees a change from dry open grassland to colder, more humid conditions (Broglio 1996, 291). In the Po valley, we see coniferous forest environments; at the foot of the Alps and the Apennines steppe and wooded grassland, with marshy conditions encountered in valley bottoms (Mussi 1990). Further south, steppe conditions are seen over relatively wide areas, with forest communities occurring in local areas at specific times, while in Liguria, between 36,000 and 32,000 BP, more humid, increasingly wooded environments replace cold, dry conditions. Steppic conditions return briefly, only to be replaced once again by more temperate conditions from 30,000 BP (Raynal & Guadelli 1990, 54). Near Naples, at Serino, the Aurignacian is associated with medium arboreal pollen frequencies (30–35 per cent) with pine, birch and alder recorded; by 31,000 BP the environment can be described as temperate and humid.

The fauna of the transition

Franco-Cantabria

The Châtelperronian (Fig. 12.1) has long been considered the classic Middle–Upper Palaeolithic transitional industry in western Europe (see Harrold 1981; Mellars & Gravina 2008; Zilhão 2006). Most of the fifty or so sites at which the industry is recognized are located in the western half of France, although some (e.g. Grotte de la Verpillière (Germolles), Solutré, Chenoves and La Mère Clochette are now being recorded further east (Floass 2003, 276).

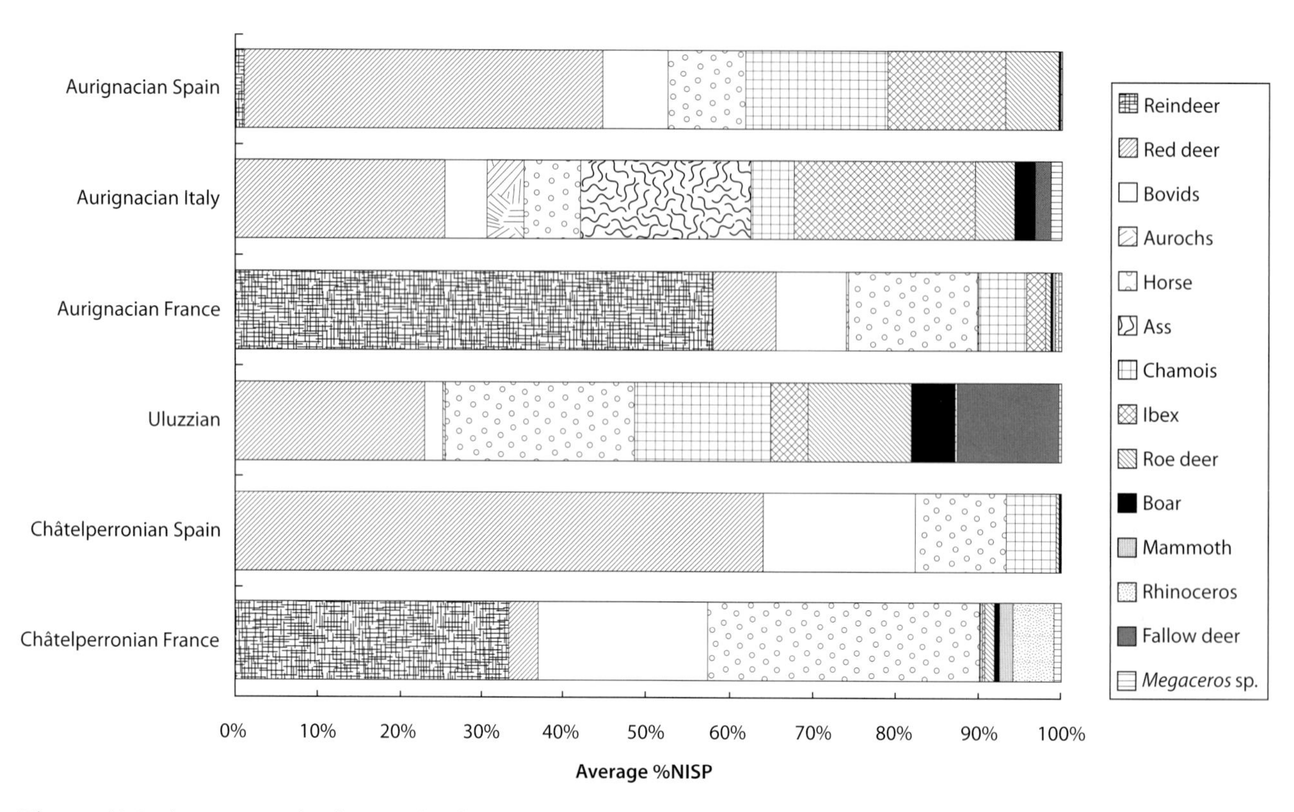

Figure 12.2. *Average species frequencies through the Middle–Upper Palaeolithic transition.*

Table 12.1. *Average large herbivore %NISP frequencies attributable to the 'Transitional' phase (see Fig. 12.2).*

	Châtel-perronian	Uluzzian	Aurignacian		
			France	Italy	Spain
Reindeer	28.51*	0.00	50.10	0.00	0.99
Red deer	8.23	21.20	6.50	24.83	41.92
Bovids	17.34	2.33	7.49	5.06	7.64
Aurochs	0.86	0.20	0.13	4.32	0.00
Horse	28.89	21.32	13.48	6.76	8.78
Ass	0.14	0.00	0.01	20.00	0.00
Chamois	3.35	15.00	5.12	5.03	16.43
Ibex	0.72	4.07	2.01	21.32	13.68
Roe deer	1.38	11.70	0.55	4.76	6.22
Boar	0.41	4.72	0.45	1.97	0.47
Mammoth	2.58	0.00	0.02	0.00	0.00
Rhinoceros	2.88	0.22	0.40	0.00	0.04
Fallow deer	0.00	11.56	0.16	2.18	0.00
Megaceros sp.	0.42	0.23	0.22	1.50	0.00
Average $NTAXA_{herb}$	**5.25**	**6.63**	**5.15**	**5.55**	**5.55**
Average $NTAXA_{carn}$	**2.17**	**4.00**	**2.00**	**3.09**	**1.29**

* France: 33.16, Cantabria .01

There are also Châtelperronian levels at a few sites in northern Spain including, but not exclusively, Cueva Morín, Labeko Koba, Amalda, Ekain and La Güelga. Added to these are broadly contemporary sites such as La Valiña (Sierras Orientais, northwest Spain), Camiac (Gironde) and Grotte des Fées[1] (Châtelperron, Allier), at which carnivore activity clearly plays an important, if not predominant, role in assemblage formation (Arrizabalaga Valbuena & Iriarte Chiapusso 2006; Fernández Rodríguez 2006; Raynal & Guadelli 1990; Zilhão *et al.* 2007).

Recent dating or redating of levels from sites, plus newly discovered occurrences, means that the chronology of the period is constantly changing especially as questions of Châtelperronian/early Aurignacian interstratification are clarified (Bordes 2002; Mellars & Gravina 2008; Zilhão *et al.* 2007). In France, dates range from 40,650± 600 BP (OxA-13621) at Grotte des Fées, Châtelperron, or even 45,100±2100 BP (Gif-101266) at Roc de Combe, to 35,000±1200 BP (GifA-95581) at Grotte XVI (Zilhão *et al.* 2007, 410).

Unfortunately the faunal record for the Châtelperronian in France and Spain is not a rich one. Quantitative faunal material (bone counts [NISP] and/or number of individuals [MNI]) are available from only a few. Sites with very poor faunal assemblages such as Cueva Morín (ΣMNI = 3: Freeman 1973, 25), are included but their significance is minimal other than as an indication of presence/absence and taxonomic richness (NTAXA).

The main taxa reported from French Châtelperronian levels are horse and reindeer (see Fig. 12.2) These occur at almost every site, sometimes in high relative frequencies, even to the degree that there is a general trend for frequencies of one species to fall as the other rises, producing a negative Pearson Correlation Coefficient value of $r = -0.699$. To these species may be added the rarer red deer and bovids, while a few large carnivores, especially the hyaena, also occur in some abundance, representing not only predators but also formidable competitors for resources (Boyle 2007, 306).

In terms of spatial patterning, reindeer frequencies peak in the Dordogne valley. Frequencies average 46 per cent NISP — well in excess of the overall regional/cultural mean of 28.5 per cent. Horse occurs at more sites (Boyle 2007), and its overall 'cultural' mean of 28.89 per cent (see Table 12.1) places it slightly above the reindeer. However the difference between the species averages is not statistically significant, although in the Dordogne river catchment its average of 7.5 per cent (Boyle 2007) makes it a secondary species. In truth horse and reindeer should be regarded as largely equal in importance. The presence of horse in any abundance might, however, be at least partly attributable to carnivore rather than just human activity (Boyle 2007, 307). Pearson Product Moment Correlation values for reindeer, horse and hyaena produce interesting results which, although not 'statistically significant' point to a possible trend in the data: reindeer is negatively correlated with *all* large and medium-sized carnivores including hyaena ($r = -0.382$). Horse, on the other hand, is positively correlated with hyaena ($r = 0.433$), wolf, cave and brown bear, fox and arctic fox, lynx *and* cave lion. Furthermore, as already stated, the relationship between horse and reindeer is a negative one. These observations lead us to ask if we are indeed looking at different faunal assemblage formation agencies. It is an issue which needs to be investigated in detail elsewhere, although it has been touched upon in discussions of sites such as Camiac (Gironde) (Delpech 1999, 70) where the presence of hyaena and associated coprolites suggests a hyaena den with horse (32.43 per cent) and bovids (32.73 per cent) the main prey. Interestingly, it is a pattern observed by Mary Stiner (1994; 2004) in her analysis of Italian Mousterian assemblages:

> Spotted hyaenas consumed horses somewhat more often than did Paleolithic humans, but there are no very significant statistical differences in prey species of proportions eaten by these two predators where they co-existed (Stiner 2004, 773).

When we look at patterns in taxonomic richness[2] and ecological diversity,[3] rather than individual species frequencies, we can see patterning among Châtelperronian assemblages which has been described in detail elsewhere (Boyle 2007). However, to summarize: low taxonomic richness (NTAXA) seen at Châtelperronian sites in the Dordogne/Vézère catchment (Boyle 2007, 307) occurs in meadow and grassland — environments which Shugart (1998) describes as being of low or relatively low primary productivity. Further west and north, primary productivity and associated NTAXA figures are higher — in areas of mixed meadow/grassland and deciduous/evergreen woodland. Where assemblage diversity is concerned, only a small number of locations yield other than diverse assemblages (Boyle 2007, 309). Mirroring the higher taxonomic richness seen in the northern half of the region, sites in this area also yield more diverse assemblages, while more specialized fauna are largely confined to the Dordogne catchment, but even here values remain relatively high (>0.50). As has already been made clear, during the Châtelperronian in France at least there is little evidence for specialization (Boyle 2007; Grayson & Delpech 2002), something which begins to emerge only later during the Upper Palaeolithic in the region (Boyle 1990).

Table 12.2. *Mean regional diversity values.*[4]

		Simpson Diversity Index
Châtelperronian		
	France	0.638
	Spain	0.648
Uluzzian		0.681
Aurignacian		
	France	0.545
	Spain	0.558
	Italy	0.588

In northern Spain faunal assemblages dating to the Châtelperronian period are rare and once again assemblages result from both carnivore and human activity. However, those that have been recorded quantitatively show a tendency for red deer to be the major species, followed by large bovids and horse. At Labeko Koba red deer accounts for 65 per cent of the faunal assemblage in layer IX_{in} (Altuna & Mariezkurrena 2000, 113), about average for the period in the region. At Ekain (10a) red deer also dominates (45.45 per cent) followed by bovids (36.36 per cent) and chamois (18.18 per cent). Even at Cueva Morín where, as already indicated, the Châtelperronian faunal assemblage is small (Freeman 1973; 1981; Straus 1982; 1992) red deer is present, but in this case it does not dominate: Freeman's (1973, 25) MNI values of just one each for horse, bovids and red deer are of limited value but NISP values tell us that the assemblage is dominated by large bovids.

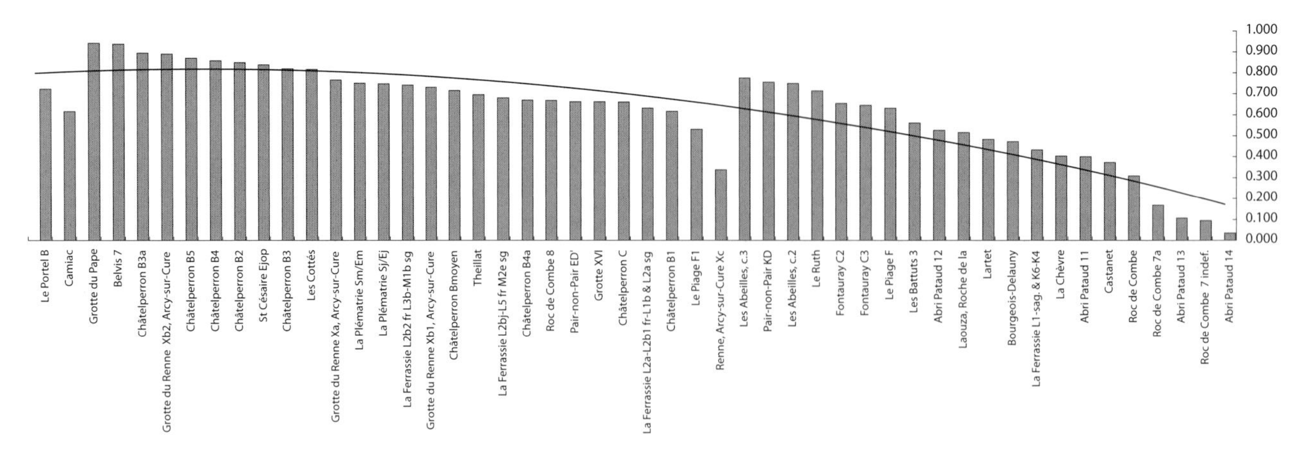

Figure 12.3. *Chart showing change and trend in assemblage diversity during the Middle–Upper Palaeolithic transition (Châtelperronian to Aurignacian) in southwest France.*

In general terms Cantabrian Châtelperronian assemblages are a little richer and more diverse than are those located north of the Pyrenees. We see a slightly higher regional Simpson Index average value (0.648; see Table 12.2) which may be due to any combination of a number of factors including:

1. latitudinal difference, namely that environmental diversity tends to increase with decreasing latitude (see Rosenweig 2003, 102);
2. greater topographic variation in northern Spain, with sites located in areas from narrow coastal belt, mountain and valley;
3. increased contemporary carnivore activity, meaning that at least two sets of top predators were actively forming faunal assemblages;
4. differing human subsistence strategies, neither of which were specialized.

Irrespective of whether one accepts arguments concerning the possible degree of chronological and geographical overlap between the Châtelperronian and early Aurignacian, it is an established fact that, by and large, the Aurignacian follows and/or outlasts the Châtelperronian. Given the possible degree of overlap it is perhaps not surprising that average faunal species frequencies during the Châtelperronian and earliest Aurignacian in southwest France are similar in many ways. However, as the early Aurignacian develops reindeer frequencies frequently rise (Boyle 1990; 2007; Grayson & Delpech 2002; Mellars 2004), with a mean value of 50.10 per cent in contrast to the Châtelperronian 28.51 per cent (see Table 12.1). It becomes the most frequently occurring dominant taxon. Secondary species, red deer, bovids and horse, on the other hand, all display a reverse pattern, i.e. frequencies fall during the earliest Aurignacian rising again in the Later Aurignacian (see Table 12.1). Where the geographic distribution of species is concerned, the difference between the Châtelperronian and early Aurignacian is less clear-cut. Reindeer continues to peak in the Dordogne catchment. Above average horse values are concentrated in the north and west, areas where the species is an important feature throughout the course of the Upper Palaeolithic (Boyle 1990). Woodland species such as roe deer and boar tend towards a more southerly distribution, as does the red deer, traditionally assigned to the woodland component despite being found in a wide range of ecological conditions today (Clutton-Brock *et al.* 1982; Flerov 1960; Straus 1981). This group of taxa occurs at valley sites in much of the region, in areas offering refuge from the increasingly cold conditions prevailing during the early Aurignacian.

Patterning in NTAXA is also not particularly different between the Châtelperronian and earliest Aurignacian. However, while average $\text{NTAXA}_{\text{herbs}}$ is slightly higher in the earlier period, the reverse is true of carnivores.[5] In both cases the difference is minimal. Of greater significance is the chronological trend which can be traced in diversity values as we move from Châtelperronian to Aurignacian (Fig. 12.3).

Although there is a significant amount of variability through time, there is a general trend of decreasing assemblage diversity and evenness (increased specialism?) in southwest France which reaches lowest values during the final stages of the late Upper Palaeolithic – Magdalenian V and VI (Boyle 1990; 2007, 311). Values peak between approximately 37,000 and 35,000 BP.

Similar patterns can be seen in Aurignacian northern Spain, although the range of species with which we are dealing is somewhat different. The most notable difference is the almost complete lack of

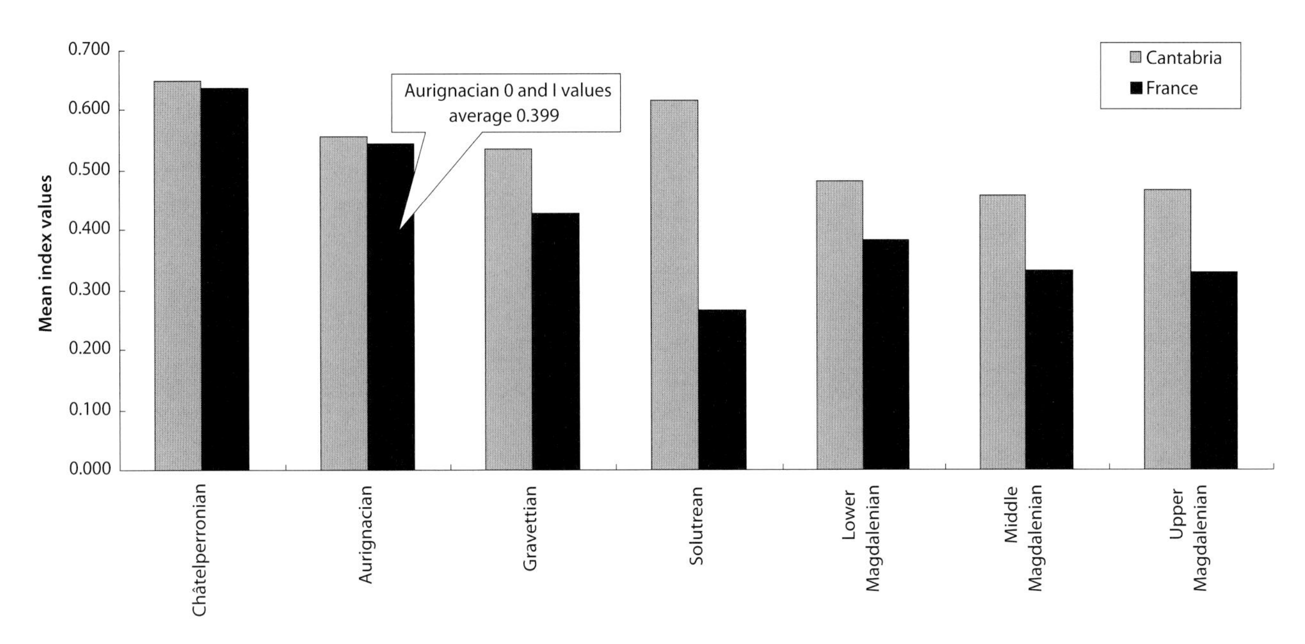

Figure 12.4. *Chart showing chronological change in mean diversity values through the Upper Palaeolithic in the Châtelperronian region.*

reindeer from the faunal spectrum. Instead of reindeer, the most frequently recorded species in the region is red deer. For example, with the exception of level 14 all the Aurignacian levels at El Castillo are dominated by this species (Landry & Burke 2006), although to differing degrees. Based on this observation and the other species present (NTAXA = 12, including three carnivores), Landry & Burke identify three standard Mediterranean faunal components — open, alpine and woodland. At Cueva Morín red deer makes up approximately 40 per cent, followed by roe deer at *c.* 24 per cent of the assemblage identifying a woodland group. Horse and bovids trail at 12 per cent each. As we move through the early Aurignacian into 'fully' Aurignacian periods and even onto the Gravettian we see a greater role for horse and ibex, e.g. at Cueto de la Miña and La Riera (Quesada López 2006, 409), so that as time progresses cold faunas become more significant (Álvarez-Lao & García 2009). Even reindeer is seen at El Castillo (Aurignacian), Santimamiñe (Aurignacian) and Lezetxiki (both Protoaurignacian, 42,500–40,000 BP and Aurignacian), as is the mammoth, at L'Arbreda (38,500 BP), and Figueira Brava (30,930 BP) (Álvarez-Lao & García 2009).

Meanwhile diversity figures fall slightly in north Spain from Châtelperronian levels, although figures remain relatively high throughout the Upper Palaeolithic, there being no real evidence for a trend of decreasing diversity such as we see in southwest France (Fig. 12.4). The average during the Aurignacian is 0.558.

Italy

During the Italian Middle–Upper Palaeolithic transition we see the Uluzzian (see Fig. 12.1). It is a flake-based industry with a low blade index (d'Errico *et al.* 1998) and crescent-shaped geometric microlith *fossiles directeurs* (Riel-Salvatore 2009, 379). It occurs in a region which covers much of the country, although with a southern bias, and extends into Greece at Klisoura Cave (Mussi 2001; Palma di Cesnola 2001; Peresani 2008). It has been dated to between >31,000 and *c.* 35,000 BP at Castelcività and Grotta del Cavallo and more recently, although controversially, to 43,000 to 40,000 BP at Fumane (Higham *et al.* 2009). Given that the Italian faunal sample size with which we are dealing is even smaller than the Châtelperronian data set, more detailed work on the period is required and it can only be hoped that future effort will increase the size of our sample.

Based on the data available however, some observations may be made. Primary amongst these is the fact that there is no sign of general predominance by a single species across the Uluzzian region. Instead we see different species dominating at different sites. This is a pattern which continues that of the late Mousterian, although prior to the Uluzzian there is a tendency for red deer to occur at slightly more sites — albeit in relatively low frequencies. For example, at Grotta Barbara, during the Late Mousterian of MIS-3, 12 species are recorded (8 herbivores, 4 carnivores) of which the predominant red deer makes up just 32 per cent (Caloi & Palombo 1989). The averages listed in Table 12.1 reflect similar observations during the Uluzzian, when frequencies are more evenly distributed than are those for the other periods considered here. Nevertheless, our current data show that, over all, horse takes first place followed closely by red deer and then chamois and roe deer. Even at La Cala, where the faunal spectrum is rather different in that the major species is fallow deer (*Dama dama*: 43.8 per cent), the temperate woodland group makes up the major component of the fauna, emphasizing the fact that the difference between the Châtelperronian and Uluzzian is primarily one of open *versus* woodland, cold *versus* temperate. In sum, three groups of species are seen: (1) woodland (of roe deer, boar, red deer and chamois); (2) open grassland (horse); and (3) valley side and bottom where we see ibex and some woodland-group species — mainly boar and chamois.

Diversity, evenness and richness figures all reflect the increased temperate nature of the faunal spectrum. In general average taxonomic richness is higher than it is further west: site values range from 2 to 11, with a period mean value of 10.63 which is significantly higher than across the Châtelperronian range (7.42). However, the Uluzzian as a whole sees fewer large mammal species than does the Châtelperronian: 11 are listed in Table 12.1 as opposed to the 13 for Franco-Cantabria.

Higher NTAXA values occur in the southern part of the range, as do the higher values of measures of assemblage diversity. More open conditions are seen towards the west and further north, and here figures are somewhat lower. However, the environment is not one of steppe or tundra conditions of the sort invoked by much of the Châtelperronian record. Indeed it can be argued that the faunal spectrum typical of much of western Europe north and west of the Alps is lacking. Instead the key feature of the Uluzzian is the degree to which temperate woodland species as a group rather than individual taxa predominate, with horse indicating a second cluster of more open but not steppe meadow environments. Indeed in a quarter of the assemblages from Castelcività, La Cala and La Fabbrica the major species make up less than 40%NISP, in contrast to the Châtelperronian for which only one tenth of assemblages have a dominant species making up less than 40 per cent. Such relatively evenly distributed species frequencies account for a mean diversity value of 0.681, with extremes ranging from 0.218 (La Fabbrica[6]) to 0.828 (Castelcività).

During the Aurignacian, as in Cantabrian Spain, the red deer dominates at several sites although the average value of just 24.83 per cent is not high. Unlike Cantabria and France, horse and bovids do not form the major secondary component. Instead we see relatively high regional frequencies of ass (e.g. at Grotta Paglicci where it is followed at a distant second by aurochs and then horse and ibex: Palma di Cesnola 2006, 356), and ibex. An ibex- and chamois-rich pocket can be found in the north at the Grotta Fumane; woodland conditions are more frequently encountered in the west and open conditions in the centre. Meanwhile the site of Fossellone displays high ass values (>53 per cent), with red deer in second place at almost 37 per cent. At La Cala the fallow deer-dominated Uluzzian level (14) is replaced by red deer, with fallow taking on a secondary role.

Meanwhile, during the earliest Aurignacian, average taxonomic richness declines — to a value of 8.64, which is still higher than in France and Spain. During this period the trends observed are a reversal of the Uluzzian situation, with higher taxonomic richness towards the north, this time reaching 11 at the La Fabbrica 3–4. Diversity values also change from earlier, the period mean falling to 0.588. However the decrease is not sufficient to justify a claim of increasing assemblage specialization through the period. Only at La Grotta Paglicci do we have assemblage diversity values which might point to 'specialization', for here in level 24 BA we see only ass (100%NISP) and in 24 A3 ass and horse (75%NISP and 25%NISP respectively). Elsewhere diversity and evenness values reflect the higher NTXA values: at Fumane Simpson diversity values range from 0.825 (A3) to 0.747 (A1–2 and D2–3), while at Grotta della Cala values fluctuate between 0.572 (level 10) and 0.761 (level 12).

Regional and chronological contrasts

When it comes to regional and chronological patterning observed in the faunal resource base of the Transitional Period, several things emerge.

The difference between the Châtelperronian and early Aurignacian in France is primarily one of scale rather than type since most of the same species are represented at sites yielding faunal assemblages from the periods in question (Fig. 12.2). Horse and reindeer are the major species in both periods and are the controlling species — taxa whose frequencies appear to determine, at least in part, the composition of the faunal assemblages in the region. However, a chronological trend of increasing reindeer values is seen overall, with gradually declining horse frequencies. An example can be seen at St Césaire where frequencies rise from less that 20 per cent in the Châtelperronian to over 80 per cent during the earliest Aurignacian (Morin 2004, 369). However, pockets in which 'key' species are of great abundance, or are rare or even absent must be treated as what they are — not necessarily typical of the macroregion as a whole.

The mosaic nature of the faunal landscape and environment of the Châtelperronian has already been established (Boyle 1990; 2007): reindeer frequencies are highest in the centre of the Châtelperronian range while the horse is higher in the north and west. Also established is the fact that during the transition primary productivity was relatively high across much of the area under consideration. Pockets of low productivity are restricted to localized areas of montane/rock-face communities, while richer woodland conditions range from open areas with scattered pockets of tree cover through open woodland to deciduous and 'mature' mixed coniferous and deciduous woodland, and are more widely distributed. Detailed analysis of the quantitative data available from Châtelperronian sites (see Boyle 1990; 2007) yields these same two groups, namely (1) the woodland and the rock-face/valley community of red deer, roe deer, boar, chamois and ibex, and (2) the geographically more widespread but taxonomically and ecologically poorer drier, open developing steppe and tundra of horse, bison and reindeer. Meanwhile, in Italy Uluzzian primary productivity appears always to have been even higher — something perhaps to be expected given its more southerly latitude. Average NTAXA values are consistently higher than at Franco-Cantabrian Châtelperronian sites; regional diversity and evenness values are also higher. During the early Aurignacian this difference between the two broadly defined areas becomes even more marked as primary productivity falls slightly in the former Châtelperronian area but remains higher in Italy. In Italy the main difference which we see between the early and later periods is the apparent disappearance of dense, as opposed to open, woodland or forest.

The security and predictability of resources to which Mellars (1985, 277) refers is perhaps one of the more notable ways in which southwest France and the Italian Uluzzian and Aurignacian differ. None of the species encountered in Italy occurs in the huge herds which we associate with the reindeer of southwest France. Similarly, none is as potentially predictable — none follows as predictable and regular a seasonal migration patterns as does the 'French' reindeer, irrespective of whether that pattern relates to long distances or altitude. Instead, the Palaeolithic sites 'astride ... major migration routes' to which Mellars refers (1985, 280) are a thing of the Palaeolithic of southwest France.

Evidence of complexity

Mellars has identified and discussed, both in 1985 and more recently in 2006, several features of the Upper Palaeolithic archaeological record of Franco-Cantabria which he considered to be signs of social complexity not apparent in the Middle Palaeolithic, and the ecological basis of this complexity. Today some of these Upper Palaeolithic, even Late Upper Palaeolithic, characteristics are recognized significantly earlier during either the Initial Upper Palaeolithic or Middle Palaeolithic.

For example, one of these features is the regional abundance of cave and rockshelter art. However the cave art which was once thought of as unique to the region is now known elsewhere, some of it relatively close to hand, e.g. Balzi Rossi (Liguria, northern Italy; Côa Valley, Portugal), while other is further afield — even as far as Kapova Cave, Russia: Solodeynikov 2005). Some of it is also significantly earlier than had been believed possible in the past. Thus the early dates obtained from the Grotte Chauvet (Pont-d'Arc, Ardèche) for the parietal art provide additional intriguing information concerning the ecological complexity of the Early Upper Palaeolithic. Although not providing any information regarding species frequency, nor any direct data of ecological nature, the presence of rhinoceros and bison dated to 30,790±600 BP (GifA-95133) and 30,940±610 BP (GifA-95126) respectively (Onoratini 1996, 264), and the predominance of the lion–mammoth–rhinoceros triad among representations (Feruglio 2006, 216), point to an environment with a significant large or mega-herbivore component. The dates obtained at the site show that at least some of the arguments put forward by Mellars in 1985 concerning Late Upper Palaeolithic ecological complexity apply equally to the Early Upper Palaeolithic and other areas away from the Classic Périgord region.

Another feature of Upper Palaeolithic complexity is gradually increasing evidence for the exploitation of marine and freshwater resources, including the use of shell as a raw material — as shell beads in particular. During the Aurignacian at Cueva Morín at least 10 species of marine molluscs are reported by Freeman (1981, 148). The existence of an even richer marine shellfish collection at Fumane (Broglio 1996), where more than 40 species are reported (>480 MNI: 80 per cent of Adriatic origin, 20 per cent Tyrrhenian), illustrates a degree of complexity, be it economic or social, frequently not expected during the earliest Upper Palaeolithic. Use of marine shell in the manufacturing of shell beads is a regular phenomenon of the Protoaurignacian in the north Mediterranean region (Joris

& Street 2008). Again this is particularly noticeable at Fumane Cave since the site is some 200 km from the coast (Joris & Street 2008, 785), thereby implying that there was some significance to the shells and that travel, or perhaps even trade/exchange, might have been involved in their procurement. At the coastal site of Mocchi in Liguria, northern Italy, we see both shell jewellery and the use of red ochre in graves. Meanwhile by 30,000 BP marine fish appear to have been exploited to a limited extent at the site of Cueto de la Miña (Adán *et al.* 2009). The possibility of such practices at this or a slightly earlier date elsewhere should therefore not be discounted.

Where exploitation of freshwater resources is concerned, as early as the mid 1920s these were being invoked when explaining site location. In 1925 Henri Martin referred to fishing as one factor explaining the location of the Aurignacian deposits at La Quina, where reindeer dominates the assemblage (Martin 1925, 13). Much more recently, significant fish faunas have been identified at Mousterian sites in Spain — well before the Late Upper Palaeolithic with which these resources are more commonly associated. In 1992 Rosello Izquierdo reported on a Mousterian fish assemblage from Cueva Millán dating to 37,600 BP and totalling, among the six species present, 279 bones of trout (*Salmo truta fario*, *N* = 198), Iberian nase (*Chondrostoma polylepsis*, *N* = 52) and eel (*Anguilla anguilla*, *N* = 29).

However, and despite cases such as Cueva Millán, it remains the case that there is little evidence of fish exploitation, particularly salmon (*Salmo* sp.), a species which features highly in Mellars's ecological complexity paper. Such fish remains are exceptionally rare during the Early Upper Palaeolithic and are as good as absent at Uluzzian and Châtelperronian sites.[7] For example, Delpech's data tables (1983) provide evidence of fish at only Roc de Combe where 28 bones are recorded throughout the Aurignacian (none in the Châtelperronian). This represents just 0.174 per cent of the complete faunal record from all the levels at the site. At Le Flageolet I the record is slightly richer. Of the 39 fish bones recorded, 13 are from the Aurignacian (0.263 per cent of the complete site faunal assemblage). To these we can add, from the list of Palaeolithic sites yielding freshwater fish provided by Cleyet-Merle (1990), the Typical Aurignacian at Grotte du Pape (Brassempouy) and Abri Pataud. Cleyet-Merle's detailed account emphasizes that, during the Upper Palaeolithic, fish do not really appear in any significant amount until the Magdalenian. However, although fish exploitation has been invoked as just one of the adaptive advantages enjoyed by anatomically modern humans (Adán *et al.* 2009), we now know, based on isotope analysis of Neanderthal bone at St Césaire, that the Châtelperronian diet included fish (Balter & Simon 2006).

Possible evidence of bird hunting at sites such as La Quina (Martin 1925) also broadens the subsistence base beyond what is commonly associated with the Middle–Upper Palaeolithic transition. Use of bird bones as a possible raw material is suggested for the Châtelperronian at Grotte du Renne, Arcy. At Isturitz, during the Aurignacian, bird bones are used as a raw material in the manufacturing of a flute. At Gatzarria bird long bones are also used, and are incised. A decorated example has also been recovered from Tuto de Camalhot (Laroulandie 2004, 167). However these are far outweighed by Later Upper Palaeolithic examples; most evidence for use of bird bones comes from Magdalenian contexts.

One difference between the Middle and Upper Palaeolithic (especially Late Upper Palaeolithic) which in all probability reflects a development in the social system of these hunter-gatherers is the apparent lack of clearly defined 'home bases' or residential sites during the earlier period. Although several sites which provide evidence of quite sophisticated use of space have been reported (see Yar & Dubois 1999 for a useful compendium), the degree to which these arrangements, including those at Lazaret (Alpes-Maritimes), Baume Bonne (Alpes-de-Haute-Provence) and Rigabe (Var), are the result of human activity remains unclear, while the presence of hearths such as we see at Pech-de-L'Azé II (Dordogne) and La Ferrassie (Dordogne) does not prove long-term residential use of a site. Regular indications of this sort of patterning are not seen until later, during the Upper Palaeolithic. A prime example of clear-cut habitation site structure is the series of well-documented Magdalenian open-air 'pavement sites' in the Isle Valley, Dordogne (Gaussen 1980; Koetje 1987; Yar & Dubois 1999). By this point the complexity of society can not be denied. The large size of some Late Upper Palaeolithic sites is also invoked by Mellars as reflecting social complexity, especially when coupled with the significant cave art at several sites in the region. Large sites are, however, known from much earlier periods. The Mousterian open-air site of Mauran (Haute-Garonne), for example, is estimated to have been over 1000 m^2, while the Grotte Figuier in the Ardèche extends over more than 1500 m^2 (Boyle 1998).

One of the main differences between the two periods considered here which remains a topic of sometimes heated debate is the degree to which subsistence specialization can be assigned to the Upper Palaeolithic in particular. In southwest France this discussion of specialization focuses almost entirely on the reindeer.

Today the fact that Neanderthals were capable of and practised comparable systematic hunting strategies is generally accepted (Adler *et al.* 2006, 104; Burke 2000 and papers therein). Indeed such practices, including 'selective hunting', are now being traced even earlier, e.g at 'Ubeidiya in Israel (Gaudzinski 2004), at Arago, France (de Lumley *et al.* 2004) and Isernia la Pineta, Italy (Thun Hohenstein *et al.* 2009) to name just three examples. As a result we can not invoke the appearance of systematic hunting to explain or characterize the Middle–Upper Palaeolithic transition. As Gaudzinski-Windhauser & Niven (2009) have recently stated quite explicitly

> one pattern that does emerge among Middle Paleolithic faunal assemblages from numerous cave and open-air sites is the predominance of a single prey species, often represented by a minimum number of individuals ... of up to 100 animals or more (2009, 100).

The site of Mauran, mentioned above, is a case in point, as is that of Salpêtre-de-Pompignan, in the Gard, where we see 80 per cent ibex, plus a significant bird assemblage (Meignen 1979; Vilette *et al.* 1983). Therefore the numerical dominance of an assemblage by one species is not something which can be used to characterize the Upper Palaeolithic. In fact the degree of difference in faunal assemblage specialization between the Middle and Early Upper Palaeolithic is not great. Middle Palaeolithic assemblages frequently display high relative frequencies of a single species. Similarly, both periods see unspecified and diverse assemblages. At the Mousterian site of St Marcel (Ardèche), red deer makes up more than 80 per cent of the fauna (Daujeard 2004). At sites such as Balazuc, above the Ardèche river, preferential exploitation of the young, this time of ibex (*Capra ibex*) is also seen. In Aurignacian levels at Caminade Est (Dordogne), the main species (bovids) makes up only 39 per cent of the assemblage. Even at the Abri Pataud, a site most often associated with high reindeer frequencies, level 8 (dated to 31,800 BP) yields 46.2 per cent bovids and only 9.7 per cent reindeer (Chiotti 2000, 244). At Pair-non-Pair, D (Gironde), reindeer 'dominates' at only 40 per cent of the large herbivores (Boyle 1990, 316). During the Late Upper Palaeolithic there are still some 'surprisingly' diverse assemblages in southwest France: at Gabillou final Magdalenian reindeer totals approximately 49 per cent, at Rond-du-Bary only 42 per cent (Boyle 1990, 327).

Conclusion

Overall the emerging feature is one of diversity during the transition. The macroregional fauna is a rich heterogeneous one across the region as a whole, even where the number of sites and occurrences is still low. Taxonomic richness is high: there are many available herbivores; the carnivore population is still significant, and in many cases the structure of individual faunal assemblages is diverse. 'Specialized' assemblages are seen, but generalized — more diverse — assemblages are the norm. Pockets of specialized faunas do appear to exist on each side of, and during, the Middle–Upper Palaeolithic transition in Italy, Spain and France, but it is not until significantly later that we begin to see, in southwest France and Italy at least, subsistence strategies developed to a degree that frequently resulted in specialized faunal assemblages, i.e. assemblages clearly dominated by a single taxon (Boyle 1990, 218–20) and lithic tool kits. In their study of late Pleistocene stone tool assemblages from southern Italy, for example, Riel-Salvatore & Barton (2004) see no signs of techno-economic change in stone-tool assemblages during the transition and Early Upper Palaeolithic. Instead they find significant signs of change coinciding with the Late Upper Palaeolithic, including 'rather different land-use strategies' after the Last Glacial Maximum (Riel-Salvatore & Barton 2004, 268) — an observation which relates well to the Late Upper Palaeolithic signs of change seen in southwest France, for it is not until later that we can identify patterns of faunal assemblage formation processes which result in the overwhelming predominance (>90 per cent) of a certain species (in southwest France usually, but not always, the reindeer) — so high a predominance in fact that it was often not necessary to fully exploit or process the carcasses of animals killed (Boyle 1994; 1997). It is during the Late Upper Palaeolithic of southwest France, that area of the world which has always held an intense fascination for Paul Mellars, that we see evidence of true regional specialization in reindeer, horse, bison or saiga hunting (see Boyle 1990), even when faunal assemblages include a surprisingly high number of taxa. But this we see only during the world of the Solutrean and subsequent Magdalenian, a time when it is frequently tempting to view the Upper Palaeolithic as a distorted mirror image of that of the reindeer hunters of northern latitudes during historic periods — an image totally inappropriate to the other regions considered in this paper and one which itself needs to be reconsidered in detail elsewhere if we are to get to grips with what it really means to be 'specialized'. However, the fact that the degree of 'specialization' seen in southwest France appears to be largely restricted to this area (it is not seen in Spain and Italy to anything like the same extent) simply reinforces Paul Mellars's 1985 contention that southwest France is special.

Acknowledgements

I should like to thank Clive Gamble for comments and suggestions on an earlier draft of this paper, an anonymous reader for additional comments and J.M. Maíllo Fernández for helping me to track down references to the Spanish faunal material.

Notes

1. Some of Zilhão *et al.* 2007's views are supported by my own examination of the small old assemblage from the site held at the Natural History Museum in London in the 1980s.
2. NTAXA or the Number of Species (taxa) recorded in a faunal assemblage, see Grayson 1984.
3. Here measured using the Simpson Indices of Diversity and Evenness: the higher the value the greater the diversity/evenness, see Boyle 2007, 305; Grayson 1984; Magurran 2004.
4. These values are not of course definitive: as new data are added so values will change, but the trends are valid.
5. These figures exclude the taxonomically rich (NTAXA = 18, 12 herbivores, 6 carnivores) faunal assemblage from La Grotte Marie (Hérault) which is dated to 31,450±270 BP (Crochet *et al.* 2007) but has not been assigned to an industry as the tool assemblage is poor.
6. La Fabbrica yields a horse-dominated assemblage (82.35%), lacking a woodland component; it is the only site where we see such an assemblage.
7. Additional evidence for fish at French Late Mousterian sites (Cleyet-Merle 1990, 22) includes that from Vallières (Tourraine), Pair-non-Pair FF' (Gironde), Grotte Vaufrey VIII (Dordogne), Salpêtre-de-Pompignan (Gard) and La Baume (Gigny-sur-Suran, Doubs). At Cueva Morín, in Cantabrian Spain, frequencies are higher. The main species are trout, eel and salmon (Cleyet-Merle 1990, 23).

References

Adán, G.E., D. Álvarez-Lao, P. Turrero, M. Arbizu & E. García-Vásquez, 2009. Fish as diet resource in north Spain during the Upper Paleolithic. *Journal of Archaeological Science* 36(3), 895–9.

Adler, D.S., G. Bar-Oz, A. Belfer-Cohen & O. Bar-Yosef, 2006. Ahead of the game: Middle and Upper Palaeolithic hunting behaviors in the southern Caucasus. *Current Anthropology* 47(1), 89–118.

Altuna, J. & K. Mariezkurrena, 2000. Macromamíferos del yacimiento de Labeko Koba (Arrasate, País Vasco), in *Labeko Koba (País Vasco). Hienas y humanos en los albores del Paleolítico Superior*, eds. A. Arrizabalaga & J. Altuna. Aranzadi: Sociedad de Ciencias, 107–81. = issue 52 of *Munibe.*

Álvarez-Lao, D.J. & N. García, 2009. Chronological distribution of Pleistocene cold-adapted large mammal faunas in the Iberian Peninsula. *Quaternary International* 202(2), 120–28.

Arrizabalaga Valbuena, A. & M.J. Iriarte Chiapusso, 2006. El Castelperroniense y otros complejos de transición entre el Paleolítico medio y el superior en la Cornisa Cantábrica: algunas reflexiones. *Zona Arqueológica* 7, 359–70.

Balter, V. & L. Simon, 2006. Diet and behavior of the Saint-Césaire Neanderthal inferred from biogeochemical data inversion. *Journal of Human Evolution* 51(4), 329–38.

Bocquet-Apel, J.-P. & P.-Y. Demars, 2000. Population kinetics in the Upper Palaeolithic in western Europe. *Journal of Archaeological Science* 27(7), 551–70.

Bordes, J.-G., 2002. Les interstratifications Châtelperronien/Aurignacien du Roc de Combe et du Piage (Lot, France). Analyse taphonomique des industries lithiques: conséquences archéologiques. Unpublished doctoral thesis, Université de Bordeaux I.

Boyle, K.V., 1990. *Upper Palaeolithic Faunas from South-West France: a Zoogeographic Perspective.* (British Archaeological Reports, International Series 557.) Oxford: BAR.

Boyle, K.V., 1994. La Madeleine (Tursac, Dordogne). Une étude paléoéconomique du paléolithique supérieur. *Paléo* 6, 55–77.

Boyle, K.V., 1997. Late Magdalenian carcase management strategies: the Périgord data. *Anthropozoologica* 25–6, 287–94.

Boyle, K.V., 1998. *The Middle Palaeolithic Geography of Southern France: Resources and Site Location.* (British Archaeological Reports, International Series 723.) Oxford: BAR.

Boyle, K.V., 2007. Changing biodiversity and complexity across the Middle–Upper Palaeolithic transition, in *Rethinking the Human Revolution: New Behavioural and Biological Perspectives on the Origin and Dispersal of Modern Humans*, eds. P. Mellars, K. Boyle, O. Bar-Yosef & C. Stringer. (McDonald Institute Monographs.) Cambridge: McDonald Institute for Archaeological Research, 303–14.

Broglio, A., 1996. Le Paléolithique supérieur en Italie du Nord (1991–1995), in *Le Paléolithique Supérieur Européen. Bilan Quinquennal 1991–1996*, ed. M. Otte. (ERAUL 76.) Liège: ERAUL, 289–304.

Brown, J.H., 1995. *Macroecology*. Chicago (IL): University of Chicago Press.

Burke, A. (ed.), 2000. *Hunting in the Middle Palaeolithic.* (*International Journal of Osteoarchaeology*, Special Issue 10(5).) New York (NY): John Wiley & Sons.

Caloi, L. & M.R. Palombo, 1989. The würmian mammalian fauna of Grotta Barbara (Monte Circeo): palaeoeconomical and environment conditions data/La mammalofauna würmiana di Grotta Barbara (Monte Circeo): implicazioni paleoeconomiche e paleoambientali. Hystrix. *Italian Journal of Mammaology* 1, 95–105.

Chiotti, L., 2000. Lamelles Dufour et grattoirs aurignaciens (careens et à museau) de la couche 8 de l'abri Pataud, les Eyzies-de-Tayac, Dordogne. *L'Anthropologie* 104, 239–63.

Cleyet-Merle, J.-J., 1990. *La Préhistoire de la Pêche*. Paris: Errance.

Clutton-Brock, T.H., F.E. Guinness & S.D. Alban, 1982. *Red Deer: Behaviour and Ecology of Two Sexes*. Edinburgh: Edinburgh University Press.

Crochet, J.-Y., J. Gence, N. Boules *et al.*, 2007. Nouvelles donneés paléoenvironnementales dans le Sud de la France vers 30000 ans ^{14}C BP: le cas de la grotte Marie (Hérault). *Comptes Rendus Palevol* 6(4), 241–51.

Dari, A. & J. Renault-Miskovsky, 2001. Études paléoenvironnementales dans la grotte «El Castillo» (Puente Viesgo, Cantabrie, Espagne). *Espacio, Tiempo y Forma, Serie 1. Prehistoria y Arqueología* 14, 121–44.

Daujeard, C., 2004. Stratégies de chasse et modalités de traitement des carcasses par les Néanderthaliens de la grotte Saint-Marcel, Ardèche (fouilles R. Gilles, ensemble 7). *Paléo* 16, 49–70.

Delpech, F., 1983. *Les faunes du Paléolithique Supérieur dans le sud-ouest de la France.* (Cahiers du Quaternaire 6.) Paris: CNRS.

Delpech, F., 1999. La chasse au bison dans le sud-ouest de la France au cours du Würm: choix humain ou contraintes paléoenvironnementales?, in *Le Bison: Gibier et Moyen de Subsistance des Hommes du Paléolithique aux Paléoindiens des Grandes Plaines,* eds. J.-Ph. Brugal, F. David, J.G. Enloe & J. Jaubert. Antibes: Editions APDCA, 63–84.

Demars, P.-Y., 2008. Paléogéographie des chasseurs de l'Europe du Paléolithique supérieur: repartition et specialisation des sites. *L'Anthropologie* 112, 157–67.

d'Errico, F. & M.F. Sánchez Goñi, 2003. Neandertal extinction and the millennial scale climatic variability of OIS 3. *Quaternary Science Reviews* 22(8–9), 769–88.

d'Errico, F., J. Zilhão, M. Julien, D. Baffier & J. Pelegrin, 1998. Neanderthal acculturation in western Europe? A critical review of the evidence and its interpretation. *Current Anthropology* 39, S1–S44.

Fernández Rodríguez, C., 2006. De humanos y carnívoros: la fauna de macromamíferos de la cueva de A Valiña (Castroverde, Lugo). *Zona Arqueológica* 7, 291–302.

Feruglio, V., 2006. De la faune au bestiaire: la grotte Chauvet-Pont-d'Arc, aux origins de l'art parietal paléolithique. *Comptes Rendus Palevol* 5(1–2), 213–22.

Finlayson, C., 2004. *Neanderthals and Modern Humans: an Ecological and Evolutionary Perspective.* Cambridge: Cambridge University Press.

Flerov, K.K., 1960. *Fauna of the USSR. Mammals,* vol. I: *Musk Deer and Deer.* Jerusalem: Israel Program for Scientific Translation.

Floass, H., 2003. Did they meet or not? Observations on Châtelperronian and Aurignacian settlement patterns in eastern France, in *The Chronology of the Aurignacian and of the Transitional Technocomplexes: Dating, Stratigraphies, Cultural Implications. Proceedings of Symposium 6.1 of the XIVth Congress of the UISPP (University of Liège, Belgium, September 2–8 2001),* eds. J. Zilhão & F. d'Errico. Lisbon: Instituto Portugués de Arqueologia, 273–87.

Freeman, L.G., 1973. The significance of Mousterian faunas from Paleolithic occupations in Cantabrian Spain. *American Antiquity* 38, 3–44.

Freeman, L.G., 1981. The fat of the land: notes on Paleolithic diet in Iberia, in *Omnivorous Primates: Gathering and Hunting in Human Evolution,* eds. R.S.O. Harding & G. Teleki. New York (NY): Columbia University Press, 104–65.

Gaudzinski, S., 2004. Subsistence patterns of Early Pleistocene hominids in theLevant—taphonomic evidence from the 'Ubeidiya Formation (Israel). *Journal of Archaeological Science* 31(1), 65–75.

Gaudzinski-Windhauser, S. & L. Niven, 2009. Hominin subsistence patterns during the Middle and Late Paleolithic in northwestern Europe, in *The Evolution of Hominin Diets: Integrating Approaches to the Study of Palaeolithic Subsistence.* London: Springer, 99–111.

Gaussen, J., 1980. *Le Paléolithique supérieur de plein air en Périgord (industries et structures d'habitat), secteur Mussidan Saint-Astier, moyenne vallée de l'Isle.* (Gallia Préhistoire suppl. 14.) Paris: CNRS.

Gioia, P., 1990. An aspect of the transition between Middle and Upper Palaeolithic in Italy: the Uluzzian, in *Paléolithique moyen recent et Paléolithique supérieur ancien en Europe: ruptures et transitions. Examen critique des Documents archéologiques, Actes du Colloque International de Nemours 9–11 mai 1988,* ed. C. Farizy. (Mémoires du Musée de Préhistoire d'Ile-de-France 3.) Nemours: Musée de Préhistoire d'Ile-de-France, 241–50.

Grayson, D.K., 1984. *Quantitative Zooarchaeology: Topics in the Analysis of Archaeological Faunas.* London: Academic Press.

Grayson, D.K. & F. Delpech, 2002. Specialized early Upper Palaeolithic hunters in southwestern France? *Journal of Archaeological Science* 29, 1439–49.

Grayson, D.K. & F. Delpech, 2008. The large mammals of Roc de Combe (Lot, France): the Châtelperronian and Aurignacian assemblages. *Journal of Anthropological Archaeology* 27(3), 338–62.

Harrold, F.B., 1981. New perspectives on the Châtelperronian. *Ampurias* 43, 1–51.

Higham, T., F. Brock, A. Peresani, A. Broglio, R. Wood & K. Douka, 2009. Problems with radiocarbon dating during the Middle to Upper Palaeolithic transition in Italy. *Quaternary Science Reviews* 28(13–14), 1257–67.

Hutson, M.A., 1994. *Biological Diversity: the Coexistence of Species on Changing Landscapes.* Cambridge: Cambridge University Press.

Joris, O. & M. Street, 2008. At the end of the ^{14}C time scale—the Middle to Upper Paleolithic record of western Eurasia. *Journal of Human Evolution* 55(5), 782–802.

Koetje, T.A., 1987. *Spatial Patterns in Magdalenian Open Air Sites from the Isle Valley, Southwestern France.* (British Archaeological Reports, International Series 246.) Oxford: BAR.

Landry, G. & A. Burke, 2006. El Castillo: the Obermaier faunal collection. *Zona Arqueológica* 7, 704–13.

Laroulandie, V., 2004. Exploitation des ressources aviaries durant le Paléolithique en France: bilan critique et perspectives, in *Petits animaux et sociétés humaines: du complément alimentaire aux ressources utilitaires,* eds. J.-P. Brugal & J. Desse. Antibes: Éditions APDCA, 163–72.

Leroyer, C., 1990. Nouvelles données palynologiques sur le passage paléolithique moyen-paléolithique supérieur, in *Paléolithique moyen recent et Paléolithique supérieur ancien en Europe: ruptures et transitions. Examen critique des documents archéologiques. Actes du Colloque*

International de Nemours 9–11 mai 1988, ed. C. Farizy. (Mémoires du Musée de Préhistoire d'Ile-de-France 3.) Nemours: Musée de Préhistoire d'Ile-de-France, 49–52.

Lévêque, C. & J.-C. Mounolou, 2001. *Biodiversité, dynamique biologique et conservation*. Paris: Dunod.

de Lumley, H., S. Grégoire, D. Barsky *et al.*, 2004. Habitat et mode de vie des chasseurs paléolithiques de la Caune de l'Arago (600 000–400 000 ans). *L'Anthropologie* 108, 159–84.

Magurran, A.E., 2004. *Measuring Biological Diversity*. Oxford: Blackwell Publishing.

Martin, H., 1925. La station aurignacienne de La Quina (Charente). *Bulletins et Mémoires de la Société d'Anthropologie de Paris* 6(1), 10–17.

Meignen, L., 1979. Le paléolithique moyen en Languedoc oriental. *Bulletin Annual de l'École Antique de Nîmes, nouvelle série* 14, 27–39.

Mellars, P., 1985. The ecological basis of social complexity in the Upper Paleolithic of south-western France, in *Prehistoric Hunter-Gatherers: the Emergence of Cultural Complexity*, eds. T.D. Price & J.A. Brown. Orlando (FL): Academic Press, 271–97.

Mellars, P., 2004. Reindeer specialization in the early Upper Palaeolithic: the evidence from south west France. *Journal of Archaeological Science* 31(5), 613–17.

Mellars, P., 2006. The ecological basis for upper Palaeolithic cave art, in *Miscelánea en Homenaje a Victoria Cabrera*, vol. II, eds. J.M. Maíllo-Fernández & E. Baquedano. Alcala de Henares: Museo Arqueológico Regional, 2–11.

Mellars, P. & B. Gravina, 2008. Châtelperron: theoretical agendas, archaeological facts, and diversionary smoke-screens. *PaleoAnthropology* 2008, 43–64. http://paleoanthro.org/journal/contents_dynamic.asp.

Morin, E., 2004. *Late Pleistocene Population Interaction in Western Europe and Modern Human Origins: New Insights based on the Faunal Remains from Saint-Césaire, Southwestern France.* PhD dissertation. Ann Arbor: University of Michigan. http://www.paleoanthro.org/dissertations/Eugene%20Morin.pdf.

Mussi, M., 1990. Le peuplement de l'Italie à la fin du paléolithique moyen et au début du paléolithique supérieur, in *Paléolithique moyen recent et Paléolithique supérieur ancien en Europe: ruptures et transitions. Examen critique des documents archéologiques. Actes du Colloque International de Nemours 9–11 mai 1988*, ed. C. Farizy. (Mémoires du Musée de Préhistoire d'Ile-de-France 3.) Nemours: Musée de Préhistoire d'Ile-de-France, 251–62.

Mussi, M., 2001. *Earliest Italy: an Overview of the Italian Paleolithic and Mesolithic*. New York (NY): Kluwer Academic/Plenum Publishers.

Onoratini, G., 1996. Le Paléolithique Supérieur dans le basin du Rhône, dans les Alpes et en Provence (1991–1995), in *Le Paléolithique Supérieur Européen. Bilan Quinquennal 1991–1996*, ed. M. Otte. (ERAUL 76.) Liège: ERAUL, 257–68.

Palma di Cesnola, A., 2001. *Le Paléolithique Supérieur en Itale.* Aubenas d'Ardèche: Jérôme Millon.

Palma di Cesnola, A., 2006. L'Aurignacien et le Gravettien ancien de la grotte Paglicci au Mont Gargano. *L'Anthropologie* 110, 355–70.

Pärtel, M., H.H. Bruun & M. Sammul, 2005. Biodiversity in temperate European grasslands: origin and conservation, in *Integrating Efficient Grassland Farming and Biodiversity*, eds. R. Lillak, R. Viiralt, A. Linke & V. Geherman. Tartu: Estonian Grassland Society, 1–14.

Peresani, M., 2008. A new cultural frontier for the last Neanderthals: the Uluzzian in northern Italy. *Current Anthropology* 49, 725–31.

Quesada López, J.M., 2006, Faunas del Auriñaciense y Gravetiense cantábrico: Asturias y Santander. Revisión critica y periodización. *Zona Arqueológica* 7, 406–21.

Raynal, J.-P. & J.-L. Guadelli, 1990. Milieux physiques et biologiques: Quels changements entre 60 et 30 000 ans à l'ouest de l'Europe? in *Paléolithique moyen recent et Paléolithique supérieur ancien en Europe: ruptures et transitions. Examen critique des documents archéologiques. Actes du Colloque International de Nemours 9–11 mai 1988*, ed. C. Farizy. (Mémoires du Musée de Préhistoire d'Ile-de-France 3.) Nemours: Musée de Préhistoire d'Ile-de-France, 53–61.

Riel-Salvatore, J., 2009. What is a 'Transitional' industry? The Uluzzian of southern Italy as a case study, in *Sourcebook of Paleolithic Transitions*, eds. M. Camps & P. Chauhan. New York (NY): Springer Science, 377–96.

Riel-Salvatore, J. & C.M. Barton, 2004. Late Pleistocene technology, economic behavior, and land-use dynamics in Southern Italy. *American Antiquity* 69, 257–74.

Rosello Izquierdo, E., 1992. La ictiofauna musteriense de Cueva Millan (Burgos): consideraciones de indole biologica y cultural contrastados con ictiocenosis paleoliticas cantabricas. *Estudios geologicos* 48, 79–83.

Rosenweig, M.L., 2003. How to reject the area hypothesis of latitudinal gradients, in *Macroecology: Concepts and Consequencies*, eds. T.M. Blackburn & K.J. Gaston. Oxford: Blackwell Publishing, 87–106.

Sémah, A.-M. & J. Renault-Miskovsky, 2004. *L'evolution de la végétation depuis deux millions d'années.* Paris: Éditions Errance.

Serangeli, J., 2006. *Verbreitung der großen Jagdfauna in Mittel- und Westeuropa im oberen Jungpleistozän. Ein Kritischer Beitrag.* (Tübinger Arteiten zur Urgeschichte 3.) Rahden/Westf.: Verlag Marie Leidorf GmbH.

Shugart, H.H., 1998. *Terrestrial Ecosystems in Changing Environments*. Cambridge: Cambridge University Press.

Solodeynikov, A., 2005. Research on the recording of rock raintings in the Kapova Cave (Urals). *International Newsletter on Rock Art* (INORA) 43. http://www.bradshawfoundation.com/inora/techniques_43_1.html.

Stiner, M., 1994. *Honor among Thieves: a Zooarchaeological Study of Neandertal Ecology*. Princeton (NJ): Princeton University Press.

Stiner, M., 2004. Comparative ecology and taphonomy of spotted hyenas, humans, and wolves in Pleistocene Italy. *Revue de Paléobiologie* 23(2), 771–85.

Straus, L.G., 1981. On the habitat and diet of *Cervus elaphus*. *Munibe* 33, 175–82.

Straus, L.G., 1982. Carnivores and cave sites in Cantabrian

Spain. *Journal of Anthropological Research* 38, 75–96.

Straus, L.G., 1992. *Iberia before the Iberians: the Stone Age Prehistory of Cantabrian Spain.* Albuquerque (NM): University of New Mexico Press.

Teyssandier, N., 2008. Revolution or evolution: the emergence of the Upper Paleolithic in Europe. *World Archaeology* 40(4), 493–519.

Thun Hohenstein, U., A. Di Nucci & A.-M. Moigne, 2009. Mode de vie à Isernia La Pineta (Molise, Italie). Stratégie d'exploitation du Bison schoetensacki par les groupes humains au Paléolithique inférieur. *L'Anthropologie* 113, 96–110.

Uzquiano, P., M. Arbizu, J.L. Arsuaga, G. Andan, A. Aranburu & E. Iriarte, 2008. Datos paleoflorísticos en la cuenca media del Nalón entre 40–32 Ka. BP: Antracoanálisis de la Cueva del Conde (Santo Andriano, Asturias). *Revista C. & G.* 22, 121–33.

Vilette, P., C. Mourer-Chauviré & L. Meignen, 1983. Les oiseaux de la Grotte de Salpêtre-de-Pompignan (Gard). *Nouvelles Archives du Muséum d'Histoire Naturelle de Lyon* 21 (suppl.), 45–8.

Yar, B. & P. Dubois, 1999. *Les structures d'habitat au Paléolithique en France.* Montagnac: Éditions Monique Mergoil.

Zilhão, J., 2006. Chronostratigraphy of the Middle-to-Upper Paleolithic transition in the Iberian Peninsula. *PYRENAE* 37(1), 7–84.

Zilhão, J., F. d'Errico, J.-G. Bordes, A. Lenoble, J.-P. Texier & J.-Ph. Rigaud, 2007. Grotte des Fées (Châtelperron, Allier) out une interstratification «châtelperronien-aurignacien» illusoire. Histoire des fouilles, stratigraphie and datations. *Paléo* 19, 391–432.

Chapter 13

Technological Characteristics at the End of the Mousterian in Cantabria: the El Castillo and Cueva Morín (Spain)

Federico Bernaldo de Quirós, Granada Sánchez-Fernández &
José-Manuel Maíllo-Fernández

Although the Mousterian has traditionally been defined typologically as a monotonous techno-complex, this characterization has more recently been refined from both a typological and, above all, technological perspective. The study of Mousterian production systems and operational sequences has demonstrated that the industry is one characterized by a wide variety of ways to manage raw material and has identified a complex range of methods of *débitage*. Terms such as Levalloisian or Mousterian facies, as defined by Bordes and Bourgon (1951), are now meaningless and devoid of content, as many production/knapping techniques have been identified at Mousterian sites, together with broad internal variability in methods. Thus, even though the Levallois concept of *débitage* is the most widely known, and therefore that to which most attention has been directed historiographically (Kelley 1954; Bordes 1980; Boëda 1988), the more recent characterization of operational sequences whose conceptualization reveals a degree of complexity similar to the Levallois deserves mention — the Discoid (Boëda 1993; Terradas 2003), Quina (Bourguignon 1996; 1997), or Laminar methods (Révillion & Tuffreau 1994).

The Mousterian has traditionally been studied from a global perspective — in both geographic and chronological terms — at the expense of a more regional focus. In consequence it has been considered as a techno-complex in which operational sequences are poorly developed from a cognitive point of view and which expresses a limited and repetitive typological array (Klein 2003; Wynn & Coolidge 2004; Lewis-Williams 2005).

A further problem facing the study of the Mousterian is its very definition. As stated above, techno-complexes which cannot necessarily be correlated have been grouped together under the same name. Instead these should be classified differently. For example, the leaf-shaped Mousterian pieces from eastern Europe have little in common with those found in western Europe, or in the Near East or North Africa: all this, if the lithic industry itself is analysed in isolation, without considering the producers of the industry and its implications.

The question of boundaries when defining a techno-complex is a deep-seated problem common to the study of all historical periods and the Mousterian is no exception. Much like the classic debates about the beginning of the Middle Ages and the end of Ancient History, the boundary between the Acheulean and the Mousterian is quite vague as regards its definition. In the Iberian Peninsula, the appearance of Levallios *débitage* methods is taken to indicate the Middle Palaeolithic. More problematically, no characterization exists for an industry that is specific to the end of the Mousterian, this being merely a matter of chronological convention. Thus, Mousterian levels later than 60,000–50,000 BP are taken as belonging to the end of the Mousterian, with all the problems of definition that this implies. Therefore, we should not speak of the Late Mousterian but rather of the End of the Mousterian. In addition it should be noted that the Middle Palaeolithic and the Mousterian are frequently considered to be synonymous, when this is not the case.

Starting from the search for a regional perspective in the study of Mousterian techno-complexes, this paper aims to define the assemblages corresponding to the end of the Mousterian in the Cantabrian region as represented at Cueva Morín (levels 11 and 12) and Level 20e at El Castillo (modern excavations), and then to examine relationships between these and other sites in the region.

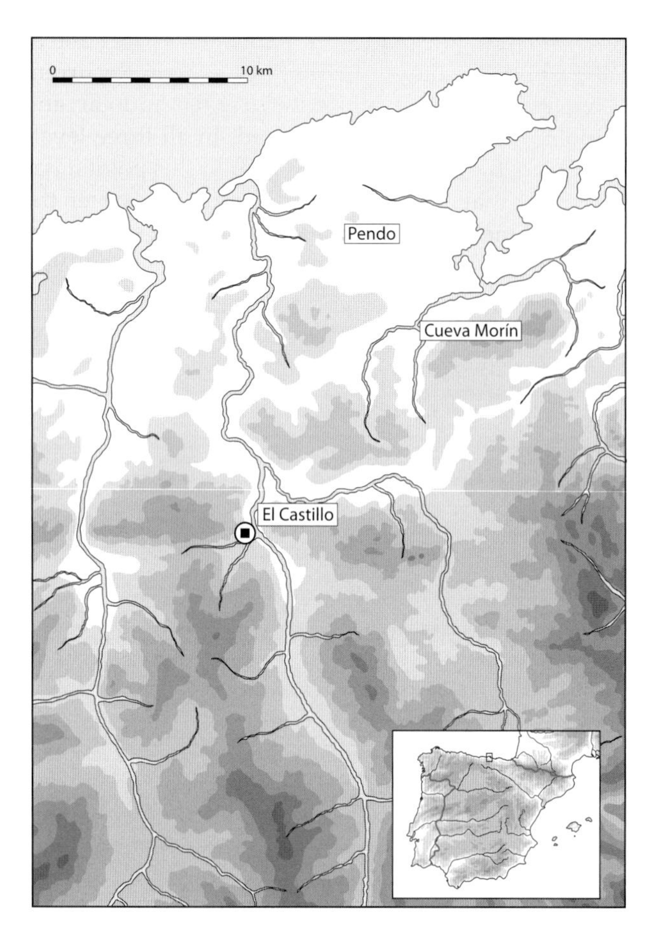

Figure 13.1. *Central Cantabrian region showing locations mentioned in the text.*

El Castillo and Cueva Morín

The sites of El Castillo and Cueva Morín are located in the central part of the Cantabrian region of northern Spain (Fig. 13.1). El Castillo is located in Puente Viesgo, some 30 km south of the present coastline, on the hill of the same name. It was discovered in 1903, and excavated between 1910 and 1914 by H. Obermaier and P. Wernet under the patronage of the Institut de Paléontologie Humaine (Paris). The site shows the most complete stratigraphy in Palaeolithic Europe with 26 stratigraphic units recording occupations from the Acheulean to the Azilian. Further excavations were conducted in the 1980s after a study of the material excavated by Obermaier (Cabrera-Valdés 1984). Recent work focuses on the Mousterian and the transition from the Middle to the Upper Palaeolithic. This work has divided Obermaier's Unit 18 (Aurignacian Delta) into 18b and 18c and identified an industry named the 'Transitional Aurignacian' (Cabrera-Valdés *et al.* 2001; 2006). It has also divided the Mousterian levels of Unit 20 (Obermaier's Alpha Mousterian) into levels 20 a/b, 20c, 20d and 20e.

Level 20a/b has been AMS radiocarbon dated to 39,300±1900 (GifA-89144) and 43,300±2900 (GifA-92506) (Cabrera-Valdés *et al.* 1996), Unit 21 has produced mean ESR dates of 69,300±9100 BP, and Unit 22 gave mean ESR dates of 70,400±9600 BP (Rink *et al.* 1997). More recently, AMS radiocarbon dates on palaeontological remains unearthed during Obermaier's excavations of Unit 20 have produced the following results: 42,100±1500 BP (OxA-10233), 45,700±1700 BP (OxA-10328), >45,700 BP (OxA-10327), >43,800 BP (OxA-10329), 42,900±1400 BP (OxA-10187) and >47,300 BP (OxA-10188). Bracketting their large standard deviations, these dates are consistent with those already available for the site (Stuart 2005; Bernaldo de Quirós *et al.* 2006).

The lithic industry from level 20e is characterized by a medium percentage of sidescrapers, with only a few denticulates and little evidence of Quina-type retouch. It can be classified as Typical Mousterian (Sánchez-Fernández 2005). A characteristic feature is the appearance of cleavers, which led Bordes to propose the Vasconian facies (Bordes 1953). The existence of this facies has subsequently been refuted (Cabrera-Valdés 1983).

Cueva Morín, on the other hand, is located on a small hill 6 km from the present-day coastline, in Villanueva (Cantabria). It was discovered in 1912 by H. Obermaier and P. Wernet. Following a series of early excavations undertaken by J. Carballo and Count de la Vega del Sella (Vega del Sella 1921; Carballo 1923), the most important investigations were carried out by J. González-Echegaray and L.G. Freeman between 1966 and 1969 (González-Echegaray & Freeman 1971; 1973). These excavations revealed 22 stratigraphic levels that cover the Middle and Upper Palaeolithic. Particularly important within these are several burials with pseudomorphs (Freeman & González-Echegaray 1970). The site is currently being re-examined by one of the authors (Maíllo-Fernández 2003).

The lithic industry found in these levels is characterized by an abundance of denticulate tools, a fact which led to its classification within the Denticulate Mousterian facies (Freeman 1971; 1973). There is little evidence of Quina retouch and a medium percentage of sidescrapers.

The Mousterian levels were not subject to radiometric dating during the 1966 to 1969 excavations (Stuckenrath 1978). Recent dates have produced the following results: a date of 39,770±730 BP (GifA-96264) has been obtained from level 11 (Maíllo-Fernández *et al.* 2001) and a date of 42,850+960/–850 BP (KIA-25883) has been obtained from level 13.

Lithic technology

The stone-tool assemblages from levels 11 and 12 of Cueva Morín and level 20e of El Castillo show important technological similarities (Sánchez-Fernández 2005; Maíllo-Fernández 2007): the predominant operational scheme is discoid and, at least from a qualitative perspective, the production of prismatic-type blades is paramount. However, differences are also observed: quantities of lithic debris resulting from secondary (non-dominant) operational schemes are different and, despite a similar overall range of raw material (principally quartzite, flint, ophite and sandstone), flint predominates in levels 11 and 12 of Cueva Morín (Maíllo-Fernández 2005) whilst fine-grained quartzite dominates and flint is a minority in level 20e of El Castillo (Sánchez-Fernández 2005).

Discoid operational scheme

This is the main operational scheme found in the three levels under discussion (Fig. 13.2). It is possible that different techniques were employed depending on raw material type (Sánchez-Fernández & Maíllo-Fernández 2006): coarse-grained raw materials (ophite and coarse-grained quartzite in the case of El Castillo, and ophite and sandstone in the case of Cueva Morín), which are present in the form of large pebbles and were reduced using a bifacial technique. Fine-grained raw materials (quartzite and flint) are generally present in the form of smaller cobbles and are reduced using a hierarchical unifacial technique. The initial shaping (*mise en forme*) of discoid cores is simple: none of the cores shows signs of previous reduction, and a striking platform, usually obtained using a unifacial technique, is prepared only in cases where an angular surface is lacking. Thus core preparation is not always peripheral; it might be partial and can be altogether absent. In the case of bifacial cores, on the other hand, reduction begins with alternating *débitage*. In the full scale *débitage* phase two flaking directions can be observed — centripetal and cordal . This is confirmed by the presence of characteristic blanks, which in all three levels show clear predominance of cordal over centripetal orientations (Sánchez-Fernández 2005; Maíllo-Fernández 2005). In all cases cores were discarded once flaking planes were exhausted.

Levallois operational scheme

This operational scheme is less well represented at Cueva Morín (Fig. 13.2) than at El Castillo: only the Levallois Recurrent Bipolar technique has been found in level 11 of Cueva Morín, whilst both the Levallois Preferential and Levallois Recurrent Unipolar techniques have been found in level 12 of the site (Maíllo-Fernández 2003). In contrast, in level 20e at El Castillo, Levallois Preferential, Levallois Recurrent Centripetal and Levallois Recurrent Unipolar and Bipolar methods are all observed. In all three levels the characteristic flakes are scarce. At this point it has to be said that it is difficult to distinguish between the different methods given the similarity that some cores show in the final stages of production (Meignen 1993; Delagnes 1992).

Laminar operational scheme

Despite the scarcity of blade cores at both sites (Fig. 13.2), four prismatic cores showing unipolar reduction were found in two levels of Cueva Morín (Maíllo-Fernández *et al.* 2004) and another four prismatic unipolar cores were recovered from level 20e of El Castillo. To these finds should be added another three recurrent unipolar cores that also show the same prismatic tendency. The cores are restricted to fine-grained quartzite and flint (the latter only at Cueva Morín) and in the majority of cases were abandoned due to accidents during *débitage.* In all cases, core preparation is simple. It begins with extraction of *entame*-type laminar flakes (in a single case from level 20e a postero-lateral crest is recorded). Striking platforms are not generally prepared. Dorsal surfaces (single in some cores or compound adjacent ones in others) show negative scars of bladelets or laminar flakes. In most cases cores were abandoned as a result of accidents during *débitage.* Laminar blanks are rare and to judge from the presence of negative flakes on some Levallois cores, some may have been produced using the Levallois production system. Some bladelets have been retouched. A semi-abrupt, laterally retouched piece and a backed piece have been found at Cueva Morín (Maíllo-Fernández 2005). In addition, three Dufour-like bladelets have been found in level of 20e of El Castillo (Sánchez-Fernández & Maíllo-Fernández 2006).

Other operational schemes

Cleavers

The production of cleavers is one of the most characteristic operational schemes in Unit 20 of El Castillo (Cabrera-Valdés 1983; 1984). In level 20e, three cleavers, two made of ophite and one of coarse-grained quartzite, have been found. All three can be classified technologically as belonging to Tixier's (1956) type 0, and in all cases the cutting edge shows clear signs of fracture from use. Two operational schemes have been identified in the production of these cleavers, one geared towards the production of flakes wider than they are long and another geared towards the production of longer flakes. Whilst the lack of cores

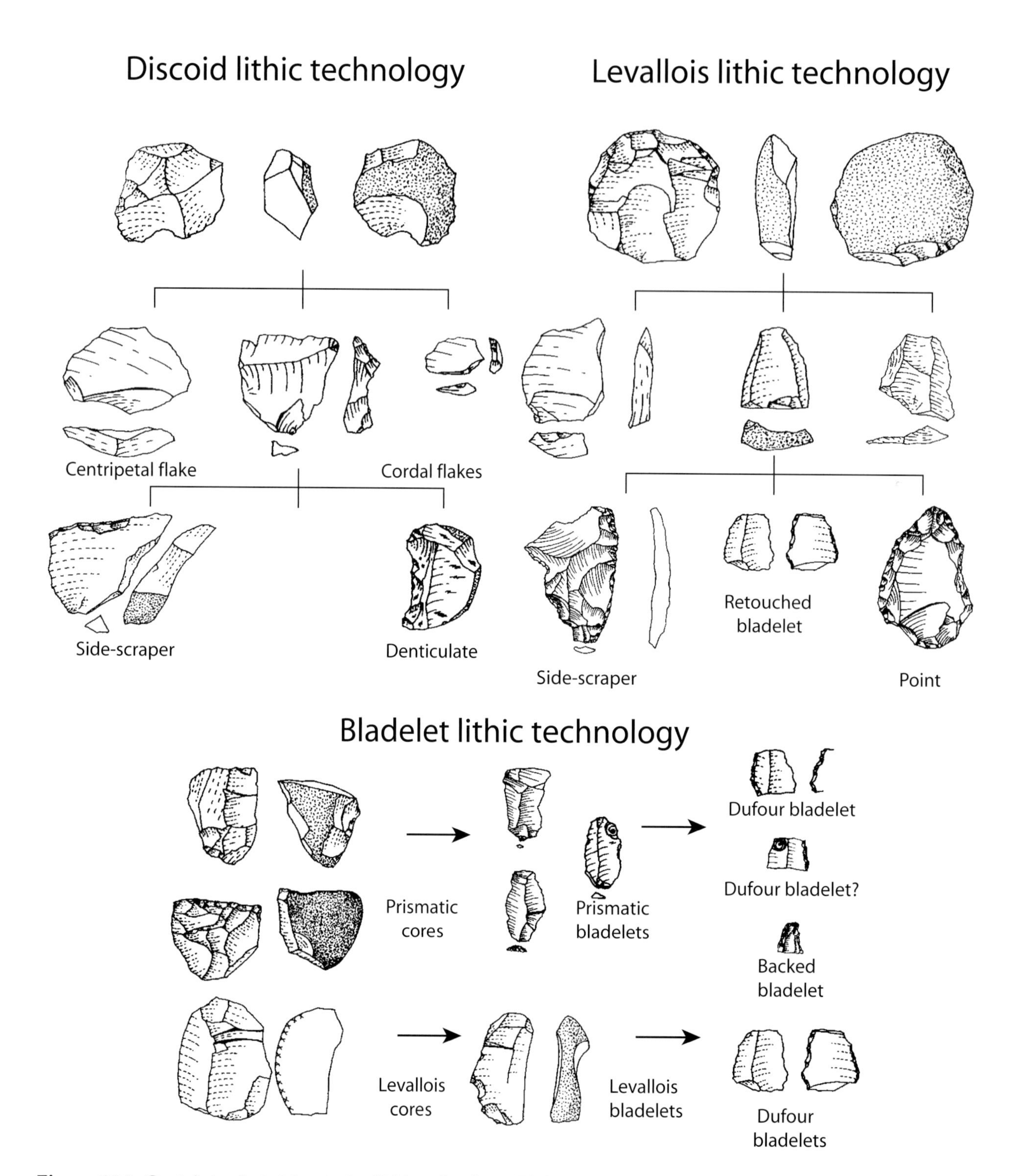

Figure 13.2. *Cantabrian Late Mousterian lithic technology: El Castillo 20e and Cueva Morín 11 and 12.*

and characteristic waste products suggests that cleavers were produced off-site, the assemblage does include large unretouched flakes in the same raw material that show signs of having been used for tasks similar to those for which cleavers were used. In addition, some smaller flakes, which could be classified as pseudo-cleavers, have been found (Sánchez-Fernández 2005).

Quina

Level 20e at El Castillo has yielded two cores which have been classified as Quina type, because their

Table 13.1. *Cantabrian Late Mousterian sites.*

Site	Principal operational scheme	Secondary operational scheme	Dominant raw material	Facies	Source
Morín 11	Discoid	Laminar Levallois?	flint	Denticulate	Maíllo-Fernández 2003
Morín 12	Discoid	Laminar Levallois?	flint	Denticulate	Maíllo-Fernández 2003
Castillo 20e	Discoid	Levallois Laminar *Façonnage* Quina?	quartzite	Typical	Sánchez-Fernández 2005
Axlor N	Levallois	Discoid	flint, lutite		González Urquijo *et al.* 2006
Axlor B, C, D	Quina	Recurrent Unipolar	flint, lutite		González Urquijo *et al.* 2006
Esquilleu XI	Quina	Levallois	quartzite	Charentian	Baena *et al.* 2005
Esquilleu IX	Levallois	Discoid	quartzite	Typical	Baena *et al.* 2005
Esquilleu III	Discoid	Laminar	quartzite	Cantabrian Mousterian	Baena *et al.* 2005
La Flecha	Discoid	Recurrent Unipolar	quartzite	Denticulate	Castanedo 2001
El Conde E	Discoid		quartzite	Denticulate	Carrión 2002
El Conde D	Discoid		quartzite	Denticulate	Carrión 2002

volumetric design corresponds to that used as a definition for cores of this *débitage* style (Bourguignon 1996; 1997). Some blanks have also been identified with this production-system style.

Kostenki
Three flakes from Level 20e at El Castillo show inverse truncations followed by one or two extractions on the dorsal side. These characteristics are similar to those described previously at other sites (Delagnes 1992; Bernard-Guelle & Porraz 2001) and can be grouped within what could be called the 'Kostenki phenomenon'.

Discussion

Techno-typological characterization of the end of the Mousterian in Cantabria suffers from problems of definition. The very concept of the Mousterian is often a 'catch-all' term that causes more uncertainty than certainty with regard to knowledge of this prehistoric period. Assemblages dated later than 60,000 BP are usually considered to belong to the end of the Mousterian or Late Mousterian. However we have already made clear that this boundary is absolutely arbitrary.

In recent years, archaeological and typological studies of sites with levels that can be assigned to the end of the Mousterian have increased considerably. In addition to levels 11 and 12 of Cueva Morín and level 20e of El Castillo there have been numerous recent studies on assemblages such as those from El Esquilleu and La Flecha in Cantabria, and Axlor in the Basque Country. In addition, more provisional reports from La Viña (Fortea 1995; 1999), Sopeña (Pinto-Llona *et al.* 2006), La Güelga (Menéndez *et al.* 2006), El Sidrón (Fortea *et al.* 2003) and El Conde (Arbizu *et al.* 2005) in Asturias, Covalejos (Martín *et al.* 2006) in Cantabria, and Arrillor (Hoyos *et al.* 1999) and Lezetxiki (Arrizabalaga 2006) in the Basque Country, provide a basis for the discussion of technological and typological variability in the region (Table 13.1).

At El Esquilleu, there is marked internal variability between levels (Baena *et al.* 2006): the main operational scheme in level XI is Quina-type, carried out on large pebbles which are not used up in a metrical way. Other methods (Levallois and Discoid) are not common. In level IX, a recurrent unipolar Levallois technique predominates. This variation is also evident at the Axlor site (González Urquijo *et al.* 2006): in level N the main operational scheme is Levallois and is geared towards the production of flakes and projectile points, whilst in levels B, C and D the main scheme is Quina type and is geared towards the production of thick backed flakes that are eventually used to manufacture Quina sidescrapers.

On the other hand, in central Cantabria the main operational sequence is discoid at El Castillo, Cueva Morín and La Flecha (Castanedo 2001). Two techniques are employed: a unifacial one that organizes surfaces hierarchically and a bifacial and non-hierarchial one. At Cueva Morín and El Castillo, the latter technique is used with coarse-grained raw materials such as ophite and sandstone. These raw materials, which are present as large pebbles, lend themselves to the extraction of longer, thicker blanks (Maíllo-Fernández 2003) and are employed in the production of elaborate cleavers at El Castillo (Sánchez-Fernández 2005).

One of the most interesting operational schemes identified at the end of the Mousterian in Cantabria is the production of bladelets at El Castillo and Cueva Morín (Cabrera-Valdés *et al.* 2000; Maíllo-Fernández *et al.* 2004; Sánchez-Fernández & Maíllo-Fernández 2006) and at Covalejos (level H), where bladelets are produced opportunistically from waste blanks (Martín *et al.* 2006). Using analytical and structural analysis the same laminar processes can be observed in the assemblages at Lezetxiki IV (Arrizabalaga 1995), Arrillor (Hoyos *et al.* 1999) and at the neighbouring French

Basque Country sites of Abri Olha 2 and Gatzarria (Laplace & Sáenz de Buruaga 2000; 2002–3).

Although a techno-complex cannot be defined solely on the basis of its operational scheme, we believe it necessary to bring knowledge of its technological characteristics up to date. In order to advance archaeological reconstruction of the end of the Mousterian in Cantabria, it is necessary to study the data presented above in conjunction with knowledge of the use of space, hunting strategies and assemblage function/use at different sites, as well as to consider evidence for symbolic behaviour at sites such as El Castillo or Lezetxiki (Cabrera-Valdés *et al.* 2004; Arrizabalaga 2006).

References

Arbizu, M., J.L. Arsuaga & G. Adán, 2005. La cueva del Forno/Conde (Tuñón, Asturias): un yacimiento del tránsito del Paleolítico medio y superior en la Cornisa Cantábrica, in *Neandertales Cantábricos: Estado de la Cuestión*, eds. R. Montes & J.A. Lasheras. (Monografías del Museo de Altamira 20.) Santander: Museo de Altamira, 425–41.

Arrizabalaga, A., 1995. La industria lítica del Paleolítico superior inicial en el Oriente Cantábrico. Unpublished PhD dissertation, Universidad del País Vasco.

Arrizabalaga, A., 2006. Lezetxiki (Arrasate, País Vasco). Nuevas preguntas acerca de un antiguo yacimiento, in *En el centenario de la cueva de El Castillo: el ocaso de los Neandertales*, eds. V. Cabrera-Valdés, F. Bernaldo de Quirós & J.M. Maíllo-Fernández. Santander: UNED-CajaCantabria, 291–309.

Baena, J., E. Carrión, B. Ruiz *et al.*, 2005. Paleoecología y comportamiento humano durante el Pleistoceno superior en la comarca de Liébana: la secuencia de la cueva de El Esquilleu (Occidente de Cantabria, España), in *Neandertales Cantábricos: Estado de la Cuestión*, eds. R. Montes & J.A. Lasheras. (Monografías del Museo de Altamira 20.) Santander: Museo de Altamira, 461–87.

Baena, J., E. Carrión & R. Velázquez, 2006. Tradición y coyuntura: claves sobre la variabilidad del musteriense occidental a partir de la cueva del Esquilleu, in *En el centenario de la Cueva de El Castillo: El Ocaso de los Neandertales*, eds. V. Cabrera-Valdés, F. Bernaldo de Quirós & J.M. Maíllo-Fernández. Santander: UNED-CajaCantabria, 249–67.

Bernaldo de Quirós, F., V. Cabrera-Valdés & A.J. Stuart, 2006. Nuevas dataciones para el Musteriense y el Magdaleniense de la cueva de El Castillo, in *En el centenario de la Cueva de El Castillo: el Ocaso de los Neandertales*, eds. V. Cabrera-Valdés, F. Bernaldo de Quirós & J.M. Maíllo-Fernández. Santander: UNED-CajaCantabria, 453–57.

Bernard-Guelle, S. & G. Porraz, 2001. Amincissement et débitage sur éclat: définitions, interprétations et discusion à partir d'industries lithiques du Paléolithique Moyen des Préalpes du Nord Françaises. *Paléo* 13, 53–72.

Boëda, E., 1988. Le concept Levallois et évaluation de son champ d'application, in *L'Homme de Néandertal*, vol. 4: *La Technique*, ed. M. Otte. Liège: ERAUL, 13–26.

Boëda, E., 1993. Le débitage discoïde et le débitage Levallois récurrent centripète. *Bulletin de la Société Préhistorique Française* 90(6), 392–404.

Bordes, F., 1953. Essai de clasification des industries moustériennes. *Bulletin de la Société Préhistorique Française* 50(7–8), 457–66.

Bordes, F., 1980. Le débitage Levallois et ses variantes. *Bulletin de la Société Préhistorique Française* 77(2), 45–9.

Bordes, F. & M. Bourgon, 1951. Le complexe Moustérien: moustériens, levalloisien et tayacien. *L'Anthropologie* 55, 1–23.

Bourguignon, L., 1996. 'La conception de débitage Quina', in Reduction processes for the European Mousterian. *Quaternaria Nova* VI, 149–66.

Bourguignon, L., 1997. Le Moustérien de Type Quina: Nouvelle Définition d'une Technique. Unpublished doctoral thesis, Université de Paris X-Nanterre.

Cabrera-Valdés, V., 1983. Notas sobre el Musteriense cantábrico: el 'Vasconiense', in *Homenaje al Prof. Martín Almagro Basch*, ed. M. Fernández-Miranda. Madrid: Ministerio de Cultura, 131–41.

Cabrera-Valdés, V., 1984. *El yacimiento de la Cueva de El Castillo (Puente Viesgo, Santander)*. Madrid: C.S.I.C.

Cabrera-Valdés, V., H. Valladas, F. Bernaldo de Quirós & M. Hoyos, 1996. La transition Paléolithique moyen-Paléolithique supérieur à El Castillo (Cantabrie): nouvelles datations par le carbone-14. *C.R. de la Academie des Sciences* 322 (série II), 1093–8.

Cabrera-Valdés, V., J.M. Maíllo-Fernández & F. Bernaldo de Quirós, 2000. Esquemas operativos laminares en el Musteriense final de la cueva del Castillo (Puente Viesgo, Cantabria). *Espacio, Tiempo y Forma* 13 (Serie I), 51–78.

Cabrera-Valdés, V., J.M. Maíllo-Fernández, M. Lloret & F. Bernaldo de Quirós, 2001. La transition vers le Paléolithique supérieur dans la grotte du Castillo (Cantabrie, Espagne): la couche 18. *L'Anthropologie* 105, 505-32.

Cabrera-Valdés, V., A. Pike-Tay & F. Bernaldo de Quirós, 2004. Trends in Middle Paleolithic settlement in Cantabrian Spain: the Late Mousterian at Castillo cave, in *Settlement Dynamics of the Middle Paleolithic and Middle Stone Age*, vol. 2, ed. N. Conard. Tübingen: Kerns Verlag, 437–60.

Cabrera-Valdés, V., J.M. Maíllo-Fernández, A. Pike-Tay, Mª.D. Garralda & F. Bernaldo de Quirós, 2006. A Cantabrian perspective on late Neanderthals, in *When Neanderthals and Modern Humans Met*, ed. N. Conard. Tübingen: Kerns Verlag, 441–65.

Carballo, J., 1923. *Excavaciones en la Cueva del Rey en Villanueva (Santander)*. Madrid: Junta Superior de Excavaciones y Antigüedades.

Carrión, E., 2002. Variabilidad Técnica en el Musteriense de Cantabria. Unpublished PhD dissertation, UAM.

Castanedo, I., 2001. Adquisición y aprovechamiento de los recursos líticos en la Cueva de La Flecha (Cantabria). San Sebastián. *Munibe* 53, 3–18.

Delagnes, A., 1992. L'organisation de la production lithique au Paléolithique moyen. Unpublished doctoral thesis, Université de Paris X.

Fortea, J., 1995. Abrigo de La Viña: informe y primera valoración de las campañas 1991–1994. *Excavaciones Arqueológicas en Asturias* 1991–94, 19–32.

Fortea, J., 1999. Abrigo de La Viña: informe y primera valoración de las campañas 1995–1998. *Excavaciones Arqueológicas en Asturias* 1995–98, 31–41.

Fortea, J., M. de la Rasilla, E. Martínez *et al.*, 2003. La cueva de El Sidron (Borines, Piloña, Asturias): primeros resultados. *Estudios Geológicos* 59, 159–79.

Freeman, L.G., 1971. Los niveles de ocupación musteriense, in *Cueva Morín*, eds. J. González Echegaray & L.G. Freeman. (Publicaciones del Patronato de las cuevas prehistóricas de la provincia de Santander VI.) Santander: Patronato de las cuevas prehistóricas, 27–161.

Freeman, L.G., 1973. El musteriense, in *Cueva Morín*, ed. J. González Echegaray & L.G. Freeman. (Publicaciones del Patronato de las cuevas prehistóricas de la provincia de Santander X.) Santander: Patronato de las cuevas prehistóricas, 13–140.

Freeman, L.G. & J. González-Echegaray, 1970. Aurignacian structural features and burials at Cueva Morín (Santander, Spain). *Nature* 226, 722–6.

González-Echegaray, J. & L.G. Freeman (eds.), 1971. *Cueva Morín*. (Publicaciones del Patronato de las cuevas prehistóricas de la provincia de Santander VI.) Santander: Patronato de las cuevas prehistóricas.

González-Echegaray, J. & L.G. Freeman (eds.), 1973. *Cueva Morín*.(Publicaciones del Patronato de las cuevas prehistóricas de la provincia de Santander X.) Santander: Patronato de las cuevas prehistóricas.

González Urquijo, J.E., J.J. Ibáñez, J. Ríos & L. Bourguignon, 2006. Aportes de las nuevas excavaciones en Axlor sobre el final del Paleolítico Medio, in *En el centenario de la Cueva de El Castillo: El Ocaso de los Neandertales*, eds. V. Cabrera-Valdés, F. Bernaldo de Quirós & J.M. Maíllo-Fernández. Santander: UNED-CajaCantabria, 269–89.

Hoyos, M., A. Sánez de Buruaga & A. Ormazabal, 1999. Cronoestratigrafía y paleoclimatología de los depósitos prehistóricos de la cueva de Arrillor (Araba, País vasco). *Munibe* 51, 137–51.

Kelley, H., 1954. Contribution à l'étude de la technique de la taille levalloisienne. *Bulletin de la Société Préhistorique Française* 51(4), 149–69.

Klein, R.G., 2003. Whither the Neanderthals? *Science* 291, 1525–7.

Laplace, G. & A. Sáenz de Buruaga, 2000. Application de la typologie analytique et structurale à l'étude de l'outillage mousteroïde de L'Abri Olha 2 à Cambo (Kanbo) en Pays Basque. *Paléo* 12, 261–324.

Laplace, G. & A. Sáenz de Buruaga, 2002–3. Typologie analytique et structurale des complexes du Moustérien de la Grotte Gatzarria (Ossas-Suhare, Pays Basque) et de leurs relations avec ceux de l'abri Olha 2 (Cambo, Pays Basque). *Pyrenae* 33–4, 81–163.

Lewis-Williams, D., 2005. New neighbours: interaction and image-making during the West European Middle to Upper Palaeolithic transition, in *From Tools to Symbols: from Early Hominids to Modern Humans*, eds. F. d'Errico & L. Backwell. Johannesburg: Witwatersrand University Press, 372–88.

Maíllo-Fernández, J.M., 2003. La Transición Paleolítico Medio-Superior en Cantabria. Análisis tecnológico de la industria lítica de Cueva Morín. Unpublished PhD dissertation, UNED, Madrid.

Maíllo-Fernández, J.M., 2005. Esquemas operativos líticos del Musteriense final de Cueva Morín (Villanueva de Villaescusa, Cantabria), in *Neandertales Cantábricos: Estado de la Cuestión*, eds. R. Montes & J.A. Lasheras. (Monografías del Museo de Altamira 20.) Santander: Museo de Altamira, 301–13.

Maíllo-Fernández, J.M., 2007. Aproximación tecnológica al Musteriense Final de Cueva Morín (Villanueva de Villaescusa, Cantabria). *Munibe* 58, 13–42.

Maíllo-Fernández, J.M., H. Valladas, V. Cabrera Valdés & F. Bernaldo de Quirós, 2001. Nuevas dataciones para el Paleolítico superior de Cueva Morín (Villanueva de Villaescusa, Cantabria). *Espacio, Tiempo y Forma* 14, 145–50.

Maíllo-Fernández, J.M., V. Cabrera-Valdés & F. Bernaldo de Quirós, 2004. Le débitage lamellaire dans le Moustérien final de Cantabria, Espagne: le cas de El Castillo et Cueva Morín. *L'Anthropologie* 108, 367–93.

Martín, P., R. Montes & J. Sanguino, 2006. La tecnología lítica del Musteriense final en la región cantábrica: los datos de Covalejos (Velo de Piélagos, Cantabria, España), in *En el centenario de la Cueva de El Castillo: El Ocaso de los Neandertales*, eds. V. Cabrera-Valdés, F. Bernaldo de Quirós & J.M. Maíllo-Fernández. Santander: UNED-CajaCantabria, 231–48.

Meignen, L., 1993. *L'Abri des Cannalettes: un habitat Moustérien sur les Grands Causses (Nant, Aveyron). Fouilles 1980–86.* (Monographie du CRA 10.) Paris: CRA.

Menéndez, M., E. García & J.M. Quesada, 2006. Excavaciones en la cueva de la Güelga (Cangas de Onís, Asturias), in *En el centenario de la Cueva de El Castillo: El Ocaso de los Neandertales*, eds. V. Cabrera-Valdés, F. Bernaldo de Quirós & J.M. Maíllo-Fernández. Santander: UNED-CajaCantabria, 209–29.

Pinto-Llona, A.C., G. Clark & A. Millar, 2006. Resultados preliminares de los trabajos en curso en el abrigo de Sopeña (Onís, Asturias), in *En el centenario de la Cueva de El Castillo: El Ocaso de los Neandertales*, eds. V. Cabrera-Valdés, F. Bernaldo de Quirós & J.M. Maíllo-Fernández. Santander: UNED-CajaCantabria, 193–207.

Révillion, S. & A. Tuffreau (eds.), 1994. *Les industries Laminaires au Paléolithique moyen.* (Dossier de Documnetation Archéologique 18.) Paris: CRA.

Rink, W.J., H.P. Schwartz, H.K. Lee, V. Cabrera, F. Bernaldo de Quirós & M. Hoyos, 1997. ESR dating of Mousterian levels at El Castillo cave, Cantabria, Spain. *Journal of Archaeological Science* 24, 593–600.

Sánchez-Fernández, G., 2005. Análisis tecnológico y tipológico del nivel 20e de la cueva de El Castillo (Puente Viesgo, Cantabria). DEA, unpublished, UNED.

Sánchez-Fernández, G. & J.M. Maíllo-Fernández, 2006.

Soportes laminares en el musteriense final cantábrico: el nivel 20e de la cueva de El Castillo (Cantabria), in *Miscelánea en Homenaje a Victoria Cabrera*, vol. I, eds. J.M. Maíllo-Fernández & E. Baquedano. (Zona Arqueológica 7.) Madrid: Museo regional de Arqueología-UNED, 264–73.

Stuart, A.J., 2005. The extinction of woolly mammoth (*Mammuthus primigenius*) and straight-tusked elephant (*Palaeoloxodon antiquus*) in Europe. *Quaternary International* 126–8, 171–7.

Stuckenrath, R., 1978. Dataciones de Carbono 14, in *Vida y Muerte en Cueva Morin*, eds. J. González Echegaray & L.G. Freeman. Santander: Institución Cultural de Cantabria, 215.

Terradas, X., 2003. Discoid flaking method: conception and technological variability, in *Discoid Lithic Technology: Advances and Implications*, ed. M. Peresani. (British Archaeological Reports, International Series 1120.) Oxford: BAR, 19–31.

Tixier, J., 1956. Le hachereau dans l'Acheuléen nord-africain: notes typologiques, in *XV Congrès préhistorique de France*. Poitiers-Angouleme: SPF, 914–23.

Vega del Sella, C. de la, 1921. *El Paleolítico de Cueva Morín (Santander) y notas para la climatología cuaternaria.* (Memoria 29.) Madrid: Comisión de Investigaciones Paleontológicas y Prehistóricas.

Wynn, T. & F.L. Coolidge, 2004. The expert Neandertal mind. *Journal of Human Evolution* 46, 467–87.

Chapter 14

The Demise of the Neanderthals and the Collapse of the Mousterian World

Jean-Philippe Rigaud

At the end of a long period of occupation in Europe and the Middle East, lasting almost 200,000 years and during which they displayed genuine creativity, the Neanderthals rapidly reached a level of technological know-how and symbolic expression which was close to that of anatomically modern humans (*Homo sapiens sapiens* = Cro-Magnon). This cultural change came about, according to some scholars, by acculturation (Mellars 1990; Hublin 1990), while for others it was the result of the technological and spiritual evolution of Neanderthals which was quite independent of the arrival of Cro-Magnons (Rigaud 1996; d'Errico *et al.* 1998). In the long debate on the nature of the Middle to Upper Palaeolithic transition the cultural capacities of Neanderthals play a key role. Assessed in many ways (see Mellars 1989; Mellars & Gibson 1995; Rigaud 1989; 1993; Hayden 1993; Zilhão & d'Errico 2003), these capacities led Neanderthals to adopt, over a relatively short period of time, technological behaviour which nothing suggests was any better or worse than that of the Cro-Magnons (Pelegrin 1995, 265, 270).

Available absolute dates and the small number of reliable archaeological sequences which we have available today tell us that the final Neanderthal industries — in western France, the Châtelperronian — appear before the earliest occupation attributable to the Aurignacian (Rigaud 2001; Bordes 2000; Zilhão & d'Errico 1999; 2003; Zilhão *et al.* 2006 *vs* Mellars 1999). It has become clear that if these occupations were contemporaneous then it was only during the later phases of the Châtelperronian (Rigaud 2000, 64) that this was the case. About 35,000 BP, by which point the Neanderthals had already advanced in a process of 'leptolithization', the earliest Aurignacians arrived and very quickly spread throughout western Europe — with the exception of a large part of the Iberian Peninsula to the south of the Pyrenees. Here, for several millennia, Neanderthals continued to occupy the mesetas and the coastal fringe of the peninsula until the later arrival of the Aurignacian (Utrilla *et al.* 2006; Zilhão 2000). Thus, for a period of 1000–2000 years, on this side of the border, Europe experienced the unusual situation in which two distinct types of humans co-existed — something which will never be seen again. But how did this situation develop so quickly prior to the extinction of the Neanderthals?

Over the last 15 years many publications have proposed different hypotheses to explain the extinction of the Neanderthals (Binford 1989; Bocquet-Appel & Demars 2000; Hublin 1990), but in reality they usually consider the extinction of *Homo sapiens neanderthalensis* on the one hand, and Mousterian techno-complexes, the Châtelperronian and other so-called transitional industries on the other, acknowledging *a priori* that Neanderthals were the creators of industries attributable to the Middle Palaeolithic while the Upper Palaeolithic was the work of Cro-Magnons. However, two important discoveries have shown that there is no direct or unequivocal relationship between fossils and stone tools, and that it is useful to consider these two aspects of the problem distinctly and independently:

1. the St Césaire Neanderthals (Lévêque & Vandermeersch 1980) associated with a Châtelperronian context considered to be part of the Upper Palaeolithic[1] (de Sonneville-Bordes 1960);
2. the presence of Middle Palaeolithic industries in the Middle East attributable to either true Neanderthals or anatomically modern humans (Bar-Yosef *et al.* 1986; Bar-Yosef & Vandermeersch 1981; Valladas *et al.* 1988; Vandermeersch 1989).

Whereas the extinction of *Homo sapiens neanderthalensis* is revealed by the study of hominin fossils, the transition from the Middle to the Upper Palaeolithic — i.e. deciphering behaviour from the analysis of technological and symbolic products — is revealed by prehistoric archaeology. However, the term extinction is not the most appropriate since hypotheses concerning the evolution of Neanderthals into Cro-Magnons and their genetic integration have been proposed (Smith 1994; Zilhão & Trinkaus 2002), and DNA analysis has recently confirmed hybridization between Neanderthals and anatomically modern humans (Green *et al.* 2010).

Several scenarios suggesting fierce competition, even if not true conflict which would have led to the exclusion of the Neanderthals (Stringer & Gamble 1993; Mellars 1999; Gat 1999), have been put forward. However, the lack of archaeological evidence for violent confrontation between populations does not appear to confirm the hypothesis — although the probability of finding evidence of such behaviour is small.

Other plausible causes of demographic replacement have been invoked and include the fact that lower birth rate, shortened lifespan, later weaning and higher mortality rates could have led to a rapid and dramatic population collapse among Neanderthals (Zubrow 1989; Flores 1998). Similarly, strong consanguinity caused by isolation of Neanderthal groups could have been a cause of their extinction, but it would have had to have manifested itself much sooner, before the arrival of the Cro-Magnons. Also invoked as a cause of demographic replacement is an immune-system deficiency among Neanderthals faced with infections introduced by the new arrivals (Pelegrin 1995). In addition, a less-varied dietary regime practised by the Neanderthals might be responsible for deficiencies not suffered by the Cro-Magnons due to the latter's better diet (Richards *et al.* 2001). These deficiencies do not seem to have seriously affected the Neanderthals during the course of the long period prior to the arrival of modern humans. Recent research, however, has shown that:

- there is no abrupt discontinuity between the faunal assemblages from the Middle to the Upper Palaeolithic, other than those related to climatic change;
- there is greater variety in taxa during the Middle Palaeolithic than the Aurignacian (Grayson & Delpech 1994; 1998; 2002; 2003; 2006); and
- there is 'no significant change in the subsistence' (according Morin 2004).

Taking into account the reservations expressed, Neanderthal extinction could have resulted, from any or all of the causes outlined above combined with other factors. Thus, the climatic impact of specific episodes of Oxygen Isotope Stage (OIS) 3 and particularly Heinrich event 4 could have played a decisive role in the extinction of Neanderthals by combining the effects of intense competition for resources with deterioration in climate (Mellars 1998, 502). This hypothesis has often been objected to on the grounds that Neanderthal populations, thought to be particularly well adapted to the cold (Steegman *et al.* 2002), had successfully survived much more extreme climatic crises and that the end of OIS-3 was, at no point, as extreme as the Last Glacial Maximum (OIS-2). The long deterioration of OIS-3 marked by intense instability was certainly an aggravating factor in a complex process which led to the Neanderthals' retreat to the Iberian Peninsula, but one always needs to consider carefully, in terms of altitude among other factors, the environmental hospitality of the meseta and sierras to the south of the Ebro depression (d'Errico & Sanchez-Goñi 2003; Utrilla *et al.* 2006).

Comparison of the social organization of the Neanderthals and that of the Cro-Magnons has often been undertaken in a general way in order to underline and accentuate the differences between these two societies (Binford 1989; Mellars 1996; Soffer 1992; White 1982). The differences between models resulting from the study of primates on the one hand and those which have been developed based on studies of modern hunter-gatherers on the other, only go to accentuate, if not exaggerate, these differences. However, can we perhaps develop a few hypotheses concerning the relationships which might have existed at the heart of the Neanderthal and Cro-Magnon populations on the basis of available archaeological data and what ethnology tells us?

One model which concerns the occupation of territory can be envisaged for western Europe between 30 and 40 kya BP in which Neanderthal populations are dispersed in small groups over a vast territory, there being a balance between human density and environmental resources. The territory occupied and exploited by these groups is determined by the abundance of available resources, while the size of the group and their mobility is caused by exhaustion, or even collapse, and renewal of food resources as a function of season and the search for available raw materials. Supporting this model, the results of excavation at Grotte Vaufrey reinforce these arguments. Throughout the occupation of the cave (OIS-9 to OIS-2), lithic raw material sources are usually located very close to the site (1–10 km, or 1–2 hours on foot). Greater distances, in the order of 60–80 km, prove environmental exploitation further afield, in an area between the foothills of the Massif Central (the Limousin plateaux) to the east and the valleys of the south and west (Rigaud 1982; 1988; Geneste 1985; 1988; Turq 2000). On the other hand, the supply of long-distance raw material, always limited in quantity, reveals clear selection (Rigaud 1988, 417; Geneste 1988, 489). Other examples of Middle Palaeolithic raw-material supply in the Dordogne reveal equivalent strategies on the same scale (Geneste 1985; 1988; Turq 2000). The origin of lithic raw material sheds light on population movements which are not necessarily motivated solely by the search for raw material. Such procurement could have occurred during subsistence activities, requiring a stay of several days and necessitating exploitation of local resources so that the immediate needs of the group could be met. Given that evidence of very long-distance movements is very rare if not completely lacking in Aquitaine (Geneste 1985), it is difficult to consider the possibility of long-distance exchange and contact. As Mellars (1996, 164) has emphasized, an important factor in determining the distances which populations travelled is the local abundance of lithic resources; whereas it is easy in the Dordogne to obtain adequate raw material without involving movement of more than 5 km, in other regions — central Europe for example — this is not possible and much more significant movement is needed in order to obtain the required raw material.

Geographical distribution of Neanderthal populations can be traced through important techo-typological variability, the causes of which have, historically, been considered to be 'ethnic' (Bordes 1961), functional (Binford & Binford 1966) or chronological (Mellars 1969). Recent studies, however, have shown that this variability also shows regional techno-typological specificities underlining strong geographical diversity (Boëda *et al.* 1990; Bourguignon 1997; Bourguignon & Turq 2003; Delagnes 1992; Delagnes & Ropars 1996; Dibble 1988; Geneste 1985; Jaubert 1993; Meignen & Bar-Yosef 1988; Mellars 1996; Roland 1988; Slimak 1999; Turq 2000), and thereby confirming what had been suspected more generally (Gamble 1982; 1986; Wobst 1976).

The occupation of northern Aquitaine by Neanderthals as proposed by Mellars (1996, 365) therefore seems to have been the result of small groups exploiting small territories beyond which incursions were rarely made. This division of space by Neanderthals, into relatively restricted areas, along with the factors responsible for strong demographic fluctuations (consanguinity, epidemics, etc.) and the effects of climatically controlled environmental change, could play a determining role.

Aurignacian populations, assuming — although sometimes with reservation (Garalda & Vandermeersch 2002) — that they are made up entirely of anatomically modern humans, were, for the same reasons as the Neanderthals, restricted as to how they occupied their environment by the same balance between available resources and demography. Allowing for difficulties inherent in analysing a population on the basis of available archaeological data (Bocquet-Appel & Demars 2000 *vs* Pettit & Pike 2001), and in the absence of any objective criteria, we can not be at all sure that the earliest waves of Aurignacians comprised significantly larger numbers of people than made up the regional Neanderthal population (Boquet-Appel & Demars 2000). And therefore the most important reservation involves suggesting demographic pressure as the reason for the driving away of the Neanderthals.

It is important to note that, from their first appearance[2] until their replacement by the Gravettian techno-complex, Aurignacian industries are remarkably well characterized as much from a technological point of view — by blade production (Bon 1996; 2002; Bordes 2002; Kozłowski 1988; Lucas 1997; 2000) and the specific bone industry (Liolios 1999; 2006, 37) — as by their typological composition (de Sonneville-Bordes 1960). What is more, mobiliary and parietal art (Clottes *et al.* 1995; Lorblanchet 1995) and numerous examples of jewellery (White 1993), are additional cultural traits which underline the homogeneity of the European Aurignacian. Geographic and chronological variability among Aurignacian industries is undeniable (Rigaud 1993; Djindjian 1993), but is limited and does not affect identifying characteristics of the Aurignacian (Bar-Yosef 2002). The distribution of technical and symbolic specificities over a vast territory stretching from Portugal to the Black Sea and to a lesser degree to the Near East is, we think, the gradual result of cultural links and close contacts between regional entities. The movement — by transport or exchange — of seashells frequently used in Aurignacian jewellery proves much more extensive contact between these communities than between Neanderthals (Taborin 1993). This evidently implies much more complex social structure during the Aurignacian, with a tight network of inter-group interactions despite their geographical distribution across various environments. This social organization is founded on essential group stability based on social and economic reciprocity between groups — according to a model close to that described and documented by Wiessner (1977; 1982a,b; 1984; 1986), the existence of inter-group links of reciprocal solidarity resulting in the rapid spread of techniques, ideas and symbols, the diffusion and development of innovations and the opening up of genetically isolated populations.

Such contrast between the social organization of Neanderthal and Aurignacian populations led in part to the progressive abandonment of some of the ecological niches occupied by the Neanderthals and their occupation by the Aurignacians, filling the empty spaces without violent confrontation. This process probably happened in fits and starts — e.g. the crossing of the Ebro depression — finishing in the south of the Iberian Peninsula towards 28–29 kya, where the population decline of the Neanderthals, reaching a point of no return, brought about the extinction or genetic absorption of the last Neanderthals.

The important role played by social organization in the scenario presented above could give the impression that we are dealing with the evaluation of intellectual aptitudes and cognitive abilities of the Palaeolithic populations concerned. However, this is not our opinion and the diversity of social organization in society today is the best proof to the contrary.

Notes

1. Although the terms Final Mousterian and Epimousterian have sometimes been used to describe Châtelperronian techno-complexes, probably because of their association with Neanderthal skeletal remains.
2. The alleged Aurignacian character of the Bachokirian (Kozłowski 1982; Delporte & Djindjian 1979; Delporte 1998; Djindjian 1993) has not stood up to several recent typological and technological revisions (Rigaud 1996; 2000; 2001; Rigaud & Lucas 2006; Teyssandier 2003; Tsanova 2006; Tsanova & Bordes 2003).

Acknowledgements

My sincere gratitude goes to Katherine Boyle for translating this paper from the original French.

References

Bar-Yosef, O., 2002. Defining the Aurignacian, in *Towards a Definition of the Aurignacian*, eds. O. Bar-Yosef & J. Zilhão. (Trabalhos de Arqueologia 45.) Lisbon: Instituto Portuguès de Arqueologia.

Bar-Yosef, O. & B. Vandermeersch, 1981. Note concerning the possible age of the Mousterian layers in Qafzeh cave, in *Préhistoire du Levant* eds. J. Cauvin & P. Sanlaville. Paris: CNRS, 281–5.

Bar-Yosef, O., B. Vandermeersch & B. Arensburg, 1986. New data on the origin of modern man in the Levant. *Current Anthropology* 27, 63–4.

Binford, L., 1989. Isolating the transition to cultural adaptations: an organizational approach, in *The Emergence of Modern Humans: Biocultural Adaptations in the Later Pleistocene*, ed. E. Trinkaus. Cambridge: Cambridge University Press, 18–41.

Binford L. & S. Binford, 1966. A preliminary analysis of functional variability in the Mousterian of Levallois facies. *American Anthropologist* 68(2), 238–95.

Bocquet-Appel, J.-P. & P.-Y. Demars, 2000. Neanderthal contraction and modern human colonization of Europe. *Antiquity* 74, 544–52.

Boëda, E., J.-M. Geneste & L. Meignen, 1990. Identification des chaînes opératoires lithiques du Paléolithique ancien et moyen. *Paléo* 2, 43–80.

Bon, F., 1996. L'industrie lithique aurignacienne de la couche 2A de le Grotte des Hyènes à Brassempouy (Landes), in *Pyrénées Préhistoriques, Arts et Sociétés*, eds. H. Delporte & J. Clottes. (Actes du 118e Congrès national des Sociétés Historiques et Scientifiques.) Paris: CTHS, 439–55.

Bon, F., 2002. *L'Aurignacien entre Mer et Océan. Réflexion sur l'Unité des Phases Anciennes de l'Aurignacien dans le Sud de la France.* (Mémoire 29.) Paris: Société Préhistorique Française.

Bordes, F., 1961. Mousterian cultures in France. *Science* 134, 803–10.

Bordes, J.-G., 2000. La séquence aurignacienne de Caminade revisitée: l'apport des raccords d'intérêt stratigraphique. *Paléo* 12, 387–407.

Bordes, J.-G., 2002. Les inter-stratifications Châtelperronien/Aurignacien du Roc de Combe et du Piage (Lot, France). Analyse taphonomique des industries lithiques, implications archéologiques. Unpublished doctoral thesis, Université de Bordeaux.

Bourguignon, L., 1997. Le Moustérien de type Quina: nouvelle définition d'une technique. Unpublished doctoral thesis, Université Paris X-Nanterre.

Bourguignon, L. & A. Turq, 2003. Une chaîne opératoire de débitage discoïde sur éclat du Moustérien à denticulés aquitain: les exemples de Champs de Bossuet et de Combe-Grenal c.14, in *Discoid Lithic Technology: Advances and Implications*, ed. M. Peresani. (British Archaeological Reports, International Series 1120.) Oxford: BAR, 131–52.

Clottes J., J.M. Chauvet, E. Brunel-Deschamps *et al.*, 1995. Les peintures paléolithiques de la grotte Chauvet-Pont d'Arc (Ardèche): datations directes et indirectes par la méthode du radiocarbone. *Comptes Rendus Académie des Sciences de Paris* Série IIa 320, 1133–40.

Delagnes, A., 1992. L'organisation de la production lithique au Paléolithique moyen, approche technologique à partir de l'étude de l'industrie de la Chaise-de-Vouthon (Charente). Unpublished doctoral thesis, Université Paris X-Nanterre.

Delagnes, A. & A. Ropars, 1996. *Le Paléolithique moyen en Pays de Caux (Haute-Normandie). Le Pucheuil, Etouville.* (Documents d'Archéologie Française 56.) Paris: Maison des Sciences de l'Homme.

Delporte, H., 1998. *Les Aurignaciens, Premiers Hommes Modernes.* Paris: La Maison des Roches.

Delporte, H. & F. Djindjian, 1979. Note à propos de l'outillage aurignacien de la couche 11 de Bacho-Kiro. *Zesztyty Naukowe Uniwersytetu Jagiellonskiego. Prace Archeologiczne* 28, 101–4.

Dibble, H., 1988. The implications of stone tool types for the presence of language during the Lower and Middle Palaeolithic, in *The Human Revolution: Behavioural and Biological Perspectives on the Origins of Modern Humans*, eds. P. Mellars & C. Stringer. Princeton (NJ): Princeton University Press, 415–32.

Djindjian, F., 1993. Les origines du peuplement aurignacien en Europe, in *Aurignacien en Europe et au Proche Orient*, eds. L. Banesz & J. Kozłowski. Nitra-Bratislava: Institut Archéologique de l'Académie Slovaque des Sciences, 136–54.

d'Errico, F. & M.-F. Sanchez-Goñi, 2003. Neanderthal extinction and the millennial scale climatic variability of OIS3. *Quaternary Science Review* 22, 769–88.

d'Errico, F., J. Zilhão, D. Baffier, M. Julien & J. Pelegrin, 1998. Neandertal acculturation in western Europe? A critical review of the evidence and its interpretation. *Current Anthropology* 39 supplement, S1–S44.

Flores, J.C., 1998. A mathematical model for Neanderthal extinction. *Journal of Theoretical Biology* 191, 295–8.

Le Gall, O., 1988. Analyse palethnologique de l'Ichtyofaune, in *La Grotte Vaufrey: Paléoenvironnements, Chronologie et Activités Humaines,* ed. J.-Ph. Rigaud. (Mémoire 19.) Paris: Société Préhistorique Française, 565–8.

Gamble, C., 1982. Interaction and alliance in Palaeolithic society. *Man* 17, 92–107.

Gamble, C., 1986. *The Palaeolithic Settlement of Europe*. Cambridge: Cambridge University Press.

Garalda, M.-D. & B.Vandermeersch, 2002. Neanderthal or modern human? The enigme of some Archaïc and early Aurignacian remains from southern Europe. *Journal of Human Evolution* 42(3), A14–A15.

Gat, A., 1999. Social organisation, group conflict and the demise of Neanderthals. *The Mankind Quarterly* 39, 435–54.

Geneste, J.-M., 1985. Analyse lithique d'industries Moustériennes du Périgord: une approche technologique du comportement des groupes humains au Paléolithique moyen. Unpublished doctoral thesis, Université de Bordeaux 1.

Geneste, J.-M., 1988. Les industries de la grotte Vaufrey; technologie du débitages, économie et circulation des matières premières lithiques, in *La Grotte Vaufrey: Paléoenvironnements, Chronologie et Activités Humaines,* ed. J.-P. Rigaud. (Mémoire 19.) Paris: Société Préhistorique Française, 441–517.

Grayson, D.K. & F. Delpech, 1994. The evidence for Middle Paleolithic scavenging from couche VIII, Grotte Vaufrey (Dordogne, France). *Journal of Archaeological Science* 21, 359–76.

Grayson, D.K. & F. Delpech, 1998. Changing diet breadth in the early Upper Paleolithic of southwestern France. *Journal of Archaeological Science* 25, 1119–30.

Grayson, D.K. & F. Delpech, 2002. Specialized early Upper Paleolithic hunters in southwestern France? *Journal of Archaeological Science* 29, 1439–49.

Grayson, D.K. & F. Delpech, 2003. Ungulates and the Middle-to-Upper Paleolithic transition at Grotte XVI (Dordogne, France). *Journal of Archaeological Science* 30, 1633–48.

Grayson, D.K. & F. Delpech, 2006. Was there increasing dietary specialization across the Middle-to-Upper Paleolithic transition in France? in *When Neanderthals and Modern Humans Met,* ed. N. Conard. (Tübingen Publications in Prehistory.) Tübingen: Kerns Verlag, 377–417.

Green, R.E., J. Krause, A.W. Briggs *et al.*, 2010. A draft sequence of the Neanderthal genome. *Science* 328(5979), 710–22.

Hayden, B., 1993. The cultural capacities of Neandertals: a review and re-evaluation. *Journal of Human Evolution* 24, 113–46.

Hublin, J.-J., 1990. Les peuplements paléolithiques de l'Europe: un point de vue géographique, in *Paléolithique moyen et récent et Paléolithique supérieur ancien en Europe,* ed. C. Farizy. (Mémoire 3.) Nemours: Musée de Préhistoire d'Île de France, 29–37.

Jaubert, J., 1993. Le gisement paléolithique moyen de Mauran (Haute-Garonne): techno-économie des industries lithiques. *Bulletin de la Société Préhistorique Française* 90, 328–35.

Kozłowski, J., 1982. *Excavation at Bacho Kiro Cave (Bulgaria): Final Report.* Warsaw: Panstwowe Wydawnictwo Naukowe.

Kozłowski, J., 1988. Transition from the Middle to the Early Upper Palaeolithic in central Europe and in the Balkans, in *The Early Upper Palaeolithic Evidence from Europe and the Near East*, eds. J.F. Hoffecker & C.A. Wolf. (British Archaeological Reports, International Series 437.) Oxford: BAR, 193–236.

Lévêque, F. & B. Vandermeersch, 1980. Découverte de restes humains dans un niveau castelperronien à Saint Césaire (Charente-Maritime). *Comptes Rendus de l'Académie des Sciences de Paris* Série 2 291, 187–9.

Liolios, D., 1999. Variabilité et caractéristiques du travail des matières osseuses au début de l'Aurignacien: approche technologique et économique. Unpublished doctoral thesis, Université Paris X – Nanterre.

Liolios D., 2006. Réflexions on the role of bone tools in the definition of the Early Aurignacian, in *Towards a Definition of the Aurignacian: Proceedings of the symposium held in Lisbon, Portugal, June 25–30, 2002*, eds. O. Bar-Yosef & J. Zilhão. (Trabalhos de Arqueologia 45.) Lisbon: Instituto Portuguès de Arqueologia, 37–51.

Lorblanchet, M., 1995. *Les Grottes Ornées de la Préhistoire. Nouveaux Regards*. Paris: Errance.

Lucas, G., 1997. Les lamelles Dufour du Flageolet 1 (Bézenac, Dordogne) dans le contexte aurignacien. *Paléo* 9, 191–219.

Lucas, G., 2000. Les industries lithiques du Flageolet 1 (Dordogne): approche économique, technologique, fonctionnelle et analyse spatiale. Unpublished doctoral thesis, Université Bordeaux 1.

Lucas, G., 2006. Re-evaluation of the principal diagnostic criteria of the Aurignacian: the example from Grotte XVI (Cénac-et-Saint-Julien, Dordogne), in *Towards a Definition of the Aurignacian: Proceedings of the symposium held in Lisbon, Portugal, June 25–30, 2002*, eds. O. Bar-Yosef & J. Zilhão. (Trabalhos de Arqueologia 45.) Lisbon: Instituto Portuguès de Arqueologia, 173–86.

Meignen, L. & O. Bar-Yosef, 1988. Variabilité technologique au Proche Orient: l'exemple de Kébara, in *L'Homme de Néandertal*, vol. 4: *La technique*, ed. M. Otte. Liège: ERAUL, 81–95.

Mellars, P., 1969. The chronology of Mousterian industries in the Périgord region of south-west of France. *Proceedings of the Prehistoric Society* 35, 134–71.

Mellars, P., 1989. Major issues in the emergence of modern human. *Current Anthropology* 30, 349–85.

Mellars, P., 1990. *The Emergence of Modern Humans: an Archaeological Perspective.* Edinburgh: Edinburgh University Press.

Mellars, P., 1996. *The Neanderthal Legacy: an Archaeological*

Perspective from Western Europe. Princeton (NJ): Princeton University Press.
Mellars, P., 1998. The impact of climatic changes on the demography of late Neandertal and early anatomically modern populations in Europe, in *Neandertals and Modern Humans in Western Asia,* eds. T. Akazawa, R. Aoki & O. Bar-Yosef. New York (NY): Plenum Press, 493–508.
Mellars, P., 1999. The Neanderthal problem continued. *Current Anthropology* 40, 341–50.
Mellars, P. & K. Gibson (eds.), 1995. *Modelling the Early Human Mind.* (McDonald Institute Monographs.) Cambridge: McDonald Institute for Archaeological Research.
Morin, E., 2004. *Late Pleistocene Population Interaction in Western Europe and Modern Human Origins: New Insights based on the Faunal Remains from Saint-Césaire, Southwestern France.* PhD dissertation. Ann Arbor: University of Michigan. http://www.paleoanthro.org/dissertations/Eugene%20Morin.pdf.
Pelegrin, J., 1995. *Technologie lithique: le Châtelperronien de Roc-de-Combe (Lot) et de la Côte (Dordogne).* (Cahiers du Quaternaire 20.) Paris: CNRS.
Pettitt, P. & A.W.G. Pike, 2001. Blind in a cloud of data: problems with the chronology of Neandertals extinction and anatomically modern human expansion (with comments by J.-P. Bocquet-Appel & P.-Y. Demars). *Antiquity* 75, 415–20.
Richards, M., P. Pettitt, M. Stiner & E. Trinkaus, 2001. Stable isotope evidence for increasing dietetic breadth in the European Mid–Upper Palaeolithic. *Proceedings of the National Academy of Sciences of the USA* 98, 6528–32.
Rigaud, J.-P., 1982. Le Paléolithique en Périgord: les données du sud-ouest Saladais et leurs implications. Unpublished doctoral thesis, Université Bordeaux 1.
Rigaud, J.-P., 1988. *La Grotte Vaufrey: paléoenvironnements, chronologie et activités humaines.* (Memoires 19.) Paris: Société Préhistorique Française.
Rigaud, J.-P., 1989. From the Middle to the Upper Palaeolithic: transition or convergence? in *The Emergence of Modern Humans: Biocultural Adaptations in the Later Pleistocene,* ed. E. Trinkaus. Cambridge: Cambridge University Press, 142–53.
Rigaud, J.-P., 1993. La transition Paléolithique moyen/Paléolithique supérieur dans le sud ouest de la France, in *El origen del hombre moderne en le suroeste de Europa,* ed. V. Cabrera-Valdés. Madrid: Universidad Nacional de Educacion a distancia, 117–26.
Rigaud, J.-P., 1996. L'émergence du Paléolithique supérieur en Europe occidentale: le rôle du Castelperronien, in *The Origin of Modern Humans: Xème colloque du XIIIème Congrés de l'UISPP,* eds. ABACO. Forli, 219–23.
Rigaud, J.-P., 2000. Late Neandertals in the south west of France and the emergence of the Upper Palaeolithic, in *Neanderthals on the Edge,* eds. C. Stringer, N. Barton & C. Finlayson. Oxford: Oxbow Books, 27–31.
Rigaud, J.-P., 2001. A propos de la contemporanéité du Castelperronien et de l'Aurignacien ancien dans le nord-est de l'Aquitaine: une révision des données et ses implications, in *Les Premiers Hommes Modernes de la Péninsule Ibérique. Colloque de la 8ème commission de l'UISPP, Vila Nova de Foz Côa, 22–24 octobre 1998,* eds. J. Zilhão, T. Aubry & A.F. Carvalho. (Trabalhos de Arqueologia 17.) Lisbon: Instituto Português de Arqueologia, 26–31.
Rigaud, J.-P. & G. Lucas, 2006. The first Aurignacian technocomplexes in Europe: a revision of the Bachokirian, in *Towards a Definition of the Aurignacian, Proceedings of the symposium held in Lisbon, Portugal, June 25–30, 2002,* eds. O. Bar-Yosef & J. Zilhão. (Trabalhos de Arqueologia 45.) Lisbon: Instituto Português de Arqueologia, 277–84.
Roland, N., 1988. Variabilité et classification: nouvelles données sur le «complexe moustérien», in *L'Homme de Néandertal,* vol. 4: *La technique,* eds. L. Binford & J.-P. Rigaud. Liège: ERAUL, 169–83.
Slimak, L., 1999. La variabilité des débitages discoïdes au Paléolithique moyen: diversité des méthodes et unité d'un concept. L'ensemble des gisements de la Baume Néron (Soyons, Ardèche) et du Champs-Grand (Saint Maurice-sur-Loire). *Préhistoire, Anthropologie Méditerranéenne* 7–8, 75–88.
Smith, F., 1994. Samples, species and speculations in the study of modern human origins, in *Origins of Anatomically Modern Humans,* eds. D.V. Nitecki & M.H. Nitecki. New York (NY): Plenum Press, 228–49.
Soffer, O., 1992. Ancestral lifeways in Eurasia: the Middle and Upper Palaeolithic records, in *Origins of Anatomically Modern Humans,* eds. D.V. Nitecki & M.H. Nitecki. New York (NY): Plenum Press, 101–19.
de Sonneville-Bordes, D., 1960. *Le Paléolithique Supérieur en Périgord.* Bordeaux: Delmas.
Steegman, A.T., F.J. Cerny & T.W. Holliday, 2002. Neandertal cold adaptation: physiological and energetic factors. *American Journal of Human Biology* 14, 566–83.
Stringer, C. & C. Gamble, 1993. *In Search of the Neanderthals: Solving the Puzzle of Human Origins.* London: Thames & Hudson.
Taborin, Y., 1993. Shells of the French Aurignacian and Perigordian, in *Before Lascaux: the Complex Record of the Early Upper Palaeolithic,* eds. H. Knecht, A. Pike-Tay & R. White. Boca Raton (FL): CRC Press, 211–29.
Teyssandier, N., 2003. Les débuts de l'Aurignacien en Europe. Discussion à partir des sites de Geissenklösterle, Willendorf II, Frems-Hundssteig et Bacho-Kiro. Unpublished doctoral thesis, Université Paris X-Nanterre.
Tsanova, T., 2006. *Les débuts du Paléolithique supérieur dans l'est des Balkans.* Unpublished doctoral thesis, Université Bordeaux 1.
Tsanova, T. & J.-G. Bordes, 2003. Contribution au débat sur l'origine de l'Aurignacien: principaux résultats d'une téude technologique de l'industrie lithique de la couche 11 de Bacho Kiro, in *The Humanized Mineral World: Toward Social and Symbolic Evaluation of Prehistoric Technologies in South Eastern Europe.* (ERAUL 103.) Liège: ERAUL, 41–50.
Turq, A., 2000. Le Paléolithique inférieur et moyen entre Dordogne et Lot. *Paléo* supplément 2.
Utrilla, P., L. Montes & P. Gonzallez-Sampériz, 2006. L'Ebre

était-elle une frontière entre 40 et 30 ka?, in *En el centenario de la Cueva de El Castillo: El Ocaso de los Neandertales*, eds. V. Cabrera- Valdés, F. Bernáldo de Quirós & J.M. Maíllo Fernández. Madrid: Centro asociado a la Universidad Nacional de Educación a Distancia en Cantabria, 165–91.

Valladas, H., J.L. Reyss, J.L. Joron, G. Valladas, O. Bar-Yosef & B. Vandermeersch, 1988. Thermoluminescence dates for the Mousterian Proto-Cro-Magnons from Qafzeh (Israel). *Nature* 331, 614–16.

Vandermeersch, B., 1989. The evolution of modern humans: recent evidence from southwest Asia, in *The Human Revolution: Behavioural and Biological Perspectives on the Origins of Modern Humans*, eds. P. Mellars & C. Stringer. Princeton (NJ): Princeton University Press, 155–64.

White, R., 1982. Rethinking the Middle/Upper Palaeolithic transition. *Current Anthropology* 23, 169–92.

White, R., 1993. Technological and social dimensions of the 'Aurignacian age' body ornaments across Europe, in *Before Lascaux: the Complex Record of the Early Upper Palaeolithic*, eds. H. Knecht, A. Pike-Tay & R. White. Boca Raton (FL): CRC Press, 277–300.

Wiessner, P., 1977. *Hxaro: a Regional System of Reciprocity for Reducing Risk among the !Kung San.* Ann Arbor (MI): University microfilm.

Wiessner, P., 1982a. Risk, reciprocity and social influence on !Kung San economics, in *Politics and History in Band Societies*, eds. E. Leacock & R.B. Lee. Cambridge: Cambridge University Press, 61–84.

Wiessner, P., 1982b. Beyond willow smoke and dog's tail: a comment on Binford analysis of hunter-gatherer settlement systems. *American Antiquity* 57, 171–8.

Wiessner, P., 1984. Reconsidering the behavioural basis for style: a case study among Kalahari San. *Journal of Anthropological Archaeology* 3, 190–234.

Wiessner, P., 1986. !Kung San network in a generational perspective, in *The Past and Future of !Kung Ethnography: Essays on Honor of Lorna Marshall*, eds. M. Biesele, R. Gordon & R. Lee. Hamburg: Helmut Buske, 103–36.

Wobst, M., 1976. Locational relationships in Palaeolithic society. *Journal of Human Evolution* 5, 49–58.

Zilhão, J., 2000. The Ebro frontier: a model for the late extinction of Iberian Neandertals, in *Neanderthals on the Edge*, eds. C. Stringer, N. Barton & C. Finlayson. Oxford: Oxbow Books, 111–21.

Zilhão, J. & F. d'Errico, 1999. The chronology and taphonomy of the earliest Aurignacian and its implications for the understanding of Neandertal extinction. *Journal of World Prehistory* 13, 1–68.

Zilhão, J. & F. d'Errico, 2003. The chronology of the Aurignacian and transitional techno-complexes. Where do we stand? in *The Chronology of the Aurignacian and of the Transitional Techno-complexes*, eds. J. Zilhão & F. d'Errico. (Trabalhos de Arqueologia 33.) Lisbon: Instituto Português de Arqueologia, 313–49.

Zilhão, J. & E. Trinkaus, 2002. *Portrait of the Artist as a Child: the Gravettian Human Skeleton from Abrigo du Lagar Velho and its Archaeological Context.* (Trabalhos de Arqueologia 22.) Lisbon: Instituto Português de Arqueologia.

Zilhão, J., F. d'Errico, J.-G. Bordes, A. Lenoble, J.-P. Texier & J.-P. Rigaud, 2006. Analysis of Aurignacian inter stratifications at the Châtelperronian-type site and implications for the behavioural modernity of neandertals. *Proceedings of the National Academy of Sciences of the USA* 103(33), 12,643–8.

Zubrow, E., 1989. The demographic modeling of Neanderthal extinction, in *The Human Revolution: Behavioral and Biological Perspectives on the Origins of Modern Humans*, eds. P. Mellars & C. Stringer. Princeton (NJ): Princeton University Press, 212–31.

Index

Note: Alphabetization is word-by-word. A page reference in italics indicates an illustration.

A

Les Abeilles (France), *142*
abraders, bone/antler, 96
Abri Antelias (Lebanon), 87
Abri Olha (Spain), 158
Abri Pataud (France), 12, *13*, 14, 15, 23, *142*, 146, 147
Abu Zif (Judean desert), *86*, 87
acculturation, 4, 27, 39, 106, 111, 161
Acheulean, 1, 15, 35, 47, 50, 73, 74, 105, 109, 153, 154
adaptability, human, 67, 68–70
Adriatic Sea, 145
adzes, 70
Afghanistan, 94
Africa, 36, 38, 39, 47, 53, 63, 68, 70, 81, 82, 104, 106, 153. *See also* South Africa
Afrolittorina africana, 57
Ahmarian artefacts, 89–90, *90*
Aiello, L., 48, 49
Aira-Tanzawa volcano, 115, 118
alder, 138, 139
Allen's Cave (Australia), *68*
Allier (France), 140
Alpes-de-Haute-Provence (France), 146
Alpes-Maritimes (France), 146
Alps, *124*, 139, 144
Altai Mountains, 2, 103, 104, 106–9, 111, 121
Altmühlian, 131
Altwies-Laagen Aker (Belgium), 128
Amalda (Spain), 140
Ambon Island, 65
Ambrona (Spain), *13*
Ambrose, S., 25
AMH. *See* anatomically modern humans
Amud (Israel), 36, 38
Anatolia, 85, 93, 97
anatomically modern humans (AMH), 2–3, 46–8, 63–70, 115, 119, 123, 133, 146, 161–3. *See also under names of individual species*
Andhra Pradesh (India), 5, 73–4, *74*
Andrews, P., 25, 37, 41
Anguilla anguilla, 146
Anti-Apartheid movement, 24
antler artefacts, 92, 93, 94
Anui (Russia), 105, 106, 109
Anui, river, 103, 104, *104*
Apennines, 139
apes, 48
Apollo 11 (Namibia), *54*, 57
ApSimon, A., 37, 38
L'Arbreda, 143
Arbutus, 138
arctic fox, 141
Arctic Sea, 132
Arcy-sur-Cure (France), 39, 138, *142*, 146
Ardèche (France), 145–7
Arjeneh. *See* Gar Arjeneh
Armstrong, L., 10
Arrillor (Spain), 157
arrowheads, 56
art, 53, 63, 69, 145, 146
Aru Islands, 63, 64, 65–7, *66*, 69
ash fall, 73, 74, *74*, *75*, *76*, *77*, 81
Ash Tree Cave (England), 11
Asia
  central, 2, 93
  east, 64
  south, 73–82
  southeast, 97
  western, 36, 47, 85–97
ass, 140, *140*, 144
Association for Cultural Exchange, 19
Aston Mills (England), *128*
Asturias (Spain), 157
Atapuerca (Spain), 38
Aterian tradition, 106
Attilly (France), *128*
Auboue (France), *128*
Auel, J., 2
Aurignacian, 27, 38–9, 87, 89, 90–93, *91*, *92*, 96, 126–9, *128*, 130, 133, 138–46, *140*, *143*, 161–4
aurochs, 140, *140*
Australasia, 41, 82
Australia, 4, 37, 40, 53, 63–70, *68*, *69*
Australian National University, 23
Australopithecines, *48*, 49
Avdeevo (Russia), 3
awls, *107*, 108
axes, 70, 96, 117
Axlor (Spain), 157
Azilian, 154

B

Bachokirian, 164
backed tools, 54, 55, 56, 58, 85, 90, 96, 106, 108, 126, 132, 133, *133*, 155, *156*, 157
Badger Hole (England), *129*
Baikal. *See* Lake Baikal
Balazuc (France), 147
Balhan-dong (Korean Peninsula), *116*
Bali, *64*, 65
Balkan–Danube axis, 123
Balla (Hungary), *127*
Balme, J., 70
Baltic countries, 124
Balve Cave (Germany), *128*
Balzi Rossi (Italy), 145
bandicoots, 66
Banganapalle (India), 74, *74*
bangle, *107*
Bar-Yosef, O., 37, 40, 96
Baradostian, 94
Barking Dog principle, 47, 49
Barton, C., 147
Basell, L., 21
Basque Country, 157, 158
bats, 67
Les Battuts (France), *142*
Batty, Ernest, *10*
Batty, Etheldred, *10*
Baume Bonne (France), 146, 148
beads, 58, 70, 94, *107*, 108, 110
  shell, 53, 67, 88, 145
bear
  brown, 141
  figurine, 110
Beaumont, P., 37, 54
becs, 117
Beedings (England), 129, *129*, *130*, 131
behaviour, 15, 26, 27, 46, 47, 85, 106, 137, 161
  complex, 63, 65, 138
  modern human, 25, 39, 68–9, 70
  symbolic, 19, 49, 53, 54, 56, 109, 110, 158
Belgium, 123, 126, 128, *128*, 129, *129*, 130, 131
Belloy-en-Santerre (France), *128*
Belvis (France), *139*, *142*
Berekhat Ram 'figurine', 49
Bergman, C., 88
Berrici Hills (Italy), 139
beta-globin gene, 40
Betamcherla (India), *74*
Betula, 138
Bhimbetka (India), 73
bifaces, 56, 78, *79*, 105
  Acheulean, 47, 50
Bilzingsleben (Germany), 49
Binford, L., 1, 2, 14, 16, 17, 19, 20, 22, 26, 40
Binford, M., 20
Binford, S., 1
biodiversity, 137
biogeography, 68–70
birch, 138, 139
birds, 110, 146, 147
birth rate, 21, 162

Bisitun (Iran), *86*, 94
bison, 108, 125, 145, 147
Biwa, 105
Biyke (Altai), 105
Black Sea, *86*, 94, 97, *124*, 163
bladelets, 90, 93, 94, 96, 138, 155, *156*, 157
blades, 4, 6, 16, 20, 54, 73, 78, *79*, 81, 85, 87, 88, 90, 96, 108, 109, 110, 111, 115, 117, *118*, 119, *120*, 121, 126, 130, *130*, 131, 132, *133*, 155, 163
  backed, 90, 106, 126
Blombos Cave (South Africa), 53, *54*, 57
blue duiker, 56
boar, 140, *140*, 142, 144, 145
Bohunician, 39, 87, 130
Boker Tachtit (Israel), *86*, 87, 88, 89, 90, 111
bone
  artefacts, 39, 92, 93, 94, 97, 106, *107*, 109, 163: beads, 108, 110; points, 66, 96
  worked, 49, 53, 56, 57
Bongmyoung-dong (Korean Peninsula), *116*, *119*
Bonsall, C., 18
Boomplaas (South Africa), *54*
Bordeaux University, 15, 16
Bordes, F., 1, 2, 12, 14, 15, 16, 17, 20, 27, 35, 38, 154
borers, 96, 106, *107*, 108, 110
Borneo, 37, 39, *66*
Bourdieu, P., 22
Bourgon, M., 14
Bourlon, J., 96
Bouyssonie, A., 96
bovids, 140, *140*, 141, 143, 144, 147
bows, 88
Boyle, K., 18
Brace, L., 37, 38
bracelet, 108
Bradshaw paintings, 69
brain size, 47, 48, 49
Brandon (England), *23*
Branigan, K., 17
Brassempouy (France), 146
Bräuer, G., 37
Bray, W., 10, *12*, 17
Breese, C., *12*
Breitenbach (Germany), *128*
Bricker, H., *10*, *13*, 16, 23
British Academy, 13, 27
British Museum, 18
Broken Hill (Australia), 35, 39
Brooks, A., 53
Brose, D., 37, 38
Brown, G., 11
Brumm, A., 53
buffalo, 56
Bui Ceri Uato (East Timor), *66*
Bükk Mountains, 127
burial(s), 37, 38, 40, 69
  dog, 67
  Neanderthal, 37, 38, 94
burins, 78, *79*, 85, 87, 88, 90, 93, 94, 96, 106, 108, 109, 110, 117, 126, 128, 130, 132
Burke, A., 143
Buru Island, 66
bushpig, 56
Byeongsan-ri (Korean Peninsula), *116*
Byika caves (Russia), *104*
Byzovaya (Russia), *131*, 132

C

La Cala (Italy). *See* Grotte della Cala
Cambridge University, 11, 22–7
Camiac (France), *139*, 140, 141, *142*
Caminade Est (France), 147
camps, 1, 46, 55, 80, 90, 93, 94, 126, 127, 129
Canberra (Australia), 23
Can Hasan 3 (Turkey), 19
Cann, B., 25
Cann, R., 37, 40, 41
Cantabria, 138, 142, *143*, 144, 148, 153–8, *154*
Capra ibex, 147
Carballo, J., 154
carnivores, 93, 137, 138, 140, 141, 142, 143, 147, 148
Carpathian region, *86*, 124, *124*, 125, 126
Carpenter's Gap (Australia), *68*, 69
Caspian Sea, *124*, 132
cassowary, 67
Castanet (France), *142*
Castelcività (Italy), *139*, 143, 144
El Castillo (Spain), 138, *139*, 143, 153–8, *154*
Caucasus Mountains, 2, 87, 94–6
cave
  art, 145, 146
  bear, 4, 141
  lion, 4, 141
  See also under names of individual cave sites
Celtis seeds, 65
chaîne opératoire, 88
chalcedony, 80
chamois, 140, *140*, 141, 144, 145
chanfreins, 87
Changnae (Korean Peninsula), *116*
Chanut, A. *See* Mellars, A.
La Chapelle-aux-Saints (France), 35
charcoal, 35, 53, 55, 65, 88, 108
Charentian, 157
Chateau de Marzac (France), *13*
Châtelperron (France), *139*, 140, *142*
Châtelperronian, 13, 27, 35, 37–9, 41, 137–42, *139*, *140*, *143*, 144–6, 161, 164
Chenoves (France), 139
Cherny Anui (Russia), 105
chert, 80
Chikhen (Mongolia), 110
Childe, G., 45, 46
chimpanzee, 49
China, 111, 121
Chommal (Korean Peninsula), *116*, *119*
Chondrostoma polylepsis, 146
Chongok-ri (Korean Peninsula), *116*
choppers, 104, 117
chronologies
  changing, 35, 38–40
  Korean Peninsula, 117–19
Chusovaya, river, 132
Cieszyn (Poland), *127*
Clark, Geoffrey, 26
Clark, Grahame, 9, 13, 26, 27
Clark, J.D., 37
Clarke, D., 14
cleavers, 154–7
Cleyet-Merle, J.-J., 146
climate, 3, 66, 85, 103, 118, 138, 162
Cnoc Coig (Scotland), 18
Côa Valley (Portugal), 145
cognition, 45, 47, 49
  complex, 53, 58
Coles, J., 17
Collins, D., 12, 15
colluvial deposits, 117
colonization, 64, 82
Combe-Grenal (France), 1, 12, 15, 16, 27
Commonwealth, 24
El Conde (Spain), 157
Conus, 70
cooking, 48
Copeland, L., 87, 89
Corbicula tibetensis, 108
cores, 78, *79*, 94, 96, 105, 110, *130*, 155
La Cotte de St Brelade (Jersey), 14
Les Cottés (France), *139*, *142*
Count de la Vega del Sella, 154
Covalejos (Spain), 157
crania, 37, 38, 39, 40
cremation, 63
Creswell Crags (England), 10, 12, *12*, 23
Crimea, 97
Cro-Magnon (France), 35
Cro-Magnons, 36, 37, 39, 96, 161, 162. *See also* anatomically modern humans
Cuddie Springs (Australia), *68*
Cueto de la Miña (Spain), 143, 146
Cueva Millán (Spain), 146
Cueva Morín (Spain), *139*, 140, 141, 143, 145, 148, 153–8, *154*
cultural ecology, 19
Cunliffe, B., 11
Cyrenaica (Libya), 87, 106
Czech Republic, *127*

D

Dabban, 87
Daejeon (Korean Peninsula), 115, *116*
Daejeong-dong (Korean Peninsula), *116*
Dama dama, 144

Dangga (Korean Peninsula), *116*, *119*
Danghasan (Korean Peninsula), *116*
Daniel, G., 11
Dansgaard/Oeschger events, 123, 124, *125*, 131
Danube, river, 97, 123, *124*, 125, 126, 128, 129, 131
dating, 2, 22, 35, 37, 38–40, 45, 88, 93, 94, 96, *119*, 140, 145
David, N., 12, *13*, 14, 16
Davidson, I., 47, 49
Davies, W., 27
Day, M., 37
Deacon, H., 25, 26
Deacon, T.W., 49
débitage, 93, 96, 121, 126, 153, 155, 157
Dederiyeh Cave (Syria), *86*, 88
deer, 93, 108. *See also* fallow deer; red deer; reindeer; roe deer
Deith, M., 18
Delpech, F., 137, 146
Denisova Cave (Russia), 103, 104, *104*, 105, 106, 108, 109
Dennell, R., 19
Dentaliidae, 70
denticulates, 88, 104, 105, 106, 108, 109, 110, 117, 154, *156*
Descartes, R., 47
deserts, 63, 67–70, *69*
Devil's Lair (Australia), 64
Diepkloof (South Africa), *54*, 57
diet, 66, 68, 162
Dili (East Timor), *66*
Dimbleby, G.W., 21
dispersal, 2, 63–70
Djebel Irhoud. *See* Jebel Irhoud
DNA, 24, 37, 40, 162
Dnepr, river, *86*, 123, *124*
Dnestr, river, *86*, 123, *124*
Dobczyce (Poland), 127
Dobra (Poland), 124
dog, 67
dolerite, 56, 80
domestication, 45
Don, river, 123, *124*, 132
Dordogne (France), 1, *13*, 138, 141, 142, 146, 147, 148, 162, 163
Doubs (France), 148
Drahany Plateau (Moravia), 127
Drinkwater, G., 11
Du Docteur Cave (Belgium), *128*
Du Prince Cave (Belgium), *128*
Dufour bladelets, 93, 94, 138, 155, *156*
Dunajec, river, 126
Dunbar, R., 48, 49
Dunsan-dong (Korean Peninsula), *116*
Durba Springs (Australia), *69*
Durban (South Africa), 56
Durubong Cave (Korean Peninsula), *116*
Dzierżysław (Poland), 126, *127*
Dzudzuana Cave (Georgia), *86*, 95, *95*, 96

E

Early Upper Palaeolithic, 93, 94, 96, 97, 131–3
East Timor, 63, 65–7, *66*, 69
Ebro depression, 162, 163
éclat debordant, 78, *79*
ecology, 137–47
eel, 146, 148
Eger (Hungary), *127*
eggshell, 57, *107*, 108, 110
Egypt, 87
Ekain (Spain), *139*, 140, 141
Elba, river (Kazakhstan), 105
Elbe, river, 128
Emireh (Israel), *86*
Emiran points, 87, 88, *89*
encephalization, 48, 49
end-scrapers, 85, 87, 88, 90, 94, 96, 106, 108, 109, 110, 117, 128, 130, 132, 133, *133*
England, 10, 11, 12, 15, 17, 18, *23*, 27, 123, 127, 128, *128*, 129, 130, 131
engravings, 93
Epimousterian, 164
equifinality, 3–4
Erica, 138
Erq el-Ahmar (Judean desert), *86*, 87, 89
Erramala Hills, *74*
El Esquilleu (Spain), 157
Ethiopia, 37, 39
EUP. *See* Early Upper Palaeolithic
Euphrates, river, *86*
Eurasia, 39, 41, 63, 106
Europe, 123, *124*, 161, 163
  blade industries, 126
  climate, 85
  eastern, 39, 96
  human evolution, 35, 36
  lithics, 153
  modern humans, 63
  population shifts, 69, 93, 97
  western, 87
European Economic Community, 24
European Lowland, 123–33
exchange, 63, 128, 146, 163
extended mind model, 45, 47, 49
external storage, 53
Les Eyzies (France), *13*

F

La Fabbrica (Italy), *139*, 144, 148
fallow deer, 140, *140*, 144
Far East, 36, 39
farmers, 46
Farrand, W., 37
fauna, 137, 139–44. *See also under individual species*
La Ferrassie (France), *142*, 146
Figueira Brava (Portugal), 143
figurines, 49, 63, 110
Finland, 124
fire ecology, 21
fish/fishing, 137, 146, 148
  hooks, 67
  pelagic, 67
Fitzwilliam House, 11
Le Flageolet I (France), 146
Flas, D., *130*
La Flecha (Spain), 157
Fleming, A., 17
flint, 38, 94, 127, 155, 157
Flint Valley (Mongolia), 110
Flores, 65, *66*
Florisbad (South Africa), 36, 39
flutes, 50, 146
FMH. *See* fully modern humans
foliates, 96, 106, 108
Font Robert type, 131
Font Yves points, 93
Fontauray (France), *142*
Fontenioux (France), *139*
foragers, 4, 5, 90, 96
Forde, C.D., 11
forests, 21, 56, 57, 66, 67, *125*, 138, 139, 145
Fossellone (Italy), 144
fox, 108, 141
France, 1, 2, 13, 23, 27, 35, 69, 126, 128, *128*, 137, 138, *139*, *143*, 144, 146, 147. *See also under names of individual sites*
Franco-Cantabria, 123, 137, 139–43
Freeman, L., 154
French, D., 19
fruitbats, 66
fully modern humans (FMH), 46, 47, 48
Fumane Cave. *See* Grotta Fumane

G

Gabillou (France), 147
Gamble, C., 24, 27, 46, 63, 69
Gangwon Province (Korean Peninsula), 118
Gar Arjeneh (Iran), *86*, 94
Garchi (Russia), *131*, 132
Gard (France), 147, 148
Gargas (France), *139*
Garrod, D., 87, 93
Gatzarria (France), *139*, 146, 158
Gaudzinski-Windhauser, S., 147
Gebe Island, 65, 66, 70
Gebel Maghara (Egypt), *86*, 90
Geloina, 66
gemstones, 108, 110
genetics, 4, 5, 24, 25, 26, 37, 40, 41, 45, 46, 73, 82, 162, 164
genome, Neanderthal, 4, 40
Georgia, *86*, 94, 95, 96
Germany, 49, 50, 128, *128*, 129, *129*, 131
Germolles (France), *139*
Gesher-Benot-Ya'aqov (Israel), 50
Geum Cave (Korean Peninsula), *116*

giant deer, 4, 140, *140*
giant rat, 56
Gibson, K., 27
Gibson Desert (Australia), *69*
Giddens, T., 22
Gigny-sur-Suran (France), 148
Gigok (Korean Peninsula), *116*, *119*
Gilead, I., 93
Gioia, P., 138
Gironde (France), 140, 147, 148
glaciers, 94
Glaston (England), 129, *129*
glues, 58
Gobi Altai (Mongolia), 110
González-Echegaray, J., 154
Góra Puławska (Poland), 128, *128*
Goring-Morris, A., 96
Gorny Altai (Russia), 109
Gorye-ri (Korean Peninsula), 115, *116*, *120*
Gostwica (Poland), *127*
Gould, R., 67
Goyet Cave (Belgium), *128*, *129*
Grande Roche de la Plématrie à Quincay (France), 138, *142*
Grange Farm (England), *129*
grassland, 54, 57, 67, 124, *125*, 138, 139, 141, 144
grave goods, 69
graves, 50, 146
Gravettian, 131, *143*, 163
Grayson, D., 137
Great Sandy Desert (Australia), *69*
Great Victoria Desert (Australia), *69*
Greater Australia. *See* Sahul
Greece, 143
Gregory Gorge (Australia), *68*
Grotta Barbara (Italy), 143
Grotta Bernardini (Italy), *139*
Grotta del Cavallo (Italy), *139*, 143
Grotta della Cala (Italy), *139*, 144
Grotta di Paina (Italy), 139
Grotta di Uluzzo (Italy), *139*
Grotta Fumane (Italy), *139*, 143, 144, 145, 146
Grotta Paglicci (Italy), 144
Grotte XVI (France), *139*, 140, *142*
Grotte Chauvet (France), 145
Grotte de la Verpillière (France), *139*
Grotte des Fées (France), *139*, 140
Grotte du Pape (France), *142*, 146
Grotte du Renne (France), *139*, *142*, 146
Grotte Figuier (France), *139*, 146
Grotte Marie (France), 148
Grotte Vaufrey (France), 148, 162
group sizes, *48*
La Güelga (Spain), 140, 157
Gumi-dong (Korean Peninsula), *116*
Gunang Cave (Korean Peninsula), *116*, *119*
Guomde (Kenya), 39
Gyeonggi Province (Korean Peninsula), 118
Gyeongsang Province (Korean Peninsula), 115

H

habitat, 4, 67, 106, 109
Haeundae (Korean Peninsula), *116*
Hahwagye-ri (Korean Peninsula), *116*, *119*
Hail Mary Hill (England), 10
Halmahera Island, 65, 66, *66*
hammer, 55
Hammersley Range (Australia), 67, *69*
hammerstones, 78
handaxes, 70, 96, 117
Hantan, river, 118
hare, 133
Harper, P., 54
Haua Fteah (Libya), *86*, 87
Haute-Garonne (France), 146
Havange (France), *128*
Hayonim Cave (Israel), *86*, 93
Hazar Merd (Iraq), *86*, 94
hearths, 55, 56, 57, 110
Hedges, R., 22, 23
Heinrich event 4, 162
hematite, 69
Henshilwood, C., 53
Hérault (France), 148
Herbeville-le-Murger (France), 129
Herto (Ethiopia), 39, 46
Hesse (Germany), 128
Higgs, E., 13, 18, 45
Higham, C., 11
Hiscock, P., 64, 67, 70
Hobbit. *See Homo floresiensis*
Hodder, I., 22
Hodson, R., 17
Holdaway, S., 70
Holocene, 4, 65, 66, 67, 68, 70, 118
Homo antecessor, *48*
Homo erectus, 39, *48*, 65, 66, 104, 105
Homo ergaster, *48*, 104
Homo floresiensis, 5, 40, 46, 63, *64*, 65, 70, 82
Homo georgicus, *48*
Homo habilis, *48*
Homo heidelbergensis, 38, 39, *48*
Homo neanderthalensis, 38, 39, 46, *48*, 49
Homo rudolfensis, *48*
Homo sapiens, 25, 35, 37, 39, 40, 46, 48, 49, 68, 81, 82, 121
Homo sapiens idaltu, 46, *48*
Homo sapiens sapiens, *48*, 119, 121, 161
Homo soloensis, *48*
hooks, jabbing, 67
Hopyeong-dong (Korean Peninsula), *116*, 118, *119*
hornfels, 56
horses, 93, 125, 133, 140, *140*, 141, 143, 144, 145, 147, 148
Houamian, 94
Howell, C., *13*, 36
Howells, B., 35
Howells, W., 37
Howiesons Poort, 54, 55, 56, 57, 58, 82
Hublin, J.-J., 37
Hungary, *127*
hunter-gatherers, 11, 12, 46, 65, 90, 146, 162
Hutchins, E., 47
Huxley's Line, *64*
Hwadae-ri (Korean Peninsula), *116*, *119*
Hwasun (Korean Peninsula), *116*
Hyakuhensentoki, 118
hybridization, 4, 162
hyaena, 4, 141
Hyaena Den (England), *128*

I

Iberian Peninsula, 69, 146, 153, 161, 162, 163
ibex, 140, *140*, 143, 144, 145, 147
Ice Age, 118
ice-wedges, 118
identity, 53, 54, 57, 69
Igeum-dong (Korean Peninsula), *116*
Ilex, 138
Ilsenhöhle-Ranis Cave (Germany), 131
Imbul-ri (Korean Peninsula), *116*
immune-system deficiency, 162
India, 5, 39, *74*
  lithic technology, 73–82
Indian Ocean, 56, 85
Indonesia, 63, 65, 82
infections, 162
Initial Upper Palaeolithic (IUP), 85, 87–9, 93, 94, 97, 106
Inskeep, R., 11
Institut de Paléontologie Humaine (Paris), 154
International Union of Pre- and Protohistoric Sciences (IUPPS), 23–4
interpleniglacial, 123–5, 133
Iran, 85, *86*, 94
Isaac, B., 12
Isaac, G., *12*
Isernia la Pineta (Italy), 147
Iskra Cave (Russia), *104*
island biogeography, 68–70
Island Southeast Asia, 67. *See also under names of individual countries and islands*
Isle Valley, Dordogne (France), 146
isotope studies, 85, 146
Israel, 37, 85, 147
Isturitz (France), *139*, 146
Italy, 38, 137, 138, *139*, 143–7
IUP. *See* Initial Upper Palaeolithic
Ivanovce (Czech Republic), *127*
ivory artefacts, 106, *107*, 108
Izquierdo, R., 146

J

Jacobi, R., 38, *130*
James Range (Australia), 67
Janggwan-ri (Korean Peninsula), *116*
Jangheung-ri (Korean Peninsula), 115, *116*
Janghung-ri (Korean Peninsula), *116*, 118, *119*
Japan, 115, 121
Jarman, M., 13
Jasna Cave (Poland), *127*
Java, 39, 65
Javakheti region (Georgia), *86*, 94
Jebel Irhoud (Morocco), 35, 36, 39
Jelinek, A., 16
Jeongjang-ri (Korean Peninsula), *116*, *119*
Jeonnam Province (Korean Peninsula), 115
Jerimalai (East Timor), 66, 67
jewellery, 50, 63, 69, 73, 97, 106, 146, 163
  bangle, *107*
  beads, 58, 70, 94, *107*, 108, 110
    shell, 53, 67, 88, 145
  bracelet, 108
  pendants, 39, 70, 93, *107*, 108, 110, 132
  pin, 56
  ring, 108
Jin Rockshelter (Korean Peninsula), *116*, *119*
Jones, D., 18
Jones, R., 21
Jordan, 88, 90, 94
Jordan Valley, 93
Judean desert, *86*, 87, 89
Juksan-ri (Korean Peninsula), *116*
Jungnae-ri (Korean Peninsula), *116*, *119*
Jurassic flint, 127
Jurreru, river, 73, 74–8, *74*, *76*, 81, 82
Jwalapuram (India), 74–81, *74*, *75*, *76*, *77*

K

Kaalpi (Australia), *69*
Kama, river, 123, *124*, 132
Kamenka, 110
Kaminnaya Cave (Russia), *104*, 105
Kanal Cave (Turkey), *86*, 88
kaolinitic agalmatolite, 108
Kapova Cave (Russia), 145
Kara-Bom (Russia), *104*, 105, 106, 109, 110, 111
Kara-Tenesh (Russia), *104*, 109
Karain E (Turkey), *86*, 93
Karakol (Russia), *104*, 106, 111
Karama (Russia), 104, *104*
Karginian period, 110
Kazakhstan, 103, 105
Kebara Cave (Israel), 26, 37, 38, 41, *86*, 88, 90, 93, 96
Kei Islands, *66*
Kent's Cavern (England), 128, *128*, 129
Khara-Khorin (Mongolia), 110
Khotyk (Russia), 110
Kimberley (Australia), 65, 69, *69*, 70
King Arthur's Cave (England), *129*
Klasies River (South Africa), *54*, 57
Klein, R., 37, 46
Klisoura Cave (Greece), 143
knapping, 153
knives, 108, 109, 110
Korean Peninsula, 111, 115–21
Kortallayar Basin (India), 73
Kostenki (Russia), 3, 130, *131*, 132, 157
el-Kowm (Syria), *86*, 88, 94
Koziarnia Cave (Poland), *129*
Kraków (Poland), 124, 125, 126, *127*, 128
Kraków-Kryspinów (Poland), 124
Kraków-Księcia Józefa Street (Poland), 125, 126, 131
Kraków-Zwierzyniec I (Poland), 127, *127*, *128*
Krapina (Croatia), 37, 38
Krems points, 93, 94
Kruszynski, R., 38
Ksar 'Akil (Lebanon), *86*, 87, 88, 89, 93, 111
Kulpi Mara (Australia), *69*
Kum, river, 115
Kumpa-ri (Korean Peninsula), *116*, *119*
Kunji Cave (Iran), *86*, 94
Kuper, A., 26
Kurnool District (India), 74
Kyrgyzstan, 103, 105

L

Labeko Koba (Spain), *139*, 140, 141
Ladybrand (South Africa), 54
Lagaman, 90
Lake Baikal region, 109–11
Lake Mungo (Australia), 37, 40, *68*, 69
Lakonis (Greece), 3
Landry, G., 143
language, 27, 46, 47, 49
Lapita, 67
Last Glacial Maximum (LGM), 2, 4, 63, 64, 65, 67, *68*, 69, 117, 123, 133, 147, 162
Later Stone Age (LSA), 36, 54
Laurent, J., 15
Laurent, P., 15
Laville, H., 16, 27
Lazaret (France), 146
Leach, E., 20
leaf points, 127, 129, 130, 131, 132, 153
Leeds University, 11
Lemdubu Woman, 69
lemming, 125
Lena, river, 110
Lene Hara Cave (East Timor), 66, *66*
leptolithic cultures, 123
leptolithization, 161
Levallois, 14, 16, 78, *79*, 81, 87, 88, 96, 105, 106, 108, 109, 110, 117, 121, 125, 131, 153, 155, *156*, 157
Levant, 2, 3, 35, 37, 38, 39, 81, 82, *86*, 87–93, 96, 97, 111
Lewin, R., 25, 40
Lezetxiki (Spain), 143, 157
LGM. *See* Last Glacial Maximum
Liang Bua (Indonesia), 70, 82
Liguria (Italy), 139, 145, 146
limestone, 80, 81, 93
Limousin plateaux, 162
limpets, 18
Lincombian-Ranisian-Jerzmanowician (L-R-J) complex, 126, 129–32, *129*, *130*
lithics. *See* stone artefacts
Liujiang (China), 39
Lombok Island, *64*, 65
Lommersum (Germany), *128*
Lorraine (France), 128
Lubbock, J., 46
luminescence techniques, 38, 39, 40, 54, 56, 64, 75
de Lumley, H., 24
Luria, A.R., 47
lutite, 157
Lydekker's Line, *64*
lynx, 141

M

Maam-ri (Korean Peninsula), *116*
Maba, 37
Macropus agilis, 66, 67
Magdalenian, 142, *143*, 146, 147
Maiserian, 131
Maisieres (Belgium), 128, *128*
Malakunanja II (Australia), 40
Malan, B., 54
Maloyalomanskaya Cave (Russia), *104*, 105, 109
Maluku Islands, 65, 66, 69
mammoth, 4, 125, 133, 140, *140*
  tusk, 132: artefacts, 106, *107*; ring, 108
Mamontovaya Kuria (Russia), *131*, 132
Mamutowa Cave (Poland), 127, *128*
Manchuria, 121
Mandu Mandu (Australia), 69
Manus Island, 67, 70
Marean, C., 53
Marine Isotope Stages
  MIS-2, 57
  MIS-3, 2, 57, 138, 143
  MIS-4, 38, 57
  MIS-5, 38
  See also Oxygen Isotope Stages
Marks, A., 93
marsupials, 67
Martin, H., 146
Massif Central, 128, 162
Matja Kuru (East Timor), 67

Mauran (France), 146, 147
McBrearty, S., 53
McBurney, C., 12, 14, 17
Mediterranean, 85, 93, 97, 123, *124*, 138, 143, 145
Megaceros sp., 4, 140, *140*
Mehlman, M.J., 25
Melanesia, 67
Mellars, A., *13*, 18, 19
Mellars, H., 9, *9*
Mellars, P., 1, 2, 9–28, *9*, *12*, *13*, *17*, *23*, *28*, 35, 37, 38, 39, 40, 41, 45, 137, 145, 146, 147
meseta, 162
Mesolithic, 1, 10, 15, 18, 19, 21, 26
Mesopotamia, 85
Mezmaiskaya Cave (Russia), *86*, 96
Micoquian, 96, 131
microblades, *79*, 115
microliths, 73, 96, 143
Microtus gregalis, 138
Middle Ages, 153
Middle East, 161
Middle Stone Age, 36, 39, 41, 53–8, 68, 73, 82
Middle–Upper Palaeolithic Transition, 20, 21, 24, 39, 46, 137, 147
  Europe and Russia, 123–33
  Siberia and Mongolia, 103–11
  Asia: south, 81–2; western, 85–97
migration, 94, 104, 105, 110, 123, 145, 163
Millak-dong (Korean Peninsula), *116*
Miller, D., 22
Mithen, S., 47
Mitochondrial Eve, 41
Mocchi (Italy), 146
Modrice (Czech Republic), *127*
molluscs, 88, *107*, 108, 145
Mongolia, 103–11, 119
monkeys, 48, 49
monsoon, 85
montane, 145
Moore, H., 22
Moore, M., 53
Moravia, 87, 127, 128
Moravian Gate, *124*, 128
Moravsky Krumlov IV (Czech Republic), *127*
Morocco, 53, 57
Morotai Island, 66
Morse, K., 70
mortality, 97, 162
Mother Grundy's Parlour (England), 12
Mount Baekdu (Korean Peninsula), 117
Mount Carmel (Israel), 87
Mount Jangbai. *See* Mount Baekdu
Mousterian, 13, 14, 20, 35, 85, 87, 88, 93, 94, 96, 106, 109, 131, 141, 146–8, 153–8, 161–4
  Denticulate, 1, 15, 154, 157
  Discoid, 153, *156*, 157
  Ferrassie, 1, 15, 16, 27
  Final, 164
  Laminar, 153, 155
  MTA-type Mousterian industries, 131
  Quina, 1, 15, 16, 27, 131, 153, 154, 156–7
  Typical, 1, 15, 154
Le Moustier (France), 27
Movius, H., 12, *13*, 14
Mugaranian, 105
Mugodjary Mountains, 105
Muluku Province (Indonesia), 68
Mungo. *See* Lake Mungo
Muralovka (Russia), 129
musical instruments, 50, 146
musk ox, 125

N

Naeheung-dong (Korean Peninsula), *116*, *119*
Nag Hamadi (Egypt), *86*, 87
Nahal Ein Gev (Israel), *86*, 93
Nahr Ibrahim (Lebanon), *86*, 88
Naples (Italy), 139
Naskalanski industry, 126
Nassarius kraussianus, 53, 57
Natchez Trace (USA), *28*
Natchex-Under-the-Hill (USA), *28*
Nawalabila (Australia), 40
Neanderthals, 1, 2–4, 25, 26, 35–41, 45–6, 82, 94, 96, 97, 125, 137, 146–7, 161–4
Near East, 93, 105, 153, 163
needles, 108
Negev highlands, 87, 90, 93
Neolithic, 3, 45, 46, 47, 49, 67
Neslovice (Czech Republic), *127*
Neuville, R., 87, 89
New Archaeology, 19, 22
Newcastle University, 17, 18
Ngandong (Indonesia), 35, 37, 39
Niah Cave (Borneo), 37, 39
Nietoperzowa Cave (Poland), *129*, *130*, 131
Nile, river, 88, 106
Niven, L., 147
Noble, W., 47, 49
Nobong (Korean Peninsula), *116*, *119*
Noeun-dong (Korean Peninsula), *116*, *119*
Nolan, R., 18
Normandy (France), 128
North Korea, 119
North Sea basin, 123
Nosan-ri (Korean Peninsula), *116*, *119*
notches, 104
Nullarbor Plain (Australia), *69*

O

O'Connor, S., 64, 65
Oase hominins, 39
Obermaier, H., 154
Obłazowa Cave (Poland), 126, 127, *127*, *128*
obsidian, 74, 94, 96
ochre, 55, 58, 69, 70
  red-, burial, 37, 40, 50, 146
Oder, river, *124*, 127
Ohnuma, K., 88
Okgwa (Korean Peninsula), *116*
Okladnikov Cave (Russia), *104*, 105
Oldowan, 104
Omo Kibish (Ethiopia), 35, *36*, 37, 39
Ondratice (Czech Republic), *127*
ophite, 155, 157
Orechov (Czech Republic), *127*
Orkhon, river, 110
ornaments, 54, 69, 73
  personal, 63, 70, 97, 106
Orok-Noor (Mongolia), 110
Oronsay (Scotland), 10, 18, 19
Ortvale Klde (Georgia), *86*, 95, 96
Ostrava (Czech Republic), *127*
ostrich(es), 58
  eggshell, 57, *107*, 108, 110
Otice (Czech Republic), *127*
Otte, M., 25, 26
Out of Africa, 37, 41, 81
Oxygen Isotope Stages
  OIS-2, 123, 162
  OIS-3, 56, 94, 123, 125, 126, 128, 131, 133, 162
  OIS-4, 56, 73, 81, 94, 123, 125, 126, 131, 133
  OIS-5, 73, 81
  OIS-9, 162
  See also Marine Isotope Stages

P

Pacific Ocean, 67
paintings, 50, 65, 69
Pair-non-Pair (France), *139*, *142*, 147, 148
palaeogeography, northern Europe, 123–5
Pan African Association, 24
Papagianni, D., 27
Pape. *See* Grotte du Pape (France)
Papua New Guinea, 64, *66*, 68
Parnkupirti (Australia), 68
Patne (India), 73
Patpara (India), 73
Paviland Cave (Wales), 128, *128*, 129
Pavlov, P., 132
Payne, S., 18, 19
pebble tools, 104, 105
Pech-de-L'Azé II (France), 146
Pechora, river, 123, 124, *124*, 131, 132
Peers Cave (South Africa), *54*, 57
pendants, 39, 70, 93, *107*, 108, 110, 132
El Pendo (Spain), *139*, *154*
Périgord (France), 137, 145
Pettitt, P., 27
Pfynnon Beuno Cave (Wales), *128*, *129*

phalangerid, 66
Phillips, P., 16
phytoliths, 57
Le Piage (France), *139*, *142*
Piekary (Poland), 125, 126, *128*, 131
Pilbara uplands (Australia), 67
pin, 56
Pin Hole (England), *129*
pine, 138, 139
Pinus sylvestris, 138
plant gum, 58
plateaux, Europe, 125–6
Pleistocene, 35, 37, 63–6, *64*, 70, 73, 94, 104, 105, 117, 118, 147
Po, river, 139
Podzvonkaya (Russia), 110
points, 54, 56, 78, 79, 85, 89, 96, 108, 109, 110, 115, 117, 118, *120*, 157
  edge-ground, 66
  Emiran, 87
  Krems, 93, 94
  leaf, 127, 129, 130, 131, 132, 153
  Levallois, 88
  Mousterian, 106
  el-Wad, 90, 93
Poland, 126–31, *127*, *128*, *129*
polar fox, 125
pollen, 138, 139
Pont-d'Arc (France), 145
Poprad, river, 126
population(s)
  African, 40
  bottlenecks, 73
  change, 36
  movements, 94. *See also* migration
Le Portel (France), *139*, *142*
Portugal, 145, 163
postprocessualism, 22
Princesse Pauline Cave (Belgium), *128*
processualism, 21
Protoaurignacian, 138, 143, 145
Protsch, R., 37
Prunus, 138
pseudomorphs, 154
Puchacza skala Cave (Poland), *129*
Puente Viesgo (Spain), 154
Puntutjarpa (Australia), 67, *69*
Punung (Indonesia), 40
Purfleet (England), 17, *69*
Puritjarra (Australia), *68*, 69
Puskaporos (Hungary), *127*
Pyeongchang-ri (Korean Peninsula), *116*
Pyrenees, 161
python, 66

Q

Qafzeh Cave (Israel), 26, 36, *36*, 37, 38, 41, *86*, 87, 90
quartz, 56, 80, 117, 118, 119, 127
quartzite, 56, 80, 117, 118, 127, 155, 157
Quaternary, 74
La Quina (France), 146

R

Raba, river, 126
rabot, 96
radiolarite, 94, 127
Radley, J., 10
Rajasthan (India), 73
Rakefet Cave (Israel), *86*, 93
Ralph, N., 18
Ranis Cave (Germany), *128*, *129*
rats, 67
red deer, 140, *140*, 141, 142, 143, 144, 145, 147
reindeer, 125, 137, 140, *140*, 141, 142, 143, 145, 146, 147
Reinhardt, S., 18
religion, 53
Renfrew, C., 2, 11, 17, 18, 19, 20, 21, 23, 46, 49
Renfrew, J., 17, 20
reptiles, 67
Reynolds, T., 11
RFLPs (Restriction Fragmentation Length Polymorphisms), 40
Rhamnus, 138
Rhine, river, *124*, 128
rhinoceros, 133, 140, *140*, 145
Riel-Salvatore, J., 137, 147
La Riera (Spain), 143
Rigabe (France), 146
ring, 108
riverine terraces, 117
Riwi (Australia), 68, *68*, 70
Robin Hood Cave (England), *129*
Roc de Combe (France), 138, *139*, 140, *142*, 146
La-Roche-à-Pierrot (France), *139*
Roche au Loup (France), *139*
Les Roches (France), *139*
rockshelters, 1, 2, 16, 40, 56, 66, 67, 68, 69, 70, 74, 89, 93, 95, *116*, 145
Roe, D., 14
roe deer, 140, *140*, 142, 144, 145
Rond-du-Bary (France), 147
Rose Cottage (South Africa), 5, 53–8, *54*
Roti Island, 65
Rouvroy (France), *128*
Rowlands, M., 47
Rozdrojovice (Czech Republic), *127*
Russia(n), *131*, 145
  Lowland, 124, 132
  Plain, 123, *124*
  See also under individual site names
Le Ruth (France), *142*
Ryokpo Cave (Korean Peninsula), 119

S

Saccopastore (Italy), 38
Sackett, J., *13*, 14, 16, 17, 19
Saemgol (Korean Peninsula), *116*
Sagi-ri (Korean Peninsula), *116*
Sahul, 5, 63, 64, *64*, 65, 66, 69
saiga, 147
Saldanha (South Africa), 35
salmon, 137, 146, 148
Salmo sp., 146
Salpêtre-de-Pompignan (France), 147, 148
Sam-ri (Korean Peninsula), *116*
Sambungmacan (Java), 39
sandstone, 155, 157
Sangmuryong-ri (Korean Peninsula), *116*
Sangsi Rockshelter (Korean Peninsula), *116*
Saône, river, *124*, 128
sapient paradox, 46, 49
Saulges Cave (France), *128*
Scandinavia, 123
Schlanger, N., 27
Schoenoplectus sp., 57
Schöningen (Germany), 50
Ściejowice (Poland), 124
scrapers, 56, 66, 78, *79*, 93, 117. *See also* end-scrapers; side-scrapers
seasonality, 14, 18, 19, 20, 22, 127, 137
sedge nutlets, 57
seeds, 53, 65
Sefunim Cave (Israel), *86*, 93
Seram Island, *66*
Serino (Italy), 139
serpentine, 108
Serpent's Glen (Australia), 68, *68*, *69*
Setzler, F.M., 14
shale, 108
Shanidar Cave (Iraq), 38, *86*, 94
Shaw, T., 24
Shawcross, W., *12*
Shea, J., 26
Sheffield Museum, 11
Sheffield University, 10, 17–22
shell(s), 53, 108, 146, 163
  adzes, 70
  artefacts, 65, 106, *107*
  beads, 67, 70, 88, 145
  jewellery, 146
  marine, 57
  middens, 18, 19
  pendant, 132
  perforated, 53, 57
  scrapers, 66
shellfish, 70, 145
Sherratt, A., 19
shrubs, 138
Shuidonggu (Mongolia), 119
Siberia, 103–11, 115
Sibudu Cave (South Africa), 5, 53–8, *54*
side-scrapers, 54, 85, 87, 94, 96, 104, 105, 106, 108, 110, 126, 132, 154, *156*, 157
El Sidrón (Spain), 157
sierras, 162
Sierras Orientais (Spain), 140

Silesian flint, 127
Sima de los Huesos (Spain), 38
Simmons, I., 21
Simpson Desert (Australia), *69*
Sinai Peninsula, 90, 93
Sinbuk (Korean Peninsula), 115, *116*, *119*
Sindae-dong (Korean Peninsula), *116*
Singa (Sudan), 39
Sinmak (Korean Peninsula), *116*
Sino-Malaysia, 106, 111
es-Skhul Cave (Israel), 26, 36, *36*, 37, 38, 41, *86*, 87
skulls, polished, 47
Sloan, D., 18
Smith, F., 25, 37
Smith, M., 67, 69
social
  archaeology, 65
  brains, 48–9
  complexity, 137–47
  organization, 162
Sokchangni (Korean Peninsula), 115, *116*, *119*
Solutré (France), *139*
Solutrean, *143*, 147
Songdu-ri (Korean Peninsula), *116*, *119*
de Sonneville-Bordes, D., 14, 16
Sopeña (Spain), 157
Soper, R., *12*
Sorbus aria, 138
Soro-ri (Korean Peninsula), *116*, *119*
South Africa, 5, 24, 53–8
Southampton University, 23
Spain, 138, 140, 142, 146–8, 153–8
Spaulding, A., 13
spears, 88
spelaeothems, 85
Spencer, F., 37
Spriggs, M., 65
Spy Cave (Belgium), 128, *128*, 129
St Césaire (France), 37–8, 39, 41, 138, *142*, 144, 146, 161
St Marcel (France), 147
Stage Three Project, 137
Star Carr (England), 15, 18, 27
steatite, 108
steppe, 123, 125, *125*, 131, 132, 138, 139, 144, 145
Sterud, G., 21
Steward, J., 14, 19
Still Bay (South Africa), 56, 57
Stiner, M., 141
stone artefacts, 14, 53, 64, 65, 70, 73–82, *79*, 89, *89*, 90, *90*, 91, *92*, 95, 103, *107*, 118, *120*, 126, 130, *130*, 133, *133*, 147, 155–7, *156*. *See also under names of individual artefact types*
Stoneking, M., 25, 37, 40
storage, 3, 45
Strashanya Cave (Russia), 109
Streletskian-Sungirian, 132, 133
Stringer, C., 24, 25, 26, 35, 37, 38, 39, 40, 41
Strzegowa (Poland), *127*
Sturdy, D., 13
subsistence, 142, 146, 147, 163
Sudetes, *124*, 125, 126
Sulawesi, 65, *66*, 68, 69
Sumatra, 74
Sumba Island, *66*
Suncheon (Korean Peninsula), *116*
Sunda, *64*, 65
Sungir (Russia), *131*
Sungirian, 132
Suyanggae (Korean Peninsula), 115, *116*, 119, *120*
Swieciechów flint, 127
symbolism, 5, 39, 46, 47, 49, 53, 54, 58
Syria, 88, 94
Syro-Arabian desert, 85
Szeleta (Hungary), *127*
Szeletian, 126, *127*, 130, 131, 133

T

et-Tabban (Judean desert), *86*, 87
et-Tabun Cave (Israel), 36, 38, *86*, 87
Tajikistan, 105
talc, 108
Les Tambourets (France), *139*
Tanami Desert, *69*
Tanimbar Island, *66*
Taramsa (Egypt), *86*
Tasmania, 64
Taurus Mountains, *86*, 93, 94, 96, 97
Tchernov, E., 37
techno-complexes, 67, 108, 111, 153, 158, 163
teeth, 38, 40
  artefacts, 92, 93, 106, *107*
  Neanderthal, 3
Telmanskaya (Russia), 132
tents, 50
tephra, 115, 118
Thar desert, 73
Theillat (France), *139*, *142*
Thorne, A., 37
Thuringia (Poland), 130
Thylogale spp., 66
Tilley, C., 22
Timor, 65, 66
Tinyuan, 39
Tiumechin (Russia), *104*, 105
Toba volcano, 5, 73, *74*, *75*, *76*, *77*
Tolbaga (Russia), 110
Tongati, river, 56
Tor Faraj (Jordan), *86*, 94
Tor Sadaf (Jordan), *86*, 88
Tourraine (France), 148
transitional industries, 87, 89, *89*, 138, 144, 154, 161
Trebom (Czech Republic), *127*
Trinkaus, E., 26, 37, 38, 40
Trou de la Mère Clochette (France), 139, *139*
Trou Magrite Cave (Belgium), *128*
trout, 146, 148
Tsagaan Agui Cave (Mongolia), 109
Tuin-Gol (Mongolia), 110
tundra, 123, 124, 125, *125*, 131, 132, 138, 144, 145
Turkey, 19, 87, 93
Turkmenistan, 105
Turner, A., 18
Tursac (France), *13*
turtle, freshwater, 67
Turville-Petre, F., 87
Tushabramishvili, D., 95
Tuto de Camalhot (France), 146
Tyrrhenian Sea, 145

U

'Ubeidiya (Israel), 147
Üçagizli (Turkey), *86*, 87, 88
Ucko, P., 20, 23, 24–5, 41
Uluzzian, 137, *139*, 140, *140*, 141, 143–6
Umhlatuzana (South Africa), 54, 56
Umm el-Tlel (Syria), *86*, 88
UNESCO, 24
United Nations, 24
University College London, 11
Uphill Quarry (England), *128*
Urban Revolution, 45
Ursul, river, 103, 109
Usa, river, *124*, 132
Ushlep (Russia), 109
Ust-Kan (Russia), *104*, 109
Ust-Karakol (Russia), *104*, 105, 106, 108, 109
Uzbekistan, 103, 105

V

Vah, river, 126, 127
La Valiña (Spain), 140
Valladas, H., 15, 26, 27
Vallières (France), 148
Vandermeersch, B., 37
Var (France), 146
Varavrina Gora (Russia), 110
Vasconian facies, 154
Vedrovice V (Czech Republic), *127*
Venetia (Italy), 138
Venus figurines, 63
Veth, P., 64, 65, 67
Vézère, river, *13*, 141
Vicencov (Czech Republic), *127*
Villanueva (Cantabria), 154
La Viña (Spain), 157
Vindija (Croatia), 41
Vistula, river, 123, *124*, 125, 126, 127, 128, 129
vole, 138
Volga, river, *86*, 123, 124
volumetric blade technology, 87, 125, 126, 130, 131, 132, 157
voyaging, 63, 69
Vygotsky, L., 46, 47

W

el-Wad Cave (Israel), *86*, 87, 90, 93, 94
Wadi Hassa (Jordan), 88
Wadley, L., 54
Wainscoat, J., 40
Wajak, 37, 40
Wales, 128
wallaby, 66
Wallace's Line, *64*, 65
Wallacea, *64*, 65, 66
Wandjina paintings, 65
Warwasi (Iran), *86*, 94
Watson, J., 19
weapons, 56, 97
Weber's Line, *64*
Wernet, P., 154
West, S., 11
Western Desert (Australia), 67, 69
White, M., 27
White, R., 21, 24, 38
Wildhaus Cave (Germany), *128*
Wildschauer Cave (Germany), *128*
Wilkinson, M., 18
Wilkinson, P., 13
Willandra Lakes (Australia), 37, 40, 69
Williams, C., 18
Wilson, A., 24, 25, 37
wolf, 141
Wolpoff, M., 25, 26, 37, 38, 40
Wolpyeong (Korean Peninsula), *116*
wooden artefacts, 106
Woodhouse school, 10
woodland, 56, 123, 124, *125*, 131, 138, 141, 142, 143, 144, 145, 148
woolly rhinoceros, 4, 110, 125
World Archaeological Congress, 15, 24–5, 41
Wu, X.Z., 37
Wylie, A., 22

Y

Yabrud II (Syria), *86*, 93
Yabrudian, 105
Yafteh Cave (Iran), *86*, 94
Yangadja-Karatengir(Turkmenistan),105
Yeonbong-2 (Korean Peninsula), *116*
Yonggok-dong (Korean Peninsula), *116*, *119*
Yongho-dong (Korean Peninsula), 115, *116*, 118, *119*
Yongsan-dong (Korean Peninsula), *116*
Younger Dryas, 4
YTT (ash deposit), 74–8, *75*, *77*
Yullyang-dong (Korean Peninsula), *116*, *119*

Z

Zagros Mountains, 2, *86*, 87, 93, 94, 96, 97
Zaozerie (Russia), *131*, 132, 133, *133*
zebra, 56
Zhoukoudian, 36, 39
Zubrow, E., 24
Zulu sleeping mats, 57
Zuttiyeh, 38
Zwergloch (Germany), *129*